My Numbers, My Friends

Springer

New York
Berlin
Heidelberg
Barcelona
Hong Kong
London
Milan
Paris
Singapore
Tokyo

Paulo Ribenboim

My Numbers, My Friends

Popular Lectures on Number Theory

Springer

Paulo Ribenboim
Department of Mathematics
  and Statistics
Queen's University
Kingston, Ontario K7L 3N6
Canada

Mathematics Subject Classification (2000): 11-06, 11Axx

Library of Congress Cataloging-in-Publication Data
Ribenboim, Paulo
  My numbers, my friends / Paulo Ribenboim
    p. cm.
  Includes bibliographical references and index.
  ISBN 0-387-98911-0 (sc. : alk. paper)
  1. Number Theory.   I. Title
QA241.R467 2000
612'.7— dc21   99-42458

Printed on acid-free paper.

Production managed by Bill Imbornoni; manufacturing supervised by Joe Quatela.
Composition by MacroTeX, Akron, OH.
Printed and bound by R.R. Donnelley and Sons, Harrisonburg, VA.
Printed in the United States of America.

9 8 7 6 5 4 3 2 1

ISBN 0-387-98911-0 Springer-Verlag New York Berlin Heidelberg   SPIN 10424971

3 3001 00831 1002

Contents

10 What Kind of Number Is $\sqrt{2}^{\sqrt{2}}$?   271

Preface

Dear Friends of Numbers:

This little book is for you. It should offer an exquisite intellectual enjoyment, which only relatively few fortunate people can experience.

May these essays stimulate your curiosity and lead you to books and articles where these matters are discussed at a more technical level.

I warn you, however, that the problems treated, in spite of being easy to state, are for the most part very difficult. Many are still unsolved. You will see how mathematicians have attacked these problems.

Brains at work! But do not blame me for sleepless nights (I have mine already).

Several of the essays grew out of lectures given over the course of years on my customary errances.

Other chapters could, but probably never will, become full-sized books.

The diversity of topics shows the many guises numbers take to tantalize* and to demand a mobility of spirit from you, my reader, who is already anxious to leave this preface.

Now go to page 1 (or 127?).

Paulo Ribenboim

---

*Tantalus, of Greek mythology, was punished by continual disappointment when he tried to eat or drink what was placed within his reach.

1

The Fibonacci Numbers and the Arctic Ocean

Introduction

There is indeed not much relation between the Fibonacci numbers and the Arctic Ocean, but I thought that this title would excite your curiosity for my lecture. You will be disappointed if you wished to hear about the Arctic Ocean, as my topic will be the sequence of Fibonacci numbers and similar sequences.

Like the icebergs in the Arctic Ocean, the sequence of Fibonacci numbers is the most visible part of a theory which goes deep: the theory of linear recurring sequences.

The so-called Fibonacci numbers appeared in the solution of a problem by FIBONACCI (also known as LEONARDO PISANO), in his book *Liber Abaci* (1202), concerning reproduction patterns of rabbits. The first significant work on the subject is by LUCAS, with his seminal paper of 1878. Subsequently, there appeared the classical papers of BANG (1886) and ZSIGMONDY (1892) concerning prime divisions of special sequences of binomials. CARMICHAEL (1913) published another fundamental paper where he extended to Lucas sequences the results previously obtained in special cases. Since then, I note the work of LEHMER, the applications of the theory in primality tests giving rise to many developments.

The subject is very rich and I shall consider here only certain aspects of it.

If, after all, your only interest is restricted to Fibonacci and Lucas numbers, I advise you to read the booklets by VOROB'EV (1963), HOGGATT (1969), and JARDEN (1958).

1   Basic definitions

A   Lucas sequences

Let P, Q be non-zero integers, let $D = P^2 - 4Q$, be called the *discriminant*, and assume that $D \neq 0$ (to exclude a degenerate case).

Consider the polynomial $X^2 - PX + Q$, called the *characteristic polynomial*, which has the roots

$$\alpha = \frac{P + \sqrt{D}}{2} \quad \text{and} \quad \beta = \frac{P - \sqrt{D}}{2}.$$

Thus, $\alpha \neq \beta$, $\alpha + \beta = P$, $\alpha \cdot \beta = Q$, and $(\alpha - \beta)^2 = D$.

For each $n \geq 0$, define $U_n = U_n(P, Q)$ and $V_n = V_n(P, Q)$ as follows:

$$U_0 = 0, \; U_1 = 1, \; U_n = P \cdot U_{n-1} - Q \cdot U_{n-2} \quad \text{(for } n \geq 2\text{)},$$
$$V_0 = 2, \; V_1 = P, \; V_n = P \cdot V_{n-1} - Q \cdot V_{n-2} \quad \text{(for } n \geq 2\text{)}.$$

The sequences $U = (U_n(P, Q))_{n \geq 0}$ and $V = (V_n(P, Q))_{n \geq 0}$ are called the (first and second) *Lucas sequences with parameters* (P, Q). $(V_n(P, Q))_{n \geq 0}$ is also called the *companion* Lucas sequence with parameters (P, Q).

It is easy to verify the following formal power series developments, for any (P, Q):

$$\frac{X}{1 - PX + QX^2} = \sum_{n=0}^{\infty} U_n X^n \quad \text{and}$$

$$\frac{2 - PX}{1 - PX + QX^2} = \sum_{n=0}^{\infty} V_n X^n.$$

The Lucas sequences are examples of sequences of numbers produced by an algorithm.

At the nth step, or at time n, the corresponding numbers are $U_n(P, Q)$, respectively, $V_n(P, Q)$. In this case, the algorithm is a linear recurrence with two parameters. Once the parameters and the initial values are given, the whole sequence—that is, its future values—is completely determined. But, also, if the parameters and two consecutive values are given, all the past (and future) values are completely determined.

B   Special Lucas sequences

I shall repeatedly consider special Lucas sequences, which are important historically and for their own sake. These are the sequences of Fibonacci numbers, of Lucas numbers, of Pell numbers, and other sequences of numbers associated to binomials.

(a) Let $P = 1$, $Q = -1$, so $D = 5$. The numbers $U_n = U_n(1, -1)$ are called the *Fibonacci numbers*, while the numbers $V_n = V_n(1, -1)$ are called the *Lucas numbers*. Here are the initial terms of these sequences:

Fibonacci numbers : 0, 1, 2, 3, 5, 8, 13, 21, 34, 55, 89, 144, ...
Lucas numbers   : 2, 1, 3, 4, 7, 11, 18, 29, 47, 76, 123, 99, 322, ...

(b) Let $P = 2$, $Q = -1$, so $D = 8$. The numbers $U_n = U_n(2, -1)$ and $V_n = V_n(2, -1)$ are the *Pell numbers* and the *companion Pell numbers*. Here are the first few terms of these sequences:

$$U_n(2, -1): 0, 1, 2, 5, 12, 29, 70, 169, \ldots$$
$$V_n(2, -1): 2, 2, 6, 14, 34, 82, 198, 478, \ldots$$

(c) Let a, b be integers such that $a > b \geq 1$. Let $P = a + b$, $Q = ab$, so $D = (a - b)^2$. For each $n \geq 0$, let $U_n = \frac{a^n - b^n}{a - b}$ and $V_n = a^n + b^n$. Then it is easy to verify that $U_0 = 0$, $U_1 = 1$, $V_0 = 2$, $V_1 = a + b = P$, and $(U_n)_{n \geq 0}$, $(V_n)_{n \geq 0}$ are the first and second Lucas sequences with parameters P, Q.

In particular, if $b = 1$, one obtains the sequences of numbers $U_n = \frac{a^n - 1}{a - 1}$, $V_n = a^n + 1$; now the parameters are $P = a + 1$, $Q = a$. Finally, if also $a = 2$, one gets $U_n = 2^n - 1$, $V_n = 2^n + 1$, and now the parameters are $P = 3$, $Q = 2$.

C   Generalizations

At this point, it is appropriate to indicate extensions of the notion of Lucas sequences which, however, will not be discussed in this lecture. Such generalizations are possible in four directions, namely, by changing the initial values, by mixing two Lucas sequences, by not demanding that the numbers in the sequences be integers, or by having more than two parameters.

Even though many results about Lucas sequences have been extended successfully to these more general sequences, and have found interesting applications, for the sake of definiteness I have opted to restrict my attention only to Lucas sequences.

(a) Let P, Q be integers, as before. Let T_0, T_1 be any integers such that T_0 or T_1 is non-zero (to exclude the trivial case). Let

$$W_0 = PT_0 + 2T_1 \quad \text{and} \quad W_1 = 2QT_0 + PT_1.$$

Let

$$T_n = P \cdot T_{n-1} - Q \cdot T_{n-2} \quad \text{and}$$
$$W_n = P \cdot W_{n-1} - Q \cdot W_{n-2} \quad (\text{for } n \geq 2).$$

The sequences $(T_n(P, Q))_{n \geq 0}$ and $W_n(P, Q))_{n \geq 0}$ are the (first and the second) *linear recurrence sequences* with parameters (P, Q) and *associated to the pair* (T_0, T_1). The Lucas sequences are special, normalized, linear recurrence sequences with the given parameters; they are associated to $(0, 1)$.

(b) LEHMER (1930) considered the following sequences. Let P, Q be non-zero integers, α, β the roots of the polynomial $X^2 - \sqrt{P} \cdot X + Q$, and define

$$L_n(P, Q) = \begin{cases} \dfrac{\alpha^n - \beta^n}{\alpha - \beta} & \text{if } n \text{ is odd,} \\[2mm] \dfrac{\alpha^n - \beta^n}{\alpha^2 - \beta^2} & \text{if } n \text{ is even.} \end{cases}$$

$L = (L_n(P, Q))_{n \geq 0}$ is the *Lehmer sequence* with parameters P, Q. Its elements are integers. These sequences have been studied by LEHMER and subsequently by SCHINZEL and STEWART in several papers which also deal with Lucas sequences and are quoted in the bibliography.

(c) Let $\mathcal{R}$ be an integral domain which need not be $\mathbb{Z}$. Let $P, Q \in \mathcal{R}$, $P, Q \neq 0$, such that $D = P^2 - 4Q \neq 0$. The sequences $(U_n(P, Q))_{n \geq 0}$, $(V_n(P, Q))_{n \geq 0}$ of elements of $\mathcal{R}$ may be defined as for the case when $\mathcal{R} = \mathbb{Z}$.

Noteworthy cases are when $\mathcal{R}$ is the ring of integers of a number field (for example, a quadratic number field), or $\mathcal{R} = \mathbb{Z}[x]$ (or other polynomial ring), or $\mathcal{R}$ is a finite field. For this latter situation, see SELMER (1966).

(d) Let P_0, P_1, ..., P_{k-1} (with $k \geq 1$) be given integers, usually subjected to some restrictions to exclude trivial cases. Let S_0, S_1, ..., S_{k-1} be given integers. For $n \geq k$, define:

$$S_n = P_0 \cdot S_{n-1} - P_1 \cdot S_{n-2} + P_2 \cdot S_{n-3} - \ldots + (-1)^{k-1} P_{k-1} \cdot S_{n-k}.$$

Then $(S_n)_{n \geq 0}$ is called a *linear recurrence sequence of order* k, *with parameters* P_0, P_1, ..., P_{k-1} *and initial values* S_0, S_1, ..., S_{k-1}. The case when $k = 2$ was seen above. For $k = 1$, one obtains the geometric progression $(S_0 \cdot P_0^n)_{n \geq 0}$.

There is great interest and still much to be done in the theory of linear recurrence sequences of order greater than 2.

2   Basic properties

The numbers in Lucas sequences satisfy many, many properties that reflect the regularity in generating these numbers.

A   Binet's formulas

BINET (1843) indicated the following expression in terms of the roots α, β of the polynomial $X^2 - PX + Q$:

(2.1) Binet's formulas:

$$U_n = \frac{\alpha^n - \beta^n}{\alpha - \beta}, \quad V_n = \alpha^n + \beta^n.$$

The proof is, of course, very easy. Note that by Binet's formulas,

$$U_n(-P, Q) = (-1)^{n-1} U_n(P, Q) \qquad \text{and}$$
$$V_n(-P, Q) = (-1)^n V_n(P, Q).$$

So, for many of the following considerations, it will be assumed that $P \geq 1$.

B   Degenerate Lucas sequences

Let (P, Q) be such that the ratio $\eta = \alpha/\beta$ of roots of $X^2 - Px + Q$ is a root of unity. Then the sequences $U(P, Q)$, $V(P, Q)$ are said to be *degenerate*.

Now I describe all degenerate sequences. Since

$$\eta + \eta^{-1} = \frac{\alpha}{\beta} + \frac{\beta}{\alpha} = \frac{P^2 - 2Q}{Q}$$

is an algebraic integer and rational, it is an integer. From $|\frac{\alpha}{\beta} + \frac{\beta}{\alpha}| \leq 2$ it follows $P^2 - 2Q = 0, \pm Q, \pm 2Q$, and this gives $P^2 = Q, 2Q, 3Q, 4Q$. If $\gcd(P, Q) = 1$, then $(P, Q) = (1, 1), (-1, 1), (2, 1)$, or $(-2, 1)$, and the sequences are

```
U(1,1)   :  0,   1,   1,   0,  -1,  -1,   0,   1,   1,   0,  ...
U(-1,1): 0,   1,  -1,   0,   1,  -1,   0,  ...
V(1,1)   :  2,   1,  -1,  -2,  -1,   1,   2,   1,  -1,  -2,  ...
V(-1,1): 2,  -1,  -1,   2,  -1,  -1,   2,  ...
U(2,1)   :  0,   1,   2,   3,   4,   5,   6,   7,  ...
U(-2,1): 0,   1,  -2,   3,  -4,   5,  -6,   7,  ...
V(2,1)   :  2,   2,   2,   2,   2,   2,   2,   2,  ...
V(-2,1): 2,  -2,   2,  -2,   2,  -2,   2,  -2,  ...
```

From the discussion, if the sequence is degenerate, then $D = 0$ or $D = -3$.

C   Growth and numerical calculations

First, I note results about the growth of the sequence $U(P, Q)$.

(2.2) If the sequences $U(P, Q)$, $V(P, Q)$ are non-degenerate, then $|U_n|, |V_n|$ tend to infinity (as n tends to ∞).

This follows from a result of MAHLER (1935) on the growth of coefficients of Taylor series. MAHLER also showed

(2.3) If $Q \geq 2$, $\gcd(P, Q) = 1$, $D < 0$, then, for every $\varepsilon > 0$ and n sufficiently large,

$$|U_n| \geq |\beta^n|^{1-\varepsilon}.$$

The calculations of U_n, V_n may be performed as follows. Let

$$M = \begin{pmatrix} P & -Q \\ 1 & 0 \end{pmatrix}.$$

Then for $n \geq 1$,

$$\begin{pmatrix} U_n \\ U_{n-1} \end{pmatrix} = M^{n-1} \begin{pmatrix} 1 \\ 0 \end{pmatrix}$$

and

$$\begin{pmatrix} V_n \\ V_{n-1} \end{pmatrix} = M^{n-1} \begin{pmatrix} 2 \\ P \end{pmatrix}.$$

To compute a power M^k of the matrix M, the quickest method is to compute successively the powers M, M^2, M^4, ..., M^{2^e} where $2^e \leq k < 2^{e+1}$; this is done by successively squaring the matrices. Next, if the 2-adic development of k is $k = k_0 + k_1 \times 2 + k_2 \times 2^2 + \ldots + k_e \times 2^e$, where $k_i = 0$ or 1, then $M^k = M^{k_0} \times (M^2)^{k_1} \times \ldots \times (M^{2^e})^{k_e}$.

Note that the only factors actually appearing are those where $k_i = 1$.

Binet's formulas allow also, in some cases, a quick calculation of U_n and V_n.

If $D \geq 5$ and $|\beta| < 1$, then

$$\left| U_n - \frac{\alpha^n}{\sqrt{D}} \right| < \frac{1}{2} \quad \text{(for } n \geq 1),$$

and $|V_n - \alpha^n| < \frac{1}{2}$ (for n such that $n \cdot (-\log|\beta|) > \log 2$). Hence, cU_n is the closest integer to $\frac{\alpha^n}{\sqrt{D}}$, and V_n is the closest integer to α^n. This applies in particular to Fibonacci and Lucas numbers for which $D = 5$, $\alpha = (1 + \sqrt{5})/2 = 1.616\ldots$, (the golden number), $\beta = (1 - \sqrt{5})/2 = -0.616\ldots$.

It follows that the Fibonacci number U_n and the Lucas number V_n have approximately $n/5$ digits.

D   Algebraic relations

The numbers in Lucas sequences satisfy many properties. A look at the issues of *The Fibonacci Quarterly* will leave the impression that there is no bound to the imagination of mathematicians whose endeavor it is to produce newer forms of these identities and properties. Thus, there are identities involving only the numbers U_n, in others only the numbers V_n appear, while others combine the numbers U_n and V_n. There are formulas for U_{m+n}, U_{m-n}, V_{m+n}, V_{m-n} (in terms of U_m, U_n, V_m, V_n); these are the addition and subtraction formulas. There are also formulas for U_{kn}, V_{kn}, and U_{n^k}, V_{n^k}, U_n^k, cV_n^k (where $k \geq 1$) and many more.

I shall select a small number of formulas that I consider most useful. Their proofs are almost always simple exercises, either by applying Binet's formulas or by induction.

It is also convenient to extend the Lucas sequences to negative indices in such a way that the same recursion (with the given parameters P, Q) still holds.

(2.4) Extension to negative indices:

$$U_{-n} = -\frac{1}{Q^n} U_n, \quad V_{-n} = \frac{1}{Q^n} V_n \quad \text{(for } n \geq 1\text{)}.$$

(2.5) U_n and V_n may be expressed in terms of P, Q. For example,

$$U_n = P^{n-1} - \binom{n-2}{1} P^{n-3}Q + \binom{n-3}{2} P^{n-5}Q^2 + \cdots$$

$$+ (-1)^k \binom{n-1-k}{k} P^{n-1-2k}Q^k + \cdots + \text{(last summand)}$$

where

$$\text{(last summand)} = \begin{cases} (-1)^{\frac{n}{2}-1} \binom{\frac{n}{2}}{\frac{n}{2}-1} PQ^{\frac{n}{2}-1} & \text{if } n \text{ is even,} \\ (-1)^{\frac{n-1}{2}} Q^{\frac{n-1}{2}} & \text{if } n \text{ is odd.} \end{cases}$$

Thus, $U_n = f_n(P, Q)$, where $f_n(X, Y) \in \mathbb{Z}[X, Y]$. The function f_n is isobaric of weight $n - 1$, where X has weight 1 and Y has weight 2.

Similarly, $V_n = g_n(P, Q)$, where $g_n \in \mathbb{Z}[X, Y]$. The function g_n is isobaric of weight n, where X has weight 1, and Y has weight 2.

(2.6) Quadratic relations:

$$V_n^2 - DU_n^2 = 4Q^n$$

for every $n \in \mathbb{Z}$.

This may also be put in the form:

$$U_{n+1}^2 - PU_{n+1}U_n + QU_n^2 = Q^n.$$

(2.7) Conversion formulas:

$$DU_n = V_{n+1} - QV_{n-1},$$
$$V_n = U_{n+1} - QU_{n-1},$$

for every $n \in \mathbb{Z}$.

(2.8) Addition of indices:

$$U_{m+n} = U_m V_n - Q^n U_{m-n},$$
$$V_{m+n} = V_m V_n - Q^n V_{m-n} = D U_m U_n + Q^n V_{m-n},$$

for all m, $n \in \mathbb{Z}$.

Other formulas of the same kind are:

$$2U_{m+n} = U_m V_n + U_n V_m,$$
$$2Q^n U_{m-n} = U_m V_n - U_n V_m,$$

for all m, $n \in \mathbb{Z}$.

(2.9) Multiplication of indices:

$$U_{2n} = U_n V_n,$$
$$V_{2n} = V_n^2 - 2Q^n,$$
$$U_{3n} = U_n(V_n^2 - Q^n) = U_n(DU_n^2 + 3Q^n),$$
$$V_{3n} = V_n(V_n^2 - 3Q^n),$$

for every $n \in \mathbb{Z}$.

More generally, if $k \geq 3$ it is possible to find by induction on k formulas for U_{kn} and V_{kn}, but I shall refrain from giving them explicitly.

E   Divisibility properties

(2.10) Let $U_m \neq 1$. Then, U_m divides U_n if and only if $m \mid n$. Let $V_m \neq 1$. Then, V_m divides V_n if and only if $m \mid n$ and n/m is odd.

For the next properties, it will be assumed that $\gcd(P, Q) = 1$.

(2.11) $\gcd(U_m, U_n) = U_d$, where $d = \gcd(m, n)$.

(2.12)

$$\gcd(V_m, V_n) = \begin{cases} V_d & \text{if } \dfrac{m}{d} \text{ and } \dfrac{n}{d} \text{ are odd,} \\ 1 \text{ or } 2 & \text{otherwise,} \end{cases}$$

where $d = \gcd(m, n)$.

(2.13)

$$\gcd(U_m, V_n) = \begin{cases} V_d & \text{if } \dfrac{m}{d} \text{ is even, } \dfrac{n}{d} \text{ is odd,} \\ 1 \text{ or } 2 & \text{otherwise,} \end{cases}$$

where $d = \gcd(m, n)$.

(2.14) If $n \geq 1$, then $\gcd(U_n, Q) = 1$ and $\gcd(V_n, Q) = 1$.

3   Prime divisors of Lucas sequences

The classical results about prime divisors of terms of Lucas sequences date back to EULER, (for numbers $\frac{a^n - b^n}{a - b}$), to LUCAS (for Fibonacci and Lucas numbers), and to CARMICHAEL (for other Lucas sequences).

A   The sets $\mathcal{P}(U)$, $\mathcal{P}(V)$, and the rank of appearance.

Let $\mathcal{P}$ denote the set of all prime numbers. Given the Lucas sequences $U = (U_n(P, Q))_{n \geq 0}$, $V = (V_n(P, Q))_{n \geq 0}$, let

$$\mathcal{P}(U) = \{p \in \mathcal{P} \mid \exists n \geq 1 \text{ such that } U_n \neq 0 \text{ and } p \mid U_n\},$$
$$\mathcal{P}(V) = \{p \in \mathcal{P} \mid \exists n \geq 1 \text{ such that } V_n \neq 0 \text{ and } p \mid V_n\}.$$

If U, V are degenerate, then $\mathcal{P}(U)$, $\mathcal{P}(V)$ are easily determined sets.

Therefore, it will be assumed henceforth that U, V are non-degenerate and thus, $U_n(P, Q) \neq 0$, $V_n(P, Q) \neq 0$ for all $n \geq 1$.

Note that if p is a prime dividing both p, q, then $p \mid U_n(P, Q)$, $p \mid V_n(P, Q)$, for all $n \geq 2$. So, for the considerations which will follow, there is no harm in assuming that $\gcd(P, Q) = 1$. So, (P, Q) belongs to the set

$$\mathcal{S} = \{(P, Q) \mid P \geq 1, \ \gcd(P, Q) = 1, \ P^2 \neq Q, \ 2Q, \ 3Q, \ 4Q\}.$$

For each prime p, define

$$\rho_U(p) = \begin{cases} n & \text{if } n \text{ is the smallest positive index where } p \mid U_n, \\ \infty & \text{if } p \nmid U_n \text{ for every } n > 0, \end{cases}$$

$$\rho_V(p) = \begin{cases} n & \text{if } n \text{ is the smallest positive index where } p \mid V_n, \\ \infty & \text{if } p \nmid V_n \text{ for every } n > 0. \end{cases}$$

We call $\rho_U(n)$ (respectively $\rho_V(p)$)) is called the *rank of appearance* of p in the Lucas sequence U (respectively V).

First, I consider the determination of even numbers in the Lucas sequences.

(3.1) Let $n \geq 0$. Then:

$$U_n \text{ even} \Longleftrightarrow \begin{cases} P \text{ even} \quad Q \text{ odd}, \quad n \text{ even}, \\ \qquad\qquad \text{or} \\ P \text{ odd} \quad Q \text{ odd}, \quad 3 \mid n, \end{cases}$$

and

$$V_n \text{ even} \Longleftrightarrow \begin{cases} P \text{ even} \quad Q \text{ odd}, \quad n \geq 0, \\ \qquad\qquad \text{or} \\ P \text{ odd} \quad Q \text{ odd}, \quad 3 \mid n. \end{cases}$$

Special Cases. For the sequences of Fibonacci and Lucas numbers $(P = 1,\, Q = -1)$, one has:
U_n is even if and only if $3 \mid n$,
V_n is even if and only if $3 \mid n$.
For the sequences of numbers $U_n = \frac{a^n - b^n}{a-b}$, $V_n = a^n + b^n$, with $a > b \geq 1$, $\gcd(a, b) = 1$, $p = a + b$, $q = ab$, one has:
If a, b are odd, then U_n is even if and only if n is even, while V_n is even for every n.
If a, b have different parity, then U_n, V_n are always odd (for $n \geq 1$).

With the notations and terminology introduced above the result **(3.1)** may be rephrased in the following way:

(3.2) $2 \in \mathcal{P}(U)$ if and only if Q is odd

$$\rho_U(2) = \begin{cases} 2 & \text{if } P \text{ even}, \quad Q \text{ odd}, \\ 3 & \text{if } P \text{ odd}, \quad Q \text{ odd}, \\ \infty & \text{if } P \text{ odd}, \quad Q \text{ even}, \end{cases}$$

$2 \in \mathcal{P}(V)$ if and only if Q is odd

$$\rho_V(2) = \begin{cases} 1 & \text{if } P \text{ even}, \quad Q \text{ odd}, \\ 3 & \text{if } P \text{ odd}, \quad Q \text{ odd}, \\ \infty & \text{if } P \text{ odd}, \quad Q \text{ even}. \end{cases}$$

Moreover, if Q is odd, then $2 \mid U_n$ (respectively $2 \mid V_n$) if and only if $\rho_U(2) \mid n$ (respectively $\rho_V(2) \mid n$).

This last result extends to odd primes:

(3.3) Let p be an odd prime.
If $p \in \mathcal{P}(U)$, then $p \mid U_n$ if and only if $\rho_U(p) \mid n$.
If $p \in \mathcal{P}(V)$, then $p \mid V_n$ if and only if $\rho_V(p) \mid n$ and $\frac{n}{\rho_V(p)}$ is odd.

Now I consider odd primes p and indicate when $p \in \mathcal{P}(U)$.

(3.4) Let p be an odd prime.
If $p \nmid P$ and $p \mid Q$, then $p \nmid U_n$ for every $n \geq 1$.
If $p \mid P$ and $p \nmid Q$, then $p \mid U_n$ if and only if n is even.
If $p \nmid PQ$ and $p \mid D$, then $p \mid U_n$ if and only if $p \mid n$.
If $p \nmid PQD$, then p divides $U_{\psi_D}(p)$ where $\psi_D(p) = p - \left(\frac{D}{p}\right)$ and $\left(\frac{D}{p}\right)$ denotes the Legendre symbol.

Thus,
$$\mathcal{P}(U) = \{p \in \mathcal{P} \mid p \nmid Q\},$$
so $\mathcal{P}(U)$ is an infinite set.

The more interesting assertion concerns the case where $p \nmid PQD$, the other ones being very easy to establish.

The result may be expressed in terms of the rank of appearance:

(3.5) Let p be an odd prime.
If $p \nmid P$, $p \mid Q$, then $\rho_U(p) = \infty$.
If $p \mid P$, $p \nmid Q$, then $\rho_U(p) = 2$.
If $p \nmid PQ$, $p \mid D$, then $\rho_U(p) = p$.
If $p \nmid PQD$, then $\rho_U(p) \mid \Psi_D(p)$.

Special Cases. For the sequences of Fibonacci numbers ($P = 1$, $Q = -1$), $D = 5$ and $5 \mid U_n$ if and only if $5 \mid n$.
If p is an odd prime, $p \neq 5$, then $p \mid U_{p-(\frac{5}{p})}$, so $\rho_U(p) \mid (p - (\frac{5}{p}))$.
Because $U_3 = 2$, it follows that $\mathcal{P}(U) = \mathcal{P}$.

Let $a > b \geq 1$, $\gcd(a, b)$, $P = a + b$, $Q = ab$, $U_n = \frac{a^n - b^n}{a - b}$.
If p divides a or b but not both a, b, then $p \nmid U_n$ for all $n \geq 1$.
If $p \nmid ab$, $p \mid a + b$, then $p \mid U_n$ if and only if n is even.
If $p \nmid ab(a + b)$ but $p \mid a - b$, then $p \mid U_n$ if and only if $p \mid n$.
If $p \nmid ab(a + b)(a - b)$, then $p \mid U_{p-1}$. (Note that $D = (a - b)^2$.)
Thus, $\mathcal{P}(U) = \{p : p \nmid ab\}$.

Taking $b = 1$, if $p \nmid a$, then $p \mid U_{p-1}$, hence $p \mid a^{p-1} - 1$ (this is Fermat's Little Theorem, which is therefore a special case of the last assertion of **(3.4)**); it is trivial if $p \mid (a+1)(a-1)$.

The result **(3.4)** is completed with the so-called *law of repetition*, first discovered by LUCAS for the Fibonacci numbers:

(3.6) Let p^e (with $e \geq 1$) be the exact power of p dividing U_n. Let $f \geq 1$, $p \nmid k$. Then, p^{e+f} divides U_{nkp^f}. Moreover, if $p \nmid Q$, $p^e \neq 2$, then p^{e+f} is the exact power of p dividing U_{nkp^e}.

It was seen above that Fermat's Little Theorem is a special case of the assertion that if p is a prime and $p \nmid PQD$, then p divides $U_{\Psi_D(p)}$. I indicate now how to reinterpret EULER's classical theorem.

If α, β are the roots of the characteristic polynomial $X^2 - PX + Q$, define the symbol

$$\left(\frac{\alpha, \beta}{2}\right) = \begin{cases} 1 & \text{if } Q \text{ is even,} \\ 0 & \text{if } Q \text{ is odd, } P \text{ is even,} \\ -1 & \text{if } Q \text{ is odd, } P \text{ is odd,} \end{cases}$$

and for any odd prime p

$$\left(\frac{\alpha, \beta}{p}\right) = \begin{cases} \left(\dfrac{D}{p}\right) & \text{if } p \nmid D, \\ 0 & \text{if } p \mid D. \end{cases}$$

Let $\Psi_{\alpha,\beta}(p) = p - \left(\frac{\alpha,\beta}{p}\right)$ for every prime p. Thus, using the previous notation, $\Psi_{\alpha,\beta}(p) = \Psi_D(p)$ when p is odd and $p \nmid D$.

For $n = \prod_p p^e$, define the *generalized Euler function*

$$\Psi_{\alpha,\beta}(n) = n \prod_r \frac{\Psi_{\alpha,\beta}(p)}{p},$$

so $\Psi_{\alpha,\beta}(p^e) = p^{e-1}\Psi_{\alpha,\beta}(p)$ for each prime p and $e \geq 1$. Define also the *Carmichael function* $\lambda_{\alpha,\beta}(n) = \text{lcm}\{\Psi_{\alpha,\beta}(p^e)\}$. Thus, $\lambda_{\alpha,\beta}(n)$ divides $\Psi_{\alpha,\beta}(n)$.

In the special case where $\alpha = a$, $\beta = 1$, and a is an integer, then $\Psi_{a,1}(p) = p - 1$ for each prime p not dividing a. Hence, if $\gcd(a, n) = 1$, then $\Psi_{a,1}(n) = \varphi(n)$, where φ denotes the classical Euler function.

The generalization of EULER's theorem by CARMICHAEL is the following:

(3.7) n divides $U_{\lambda_{\alpha,\beta}(n)}$ hence, also, $U_{\Psi_{\alpha,\beta}(n)}$.

It is an interesting question to evaluate the quotient $\frac{\Psi_D(p)}{\rho_U(p)}$. It was shown by JARDEN (1958) that for the sequence of Fibonacci numbers,

$$\sup\left\{\frac{p - \left(\frac{5}{p}\right)}{\rho_U(D)}\right\} = \infty$$

(as p tends to ∞). More generally, KISS (1978) showed:

(3.8) (a) For each Lucas sequence $U_n(P, Q)$,

$$\sup\left\{\frac{\Psi_D(p)}{\rho_U(p)}\right\} = \infty.$$

(b) There exists $C > 0$ (depending on P, Q) such that

$$\frac{\Psi_D(p)}{\rho_U(p)} < C\frac{p}{\log p}.$$

Now I turn my attention to the companion Lucas sequence $V = (V_n(P, Q))_{n \geq 0}$ and I study the set of primes $\mathcal{P}(V)$. It is not known how to describe explicitly, by means of finitely many congruences, the set $\mathcal{P}(V)$. I shall indicate partial congruence conditions that are complemented by density results.

Because $U_{2n} = U_n V_n$, it then follows that $\mathcal{P}(V) \subseteq \mathcal{P}(U)$. It was already stated that $2 = \mathcal{P}(V)$ if and only if Q is odd.

(3.9) Let p be an odd prime.
If $p \nmid P$, $p \mid Q$, then $p \nmid V_n$ for all $n \geq 1$.
If $p \mid P$, $p \nmid Q$, then $p \mid V_n$ if and only if n is odd.
If $p \nmid PQ$, $p \mid D$, then $p \nmid V_n$ for all $n \geq 1$.
If $p \nmid PQD$, then $p \mid V_{\frac{1}{2}\Psi_D(p)}$ if and only if $\left(\frac{Q}{p}\right) = -1$.
If $p \nmid PQD$ and $\left(\frac{Q}{p}\right) = 1$, $\left(\frac{P}{p}\right) = -\left(\frac{-1}{p}\right)$, then $p \nmid V_n$ for all $n \geq 1$.

The above result implies that $\mathcal{P}(V)$ is an infinite set.* One may further refine the last two assertions; however, a complete determination of $\mathcal{P}(V)$ is not known.

In terms of the rank of appearance, **(3.9)** can be rephrased as follows:

---

*This was extended by WARD (1954) for all binary linear recurrences

(3.10) Let p be an odd prime.

If $p \mid P$, $p \nmid Q$, then $\rho_V(p) = 1$.

If $p \nmid P$, $p \mid Q$, then $\rho_V(p) = \infty$.

If $p \nmid PQ$, $p \mid D$, then $\rho_V(p) = \infty$.

If $p \nmid PQD$, $(\frac{Q}{p}) = -1$, then $\rho_V(p)$ divides $\frac{1}{2}\Psi_D(p)$.

If $p \nmid PQD$, $(\frac{Q}{p}) = 1$, $(\frac{D}{p}) = -(\frac{-1}{p})$, then $\rho_V(p) = \infty$.

The following conjecture has not yet been established in general, but has been verified in special cases, described below:

Conjecture. For each companion Lucas sequence V, the limit

$$\delta(V) = \lim \frac{\pi_V(x)}{\pi(x)}$$

exists and is strictly greater than 0.

Here, $\pi(x) = \#\{p \in \mathcal{P} \mid p \leq x\}$ and $\pi_V(x) = \#\{p \in \mathcal{P}(V) \mid p \leq x\}$. The limit $\delta(V)$ is the *density* of the set of prime divisors of V among all primes.

Special Cases. Let $(P, Q) = (1, -1)$, so V is the sequence of Lucas numbers. Then the above results may be somewhat completed. Explicitly:

If $p \equiv 3, 7, 11, 19 \pmod{20}$, then $p \in \mathcal{P}(V)$.

If $p \equiv 13, 17 \pmod{20}$, then $p \notin \mathcal{P}(V)$.

If $p \equiv 1, 9 \pmod{20}$ it may happen that $p \in \mathcal{P}(V)$ or that $p \notin \mathcal{P}(V)$.

JARDEN (1958) showed that there exist infinitely many primes $p \equiv 1 \pmod{20}$ in $\mathcal{P}(V)$ and also infinitely many primes $p \equiv 1 \pmod{20}$ not in $\mathcal{P}(V)$. Further results were obtained by WARD (1961) who concluded that there is no finite set of congruences to decide if an arbitrary prime p is in $\mathcal{P}(V)$.

Inspired by a method of HASSE (1966), and the analysis of WARD (1961), LAGARIAS (1985) showed that, for the sequence V of Lucas numbers, the density is $\delta(V) = \frac{2}{3}$.

BRAUER (1960) and HASSE (1966) studied a problem of SIERPIŃSKI, namely, determine the primes p such that 2 has an even order modulo p, equivalently, determine the primes p dividing the numbers $2^n + 1 = V_n(3, 2)$. He proved that $\delta(V(3, 2)) = 17/24$. LAGARIAS pointed out that HASSE's proof shows also that if $a \geq 3$ is square-free, then $\delta(V(a + 1, a)) = 2/3$; see also a related paper of HASSE (1965).

LAXTON (1969) considered, for each $a \geq 2$, the set $\mathcal{W}(a)$ of all binary linear recurrences W with W_0, W_1 satisfying $W_1 \neq W_0$, $W_1 \neq aW_0$, and $W_n = (a+1)W_{n-1} - aW_{n-2}$, for $n \geq 2$. This set includes the Lucas sequences $U(a+1, a)$, $V(a+1, a)$. For each prime p, let

$$e_p(a) = \begin{cases} 0 & \text{if } p \mid a, \\ \text{order of } a \bmod b & \text{if } p \nmid a. \end{cases}$$

LAXTON gave a heuristic argument to the effect that if the limit, as x tends to ∞, of

$$\frac{1}{\pi(x)} \sum_{p \leq x} \frac{e_p(a)}{p-1}$$

exists, then it is the expected (or average value), for any $W \in \mathcal{W}(a)$, of the density of primes in $\mathcal{P}(\mathcal{W})$ (that is, the set of primes dividing some W_n).

STEPHENS (1976) used a method of HOOLEY (1967) who had proved, under the assumption of a generalized Riemann's hypothesis, ARTIN's conjecture that 2 is a primitive root modulo p for infinitely many primes p. Let $a \geq 2$, a not a proper power. Assume the generalized Riemann hypothesis for the Dedekind ζ function of all fields $\mathbb{Q}(a^{1/n}, \zeta_k)$, where ζ_k is a primitive kth root of 1. Then, for every $x \geq 2$,

$$\sum_{p \leq x} \frac{e_p(a)}{p-1} = c(a)\frac{x}{\log x} + O\left(\frac{x \log \log x}{(\log x)^2}\right);$$

by the Prime Number Theorem, the limit considered above exists and is equal to $c(a)$. STEPHENS evaluated $c(a)$. Let

$$C = \prod_p \left(1 - \frac{p}{p^3 - 1}\right),$$

let $a = a_1 \cdot (a_2)^2$ where a_1 is square-free, let r be the number of distinct prime factors of a_1, and let f be defined as

$$f = \begin{cases} -\dfrac{2}{5} & \text{if } a_1 \equiv 1 \pmod 4, \\ -\dfrac{1}{64} & \text{if } a_1 \equiv 2 \pmod 4, \\ -\dfrac{1}{20} & \text{if } a_1 \equiv 3 \pmod 4. \end{cases}$$

Then,

$$
c(a) = C \left[1 - (-1)^r f \prod_{\substack{q \mid a_1 \\ q \text{ prime}}} \frac{q}{q^3 - q - 1} \right].
$$

STEPHENS also showed that even without the assumption of the generalized Riemann hypothesis the above estimation holds on average. Precisely, given $a \geq 2$ (as before), $e > 1$, and $x \geq 1$, there exists $c_1 > 0$ such that if $N > \exp\{c_1 (\log x)^{\frac{1}{2}}\}$, then

$$
\sum_{x \leq N} \sum_{p \leq x} \frac{e_p(a)}{p - 1} = C \int_1^x \frac{dt}{t} + O\left(\frac{x}{(\log x)^e}\right).
$$

B Primitive factors of Lucas sequences

Let p be a prime. If $\rho_U(p) = n$ (respectively $\rho_V(p) = n$), then p is called a *primitive factor* of $U_n(P, Q)$ (respectively $V_n(P, Q)$). Denote by $\text{Prim}(U_n)$ the set of primitive factors of U_n, similarly, by $\text{Prim}(V_n)$ the set of primitive factors of V_n. Let $U_n = U_n^* \cdot U_n'$, $V_n = V_n^* \cdot V_n'$, where $\gcd(U_n^*, U_n') = 1$, $\gcd(V_n^*, V_n^1) = 1$, $p \mid U_n^*$ (respectively $p \mid V_n^*$) if and only if p is a primitive factor of U_n (respectively V_n). U_n^*, (respectively V_n^*) is called the *primitive part* of U_n (respectively V_n). From $U_{2n} = U_n \cdot V_n$ it follows that $U_{2n}^* \mid V_n^*$, hence, $\text{Prim}(U_{2n}) \subseteq \text{Prim}(V_n^*)$. It is not excluded that $U_n^* = 1$ (respectively $V_n^* = 1$); I shall discuss this question.

a Existence of primitive factors

The study of primitive factors of Lucas sequences originated with BANG and ZSIGMONDY for special Lucas sequences (see below). The first main theorem is due to CARMICHAEL (1913):

(3.11) Let $(P, Q) \in \mathcal{S}$ and assume that $D > 0$.

1. If $n \neq 1, 2, 6$, then $\text{Prim}(U_n) \neq \varnothing$, with the only exception $(P, Q) = (1, -1)$, $n = 12$ (which gives the Fibonacci number $U_{12} = 144$).

   Moreover, if D is a square and $n \neq 1$, then $\text{Prim}(U_n) \neq \varnothing$, with the only exception $(P, Q) = (3, 2)$, $n = 6$ (which gives the number $2^6 - 1 = 63$).

2. If $n \neq 1, 3$, then $\text{Prim}(V_n) \neq \varnothing$, with the only exception $(P, Q) = (1, -1)$, $n = 6$ (which gives the Lucas numbers

$V_6 = 18$).

Moreover, if D is a square and $n \neq 1$, then $\mathrm{Prim}(V_n) \neq \varnothing$, with the only exception $(P, Q) = (3, 2)$, $n = 3$ (which gives the number $2^3 + 1 = 9$).

In his paper, CARMICHAEL also proved that if p does not divide D and $p \in \mathrm{Prim}(U_n)$, then $p \equiv \pm 1 \pmod{n}$, while if $p \in \mathrm{Prim}(V_n)$, then $p \equiv \pm 1 \pmod{2n}$.

The result of CARMICHAEL was extended by LEKKERKERKER (1953):

Even without assuming that $\gcd(P, Q) = 1$, if $D > 0$, there exist only finitely many n such that $U_n(P, Q)$ (respectively $V_n(P, Q)$) does not have a primitive factor.

DURST (1961) proved:

(3.12) Let $(P, Q) \in \mathcal{S}$ and $D > 0$. Then, $U_6(P, Q)$ has no primitive factor if and only if one the following conditions holds:

1. $P = 2^{t+1} - 3r$, $Q = (2^t - r)(2^t - 3r)$ where $t \geq 1$, $2^{t+1} > 3r$, and r is odd and positive.
2. $P = 3^s k$, $Q = 3^{2s-1} k^2 - 2^t$ where $s \geq 1$, $t \geq 0$, $k \equiv \pm 1 \pmod 6$, and $3^{2s-1} k^2 < 2^{t+2}$.

Thus, there exist infinitely many (P, Q) as above with $U_6(P, Q)$ having no primitive factor. DURST dealt also with parameters (P, Q) where $\gcd(P, Q)$ may be greater than 1.

(3.13) Let I be a finite set of integers, with $1 \in I$. Then, there are infinitely many pairs (P, Q), with $P \geq 1$, $P \neq Q$, $2Q$, $3Q$, $4Q$, $P^2 - 4Q > 0$, such that $\mathrm{Prim}(U(P, Q)) = I$.

If $D < 0$, the above result does not hold without modification. For example, for $(P, Q) = (1, 2)$ and $n = 1, 2, 3, 5, 8, 12, 13, 18$, $\mathrm{Prim}(U_n) = \varnothing$.

In 1962, SCHINZEL investigated the case when $D < 0$. In 1974, he proved a general result of which the following is a corollary.

(3.14) There exists $n_0 > 0$ such that for all $n \geq n_0$, $(P, Q) \in \mathcal{S}$, $U_n(P, Q)$, $V_n(P, Q)$ have a primitive factor.

The proof involves BAKER's lower bounds for linear forms in logarithms and n_0 is effectively computable. It is important to stress that n_0 is independent of the parameters. STEWART (1977a) showed that $n_0 \leq e^{452}4^{67}$. STEWART also showed that if $4 < n$, $n \neq 6$, there exist only finitely many Lucas sequences $U(P, Q)$, $V(P, Q)$ (of the kind indicated), which may in principle be explicitly determined, and such that $U_n(P, Q)$ (respectively $V_n(P, Q)$) does not have a primitive factor.

VOUTIER (1995) used a method developed by TZANAKIS (1989) to solve Thue's equations and determined for each n, $4 < n \leq 30$, $n \neq 6$, the finite set of parameters $(P, Q) \in \mathcal{S}$ such that $U_r(P, Q)$ has no primitive factor.

The next result of GYÖRY (1981) concerns terms of Lucas sequences with prime factors in a given set. If E is a finite set of primes, let $E^{\times}$ denote the set of natural numbers, all of whose prime factors belong to E.

(3.15) Let $s > 1$ and $E = \{p \text{ prime} \mid p \leq s\}$. There exist $c_1 = c_1(s) > 0$, $c_2 = c_2(s) > 0$, effectively computable, such that if $(P, Q) \in \mathcal{S}$, $4 < n$, and $U_n(P, Q) \in E^{\times}$, then

$$n \leq \max\{s + 1, e^{452} \cdot 2^{67}\},$$

$\max\{P, |Q|\} \leq c_1$, and $|U_n(P, Q)| \leq c_2$.

In 1982, GYÖRY gave an explicit value for the constants. An interesting corollary is the following:

(3.16) Let $s > 1$ and $E = \{p \text{ prime} \mid p \leq s\}$. There exists $c_3 = c_3(s) > 0$, effectively computable, such that if $a > b \geq 1$ are integers, $\gcd(a, b) = 1$, if $3 < n$, $\frac{a^n - b^n}{a - b} = m \in E^{\times}$, then $n < s$ and $\max\{a, m\} < c_3$.

Special Cases. The following very useful theorem was proved by ZSIGMONDY (1892); the particular case where $a = 2$, $b = 1$ had been obtained earlier by BANG (1886). ZSIGMONDY's theorem was rediscovered many times (BIRKHOFF (1904), CARMICHAEL (1913), KANOLD (1950), ARTIN (1955), and LÜNEBURG (1981) who gave a simpler proof). For an accessible proof, see RIBENBOIM (1994)

Let $a > b \geq 1$, $\gcd(a, b) = 1$, and consider the sequence of binomials

$$(a^n - b^n)_{n \geq 0}.$$

If $P = a + b$, $Q = ab$, then $a^n - b^n = U_n(P, Q) \cdot (a - b)$. The prime p is called a *primitive factor* of $a^n - b^n$ if $p \mid a^n - b^n$ but $p \nmid a^m - b^m$ for all m, $1 \leq m < n$. Let $\text{Prim}(a^n - b^n)$ denote the set of all primitive factors of $a^n - b^n$. Clearly, if $n > 1$, then $\text{Prim}(a^n - b^n) = \text{Prim}(U_n(P, Q)) \setminus \{p \mid p \text{ divides } a - b\}$.

(3.17) Let $a > b \geq 1$, $\gcd(a, b) = 1$.

1. For every $n > 1$, the binomial $a^n - b^n$ has a primitive factor, except in the following cases:

$$a = 2,\ b = 1,\ n = 6 \text{ (this gives } 2^6 - 1 = 63\text{)},$$
$$a,\ b \text{ are odd, } a + b \text{ is a power of } 2,\ n = 2.$$

Moreover, each primitive factor of $a^n - b^n$ is of the form $kn + 1$.

2. For every $n > 1$, the binomial $a^n + b^n$ has a primitive factor, except for $a = 2$, $b = 1$, $n = 3$ (this gives $2^3 + 1 = 9$).

b The number of primitive factors

Now I consider the primitive part of terms of Lucas sequences and discuss the number of distinct prime factors of U_n^*, V_n^*. The following question remains open: Given $(P, Q) \in \mathcal{S}$, do there exist infinitely many $n \geq 1$ such that $\#(\text{Prim}(U_n)) = 1$, respectively $\#(\text{Prim}(V_n)) = 1$, that is, U_n^* (respectively V_n^*) is a prime power? This question is probably very difficult to answer. I shall discuss a related problem in the next subsection (c).

Now I shall indicate conditions implying

$$\#(\text{Prim}(U_n)) \geq 2 \quad \text{and} \quad \#(\text{Prim}(V_n)) \geq 2.$$

If c is any non-zero integer, let $k(c)$ denote the *square-free kernel* of c, that is, c divided by its largest square factor. If $(P, Q) \in \mathcal{S}$, let $M = \max\{P^2 - 4Q, P^2\}$, let $\kappa = \kappa(P, Q) = k(MQ)$, and define

$$\eta = \eta(P, Q) = \begin{cases} 1 & \text{if } \kappa \equiv 1 \pmod 4, \\ 2 & \text{if } \kappa \equiv 2 \text{ or } 3 \pmod 4. \end{cases}$$

SCHINZEL (1963a) proved (see also ROTKIEWICZ (1962) for the case when $Q > 0$ and $D > 0$):

(3.18) There exist effectively computable finite subsets $\mathcal{M}_0$, $\mathcal{N}_0$ of $\mathcal{S}$ and for every $(P, Q) \in \mathcal{S}$ an effectively computable integer $n_0(P, Q) > 0$ such that if $(P, Q) \in \mathcal{S}$, $\neq 1, 2, 3, 4, 6$, and $\frac{n}{\eta \kappa}$ is odd, then $\#(\text{Prim}(U_n(P, Q))) \geq 2$, with the following exceptions:

1. $D = P^2 - 4Q > 0$:

$$n = \eta \cdot |\kappa| \text{ and } (P, Q) \in \mathcal{M}_0 \,;$$
$$n = 3 \cdot \eta \cdot |\kappa| \text{ and } (P, Q) \in \mathcal{N}_0 \,;$$

$$(n, P, Q) = (2D, 1, -2), \ (2D, 3, 2)$$

2. $D = P^2 - 4Q < 0$:

$$(n, P, Q) \text{ with } n \le n_0(P, Q).$$

Thus, for each $(P, Q) \in \mathcal{S}$ there exist infinitely many n with $\#(\mathrm{Prim}(U_n(P, Q))) \ge 2$. SCHINZEL gave explicit finite sets $\mathcal{M}$, $\mathcal{N}$ containing respectively the exceptional set $\mathcal{M}_0$, $\mathcal{N}_0$, which were later completely determined by BRILLHART and SELFRIDGE, but this calculation remained unpublished. Later, I shall invoke the following corollary:

(3.19) Let $(P, Q) \in \mathcal{S}$ with Q a square and $D > 0$. If $n > 3$, then

$$\#(\mathrm{Prim}(U_n(P, Q))) \ge 2,$$

with the exception of $(n, P, Q) = (5, 3, 1)$.

Thus, in particular, $U_n(P, Q)$ is not a prime when $n > 3$ and Q is a square, except for $(n, P, Q) = (5, 3, 1)$.

Since $\mathrm{Prim}(U_n(P, Q)) \subseteq \mathrm{Prim}(V_n(P, Q))$, it is easy to deduce from **(3.16)** conditions which imply that $\#(\mathrm{Prim}(V_n(P, Q))) \ge 2$; in particular, for each $(P, Q) \in \mathcal{S}$ there are infinitely many such indices n.

These results have been strengthened in subsequent papers by SCHINZEL (1963), (1968), but it would be too technical to quote them here. It is more appropriate to consider:

Special Cases. Let $a > b \ge 1$ be relatively prime integers, let $P = a+b$, $Q = ab$, so $U_n(P, Q) = \frac{a^n - b^n}{a - b}$, $V_n(P, Q) = a^n + b^n$. Even for these special sequences it is not known if there exist infinitely many n such that $\#\mathrm{Prim}(U_n(P, Q)) = 1$, respectively $\#\mathrm{Prim}(V_n(P, Q)) = 1$.

SCHINZEL (1962b) showed the following result, which is a special case of **(3.16)**. Let $\kappa = k(a, b)$,

$$\eta = \begin{cases} 1 & \text{if } \kappa \equiv 1 \pmod{4}, \\ 2 & \text{if } \kappa \equiv 2 \text{ or } 3 \pmod{4}. \end{cases}$$

(3.20) Under the above hypotheses:

1. If $n > 20$ and $\frac{n}{\eta\kappa}$ is an odd integer, then $\# \operatorname{Prim}(\frac{a^n - b^n}{a-b}) \geq 2$.
2. If $n > 10$, κ is even, and $\frac{n}{\kappa}$ is an odd integer, then $\# \operatorname{Prim}(a^n + b^n) \geq 2$.

Thus, there exist infinitely many n such that $\# \operatorname{Prim}(\frac{a^n - b^n}{a-b}) \geq 2$, respectively $\# \operatorname{Prim}(a^n + b^n) \geq 2$. SCHINZEL also showed:

(3.21) With the above hypotheses, if $\kappa = c^h$ where $h \geq 2$ when $k(c)$ is odd, and $h \geq 3$ when $k(c)$ is even, then there exist infinitely many n such that $\# \operatorname{Prim}(\frac{a^n - b^n}{a-b}) \geq 3$.

However, for arbitrary (a, b) with $a > b \geq 1$, $\gcd(a, b) = 1$, it is not known if there exist infinitely many n with $\# \operatorname{Prim}(\frac{a^n - b^n}{a-b}) \geq 3$.

c   Powers dividing the primitive part

Nothing is known about powers dividing the primitive part, except that it is a rare occurrence. To size up the difficulty of the question, it is convenient to consider right away the very special case where $(P, Q) = (3, 2)$, so $U_n = 2^n - 1$, $V_n = 2^n + 1$. Recall that if $n = q$ is a prime, then $U_q = 2^q - 1$ is called a *Mersenne number*, usually denoted $M_q = U_q = 2^q - 1$. Also, if $n = 2^m$, then $V_{2^m} = 2^{2^m} + 1$ is called a *Fermat number* and the notation $F_m = V_{2^m} = 2^{2^m} + 1$ is used.

The following facts are easy to show: $\gcd(M_q, M_p) = 1$ when $p \neq q$, and $\gcd(F_m, F_n) = 1$ when $m \neq n$. It follows that M_q, F_m are equal to their primitive parts.

A natural number which is a product of proper powers is said to be a *powerful number*.

I indicate below several statements which are related, but have never been proved to be true.

(M)   There exist infinitely many primes p such that M_p is square-free.

(M')   There exist infinitely many primes such that M_p is not powerful.

(F)   There exist infinitely many n such that F_n is square-free.

(F')   There exist infinitely many n such that F_n is not powerful.

(B)   There exist infinitely many n such that the primitive part of $2^n - 1$ is square-free.

(B')   There exist infinitely many n such that the primitive part of $2^n - 1$ is not powerful.

(C)   There exist infinitely many n such that the primitive part of $2^n + 1$ is square-free.

(C′) There exist infinitely many n such that the primitive part of
$2^n + 1$ is not powerful.

I shall discuss these and related conjectures in Chapter 9 where
it will be explained why the proof of any of the above conjectures
should be very difficult.

d   The greatest prime factor of terms of Lucas sequences.

The problem of estimating the size of the greatest prime division of
terms of Lucas sequences has been the object of many interesting
papers.

If n is a natural number, let $P[n]$ denote the greatest prime factor
of n, and let $\nu(n)$ denote the number of distinct prime factors of n. So,
the number $q(n)$ of distinct square-free factors of n is $q(n) = 2^{\nu(n)}$.
There have also been studies to estimate the size of $Q[n]$, the largest
square-free factor of n, but I shall not consider this question.

For every $n \geq 1$, let $\Phi_n(X,Y) \in \mathbb{Z}[X,Y]$ be the nth homogenized
cyclotomic polynomial

$$\Phi_n(X,Y) = \prod_{\substack{\gcd(i,n)=1 \\ 1 \leq i \leq n}} (X - \zeta^i Y)$$

where ζ is a primitive nth root of 1; so, $\Phi_n(X,Y)$ has degree $\varphi(n)$
(the EULER totient of n).

If P, Q are non-zero integers, $D = P^2 - 4Q \neq 0$ and α, β the roots
of $X^2 - PX + Q$, then $\Phi_n(\alpha, \beta) \in \mathbb{Z}$ (for $n \geq 2$) and $\alpha^n - \beta^n = \prod_{d \mid n} \Phi_d(\alpha, \beta)$.

It follows easily that

$$P\left[\frac{\alpha^n - \beta^n}{\alpha - \beta}\right] \geq P[\Phi_n(\alpha,\beta)],$$

$$P[\alpha^n - \beta^n] \geq P[\Phi_n(\alpha,\beta)],$$

$$P[\alpha^n + \beta^n] \geq P[\Phi_{2n}(\alpha,\beta)].$$

Therefore, it suffices to find lower estimates for $P[\Phi_n(\alpha,\beta)]$.

The first result was given by ZSIGMONDY (1892) and again by
BIRKHOFF (1904): *If a, b are relatively prime integers, $a > b \geq 1$,
then $P[a^n - b^n] \geq n+1$ and $P[a^n + b^n] \geq 2n+1$ (with the exception
$2^3 + 1 = 9$).* SCHINZEL added to this result (1962): *If ab is a square*

*or the double of a square, then $P[a^n - b^n] \geq 2n+1$, except for $a = 2$,
$b = 1$, and $n = 4$, 6, 12.*

In his work on primitive factors of LUCAS sequences with $D > 0$,
CARMICHAEL (1913) showed that if $n > 12$, then $P[U_n] \geq n - 1$ and
$P[V_n] \geq 2n - 1$. ERDÖS (1965) conjectured:

$$\lim_{n \to \infty} \frac{P[2^n - 1]}{n} = \infty.$$

This problem, as well as related questions which are still unsolved,
has been extensively studied by STEWART (see STEWART (1975,
1977b); SHOREY (1981); STEWART (1982, 1985)). Several of the results which I shall describe concern the greatest prime factor when
the index n belongs to some set with asymptotic density 1.

A subset S of $\mathbb{N}$ has asymptotic density γ, $0 \leq \gamma \leq 1$, where

$$\lim_{N \to \infty} \frac{\#\{n \in S \mid n \leq N\}}{N} = \gamma.$$

For example, the set $\mathcal{P}$ of prime numbers has asymptotic density 0.

Combining the Prime Number Theorem with the fact that each
primitive factor of $\Phi_n(a, b)$ is of the form $hn + 1$ yields:

(3.22) There exists a set T of asymptotic density 1 such that

$$\lim_{\substack{n \to \infty \\ n \in T}} \frac{P[\Phi(a, b)]}{n} = \infty.$$

In particular, $\lim_{n \to \infty, n \in T} \frac{P[2^n - 1]}{n} = \infty$ where T is a set with asymptotic density 1. The above result was made more precise and extended
for sequences with arbitrary discriminant $D \neq 0$. Let $0 \leq \kappa \leq 1/\log 2$
and define the set

$$\mathcal{N}_\kappa = \{n \in \mathbb{N} \mid n \text{ has at most } \kappa \log \log n \text{ distinct prime factors}\}.$$

For example, $\mathcal{P} \subset \mathcal{N}_\kappa$, for every κ as above. A classical result
(see the book of HARDY and WRIGHT (1938)) is the following: If
$0 \leq \kappa \leq 1/\log 2$, then $\mathcal{N}_\kappa$ has asymptotic density equal to 1.

In other words, "most" natural numbers have "few" distinct prime
factors.

The following result is due to STEWART (1977b) for α, β real, and
to SHOREY (1981) for arbitrary α, β.

(3.23) Let κ, α, β be as above. If $n \in \mathcal{N}_\kappa$, $n \geq 3$, then

$$P[\Phi_n(\alpha, \beta)] \geq C\varphi(n)\frac{\log n}{q(n)}$$

where $C \geq 0$ is an effectively computable number depending only on α, β, and κ.

Recall that $q(n) = 2^{\nu(n)}$ and $\nu(n) \leq \kappa \log \log n$. It follows, with appropriate constants $C_1 > 0$ and $C_2 > 0$, that

$$P[\Phi_n(\alpha, \beta)] > C_1\frac{n \log n}{2^{\nu(n)} \log(1 + \nu(n))}$$

and

$$P[\Phi_n(\alpha, \beta)] > C_2\frac{n \log n^{1-\kappa \log 2}}{\log \log \log n}.$$

In particular, the above estimates hold for $n \in \mathcal{N}_\kappa$, $n > 3$, and each Lucas sequence $U_n(P, Q)$, $V_n(P, Q)$, and $\alpha^n - \beta^n$.

Since $\nu(p) = 1$ for each prime p, then

$$P[a^p - b^p] \geq Cp \log p,$$
$$P[a^p + b^p] \geq Cp \log p$$

(with appropriate $C > 0$). In particular, for the Mersenne numbers $M_p = 2^p - 1$,

$$P[2^p - 1] \geq Cp \log p,$$

and for the Fermat number $F_m = 2^{2^m} + 1$,

$$P[2^{2^m} + 1] \geq Cm \times 2^m,$$

but this estimate may also be obtained in a more direct way, as suggested by D. KNAYSWICK.

STEWART obtained also sharper, more technical expressions for lower bounds of $P[\Phi_n(\alpha, \beta)]$, and he conjectured that

$$P[\Phi(\alpha, \beta)] > C[\varphi(n)]^2$$

for α, β real, for all $n > 3$, where $C > 0$ is an effectively computable number (depending on α, β). This statement is true if n is square-free.

Using a more refined form of BAKER's lower bounds for linear forms in logarithms (as given by WALDSCHMIDT (1980)), STEWART (1982) proved the following result, valid for all $n > C_0$ (an absolute constant):

(3.24) For every $(P, Q) \in \mathcal{S}$ there exists an effectively computable number $C_1 = C_1(P, Q) > 0$ such that if $n > C_0$, then $P[U_n]$, $P[V_n]$ are bounded below by

$$\max\left\{n - 1, C_1 \frac{n \log n}{q(n)^{\frac{4}{3}}}\right\}.$$

The following result is non-effective, but gives sharper bounds on sets of asymptotic density 1 (STEWART (1982)):

(3.25) Let $f : \mathbb{N} \to \mathbb{R}_{>0}$ be any function such that $\lim f(n) = 0$. For each $(P, Q) \in \mathcal{S}$ there exists a set $T \subseteq \mathbb{N}$ of asymptotic density 1, such that if $n \in T$, then

$$P[U_n] \geq f(n) \frac{n(\log n)^2}{\log \log n}.$$

STEWART obtained further results about linear recurrence sequences other than Lucas sequences, and even for linear recurrence sequences of order greater than 2, but they fall beyond my scope. For a comprehensive survey, see STEWART (1985).

An interesting result related to these questions had already been obtained by MAHLER (1966):

(3.26) Let $Q \geq 2$, $D = P^2 - 4Q < 0$, and let E be a finite set of primes and denote by $E^{\times}[U_n]$ the largest factor of U_n, where prime factors all belong to E. If $0 < \epsilon < \frac{1}{2}$, there exists $n_0 > 1$ such that if $n > n_0$, then $\left|\frac{U_n}{E^{\times}[U_n]}\right| > Q^{(1/2-\epsilon)n}$. In particular, $\lim P[U_n] = \infty$.

The proof used p-adic methods.

4   Primes in Lucas sequences

Let U, V be the Lucas sequences with parameters $(P, Q) \in S$.

The main questions about primes in Lucas sequences are the following:

1. Does there exist $n > 1$ such that $U_n(P, Q)$, respectively $V_n(P, Q)$, is a prime?
2. Do there exist infinitely many $n > 1$ such that $U_n(P, Q)$, respectively $V_n(P, Q)$, is a prime?

I discuss the various possibilities, indicating what is known in the most important special cases.

The following is an example of a Lucas sequence with only one prime term, namely U_2:

$U(3,1)$: 0 1 3 8 21 55 144 377 987 ...

This was remarked after **(3.19)**. Similarly, if $a > b \geq 1$, with a, b odd, if $P = a + b$, $Q = ab$, then $V_n(P,Q) = a^n + b^n$ is even for every $n \geq 1$, so it is not a prime.

Applying CARMICHAEL's theorem **(3.11)** on the existence of primitive factors, it follows easily that:

(4.1) If $D > 0$ and $U_n(P,Q)$ is a prime, then $n = 2$, 4 or n is an odd prime. If $V_n(P,Q)$ is a prime, then n is a prime or a power of 2.

This result is not true if $D < 0$, as this example shows:
Let $(P,Q) = (1,2)$, so $D = -7$ and

$$U(1,2): 0\ 1\ 1\ -1\ -3\ -1\ 5\ 7\ -3\ -17\ -11\ 23\ 45\ -1\ -91\ -89\ \ldots$$

In this example, U_6, U_8, U_9, U_{10}, U_{15}, ..., are primes.
Similarly, in $V(1,2)$, for example, the terms $|V_9|$, $|V_{10}|$ are primes.

Special Cases. In (1999), DUBNER and KELLER indicated all the indices $n < 50000$ for which the Fibonacci number U_n, or the Lucas number V_n, are known to be prime: U_n is known to be a prime for $n = 3$, 4, 5, 7, 11, 13, 17, 23, 29, 43, 47, 83, 131, 137, 359, 431, 433, 449, 509, 569, 571, $2971^{(W)}$, $4723^{(M)}$, $5387^{(M)}$, $9311^{(DK)}$ [W: discovered by H. C. WILLIAMS; M: discovered by F. Morain; DK: discovered by H. DUBNER and W. KELLER].

Moreover, for $n < 50000$, U_n is a probable prime for $n = 9677$, 14431, 25561, 30757, 35999, 37511 (and for no other $n < 50000$). This means that these numbers were submitted to tests indicating that they are composite.

For $n \leq 50000$, V_n is known to be a prime for $n = 2$, 4, 5, 7, 8, 11, 13, 16, 17, 19, 31, 37, 41, 47, 53, 61, 71, 79, 113, 313, 353, $503^{(W)}$, $613^{(W)}$, $617^{(W)}$, $863^{(W)}$, $1097^{(DK)}$, $1361^{(DK)}$, $4787^{(DK)}$, $4793^{(DK)}$, $5851^{(DK)}$, $7741^{(DK)}$, $10691^{(DK)}$, $14449^{(DK)}$ [W: discovered by H. C. WILLIAMS; DK: discovered by H. DUBNER and W. KELLER].

Moreover, V_n is a probable prime for $n = 8467$, 12251, 13963, 19469, 35449, 36779, 44507 (and for no other $n \leq 50000$).

Due to the size of the probable primes, an actual prime certification is required to be done.

The paper of DUBNER and KELLER contains a lot more factorizations; it is a continuation of previous work of numerous other mathematicians; we call attention to JARDEN (1958), the edition of JARDEN's book by BRILLHART (1973), and the paper by BRILLHART (1988) which contains complete factorizations of U_n (for $n \leq 1000$) and of V_n (for $n \leq 500$).

If $a = 2$, $b = 1$, the associated Lucas sequences are $U_n = 2^n - 1$ and $V_n = 2^n + 1$.

Now, if U_n is a prime, then $n = q$ is a prime, and $M_q = U_q = 2^q - 1$ is a prime Mersenne number. If V_n is a prime, then $n = 2^m$, and $F_m = 2^{2^m} + 1$ is a prime Fermat number.

Up to now, only 37 Mersenne primes are known, the largest one being M_{302137}, proved prime in 1999; it has more than 2 million digits. On the other hand, the largest known Fermat prime number is F_4. For a detailed discussion of Mersenne numbers and Fermat numbers, see my book *The Little Book of Big Primes* (1991a) or the up-to-date Brazilian edition (1994).

It is believed that there exist infinitely many Mersenne primes. Concerning Fermat primes, there is insufficient information to support any conjecture.

5 Powers and powerful numbers in Lucas sequences

In this section, I deal with the following questions. Let U, V be the Lucas sequences with parameters $(P, Q) \in \mathcal{S}$. Let $k \geq 1$, $h \geq 2$, and consider the set

$$\mathcal{C}_{U,k,h} = \{U_n \mid U_n = kx^h, \text{ with } |x| \geq 2\}.$$

Let $\mathcal{C}_{U,k} = \bigcup_{h \geq 2} \mathcal{C}_{U,k,h}$, so $\mathcal{C}_{U,k}$ consists of all U_n of the form $U_n = kx^h$ for some $|x| \geq 2$ and $h \geq 2$. If $k = 1$, one obtains the set of all U_n that are proper powers.

Similarly, let

$$\mathcal{C}^*_{U,k} = \{U_n \mid U_n = kt \text{ where } t \text{ is a powerful number}\}.$$

If $k = 1$, one obtains the set of all U_n which are powerful numbers.

Corresponding definitions are made for the sets $\mathcal{C}_{V,k,h}$ and $\mathcal{C}_{V,k}*$ associated to the sequence V.

The basic question is to find out if, and when, the above sets are empty, finite, or infinite, and, whenever possible, to determine the sets explicitly.

A related problem concerns the square-classes in the sequences U, V.

U_n, U_m are said to be *square-equivalent* if there exist integers $a, b \neq 0$ such that $U_m a^2 = U_n b^2$ or, equivalently, $U_m U_n$ is a square. This is clearly an equivalence relation on the set $\{U_n \mid n \geq 1\}$ whose classes are called the *square-classes of the sequence U*. If U_n, U_m are in the same square-class, and if $d = \gcd(U_n, U_m)$, then $U_m = dx^2$, $U_n = dy^2$, and conversely.

The square-classes of the sequence V are defined in a similar manner.

Concerning square-classes, the problems are the same: to determine if there are square-classes which are not trivial, that is, having more than one element; next, to ascertain if there are only finitely many nontrivial square-classes, if a square-class may be finite and, if possible, to determine explicitly the square-classes.

If $k \geq 1$, the notation $k\square$ indicates a number of the form kx^2, with $x \geq 2$; thus, $\square$ indicates a square greater than 1.

The first results on these questions were the determinations of those Fibonacci and Lucas numbers that are squares. This was achieved using rather elementary, but clever, arguments. In my presentation, I prefer to depart from the order in which the subject unfolded, and, instead, to give first the general theorems.

A   General theorems for powers

The general theorem of SHOREY (1981, 1983) (valid for all non-degenerate binary recurrence sequences) was proved using sharp lower bounds for linear forms in logarithms by BAKER (1973), plus a p-adic version by VAN DER POORTEN (1977), assisted by another result of KOTOV (1976).

A result of SHOREY (1977) may also be used, as suggested by PETHÖ.

(5.1) Let $(P, Q) \in \mathcal{S}$, $k \geq 1$. There exists an effectively computable number $C = C(P, Q, k) > 0$ such that if $n \geq 1$, $|x| \geq 2$, $h \geq 2$ and

$U_n = kx^h$, then n, $|x|$, $h < C$. A similar statement holds for the sequence V.

In particular, in a given Lucas sequence there are only finitely many terms which are powers.

STEWART's paper (1980) contains also the following result, suggested by MIGNOTTE and WALDSCHMIDT. For $h \geq 2$, $n \geq 1$, let $[n]^h$ denote the h-power closest to n.

(5.2) If $Q = \pm 1$, then

$$\lim_{n \to \infty} |U_n - [U_r]^h| = \infty.$$

This is achieved by showing that for every d, there exists an effectively computable number $C = C(P, d) > 0$ such that if $U_n = x^h + d$ with $|x| \geq 1$, $h \geq 2$, then n, $|x|$, $h < C$.

The above general results are not sufficient to determine explicitly all the terms U_n of the form kx^h, because the bounds indicated are too big.

PETHÖ (1982) gave the following extension of **(5.1)** (valid for all non-degenerate binary recurrences):

(5.3) Let E be a finite set of primes, $E^\times$ the set of integers all of whose prime factors belong to E. Given $(P, Q) \in \mathcal{S}$, there exists an effectively computable number $C > 0$, depending only on P, Q, and E, such that if $n \geq 1$, $|x| \geq 2$, $h \geq 2$, $k \in E^\times$, and $U_n = kx^h$, then n, $|x|$, h, $k \geq C$. A similar result holds for the sequence V.

B   Explicit determination in special sequences

Now I shall consider special sequences, namely, those with parameters $(1, -1)$ (the Fibonacci and Lucas numbers), those with parameters $(2, -1)$ (the Pell numbers), and those with parameters $(a + 1, a)$, where $a > 1$, in particular with parameters $(3, 2)$.

The questions to be discussed concern squares, double squares, other multiples of squares, square-classes, cubes, and higher powers.

The results will be displayed in a table (see page 35).

a   Squares

The only squares in the sequence of Fibonacci numbers are $U_1 = U_2 = 1$ and $U_{12} = 144$. This result was proved independently in 1964 by COHN and WYLER.

The only square in the sequence of Lucas numbers is $V_3 = 4$, proved by COHN (1964a).

One proof uses only divisibility properties and algebraic identities involving the Fibonacci and Lucas numbers. Another proof is based on the solution of the equations $X^2 - 5Y^4 = \pm 4$, $X^4 - 5Y^2 = \pm 4$.

For the parameters $(P, Q) = (2, -1)$, which give the sequences of Pell numbers, it is easy to see that V_n is never a square. The only U_n (with $n > 1$) which is a square is $U_7 = 169$. The proof follows from a study of the equation $X^2 - 2Y^4 = -1$, which was the object of a long paper by LJUNGGREN (1942c). ROBBINS reported this result in (1984) and it was again discovered by PETHÖ (1991) using a method of Diophantic approximation and computer calculations.

Let $a \geq 2$, $P = a + 1$, and $Q = a$. NAGELL (1921a) (and LJUNG-GREN (1942c), who completed the work) proved: If $\frac{a^n - 1}{a - 1}$ is a square, and $n > 1$, then $(a, n) = (3, 5)$ or $(7, 4)$.

KO (1960, 1964) proved: If $a^n + 1$ is a square, then $(a, n) = (2, 3)$. This result answered a long-standing problem.

A short proof of Ko's theorem is due to CHEIN (1976); another one was given by ROTKIEWICZ (1983) involving the computation of Jacobi symbols.

Detailed proofs of the above results are given in my book *Catalan's Conjecture* (1994).

The special case of parameters $(3, 2)$ gives the numbers $U_n = 2^n - 1$, $V_n = 2^n + 1$, and it is very easy to see that $2^n - 1 = \square$ only for $n = 1$, and $2^n + 1 = \square$ only for $n = 3$.

b Double squares

COHN (1964b) showed for Fibonacci numbers U_n and Lucas numbers V_n:

If $U_n = 2\square$, then $n = 3$ or 6, giving $U_3 = 2$, $U_6 = 8$.

If $V_n = 2\square$, then $n = 0$ or 6, giving $V_0 = 2$, $V_6 = 18$.

I have not found in the literature the determination of the Pell numbers $U_n(2, -1)$, $V_n(2, -1)$, $\frac{a^n - 1}{a - 1}$, $a^n + 1$ which are double squares (apart from the trivial cases).

c Square-classes

COHN (1972) determined the square-classes of Fibonacci and Lucas numbers (and even of more general sequences). In (1989a), I used another method to solve this problem:

The square-classes of Fibonacci numbers consist all of one number, except $\{U_1, U_2, U_{12}\}$ and $\{U_3, U_6\}$.

The square-classes of Lucas numbers consist only of one number, except $\{V_1, V_3\}$, $\{V_0, V_6\}$.

The determination of the square-classes of sequences of Pell numbers remains to be done.

For the square-classes of the sequences $U_n = \frac{a^n - 1}{a - 1}$, $V_n = a^n + 1$ $(n \geq 1)$, see RIBENBOIM (1989b).

The square-classes of the sequence U consist all of only one number. If a is even, the square-classes of V are also reduced to one element. Furthermore, there is an effectively computable number $C > 0$ such that if

$$(a^n + 1)(a^m + 1) = \square$$

with $m \neq n$, a odd, then a, m, $n < C$. So, only finitely many square-classes are not trivial and they are all finite.

d   Numbers of the form $k\square$ with $k \geq 3$

Let $k \geq 3$, assumed without loss of generality to be square-free. Often, k is taken to be an odd prime.

I have mentioned some papers concerning the special Lucas sequences with terms of the form $k\square$. On this matter, it is unavoidable to be incomplete and I wish to apologize to any author whose work I did not report.

On Fibonacci numbers, respectively Lucas numbers, of the form $p\square$ (where p is an odd prime) there are papers by STEINER (1980), ROBBINS (1983a), and GOLDMAN (1988).

STEINER showed that if $U_n = 3\square$, then $n = 4$. ROBBINS proved that if $U_n = p\square$, where p is a prime, $p \equiv 3 \pmod{4}$ or $3 < p < 10000$, then $p = 3001$. GOLDMAN showed that if $p = 3$, 7, 47 or 2207, and the Lucas number $V_n = p\square$, then $V_n = p$; note that then $n = 2^e$ (with $e = 1, 2, 3, 4$).

For the sequence $\frac{a^n - 1}{a - 1}$, $(n \geq 0, a \geq 2)$, there is also a partial result by ROTKIEWICZ (1983): if $a \equiv 0$ or $3 \pmod{4}$ and $n > 1$, n odd, then $\frac{a^n - 1}{a - 1} \neq n\square$. This is obtained using the calculation of Jacobi symbols.

e   Cubes

LONDON and FINKELSTEIN (1969) showed that the only Fibonacci cubes are $U_1 = U_2 = 1$ and $U_6 = 8$, while the only Lucas number which is a cube is $V_1 = 1$. The proof by LONDON and FINKELSTEIN requires the explicit solution of the cubic diophantine equations $x^2 \pm 100 = y^3$, subject to certain conditions. The latter result was obtained by LAGARIAS (1981) as well as by PETHÖ (1983) with a different proof using WALDSCHMIDT's form (1980) of the lower bound for linear forms in logarithms, followed by computer calculations. PETHÖ also gave results about Fibonacci numbers of the form px^3 or p^2x^3. For Pell numbers, PETHÖ (1991) showed that for $n > 1$, $U_n(2, -1)$ is never a cube.

NAGELL (1920, 1921b) (work completed by LJUNGGREN (1942a, 1943)) showed that if $\frac{a^n-1}{a-1}$ is a cube, with $n = 3$, then $a = 18$; moreover, if $n > 3$, then $n \not\equiv -1 \pmod 6$, which is just a partial result.

The work of NAGELL and LJUNGGREN also showed that $a^n + 1$ is a cube only in trivial cases.

These results are of course trivial for the numbers $2^n - 1$, $2^n + 1$, which cannot be cubes. They were given by GÉRONO (1870).

f   Higher powers

Nobody has as yet found any power higher than a cube among Fibonacci or Lucas numbers (except, trivially, 1).

In (1978) and (1983b), ROBBINS showed, if $q \geq 5$, q a prime, and if n is the smallest index such that the Fibonacci number U_n is a qth power, then n is a prime. Thus, if p is a prime dividing U_n, then $n = \rho_U(p)$, but also $p^q \mid U_n$, a fact which seems very unlikely to happen. The same result was also obtained in (1983) by PETHÖ.

PETHÖ (1991) showed also that a Pell number $U_n(2, -1)$ (with $n > 1$) is not a power (higher than a square).

The work of NAGELL and LJUNGGREN already quoted gives: If $\frac{a^n-1}{a-1} = y^m$ where $m > 3$, $n \geq 3$, then $n \neq 3$. Moreover, from NAGELL (1920) and LJUNGGREN (1943), necessarily 3 and 4 do not divide n when $m > 3$ (this is a partial result only).

INKERI communicated to me: If $\frac{a^n-1}{a-1}$ is a pth power (with $a > 1$, $n > 1$ and p a prime), then the p-adic value $v_p(a) \neq 1$ (see the proof in my book *Catalan's Conjecture* (1994), page 120).

The problem to determine if $a^n + 1$ can be equal to a higher power, or the similar problem for $a^n - 1$, amounts to the determination of all consecutive powers of integers. CATALAN (1844) conjectured that 8 and 9 are the only consecutive powers. This problem remains open, and my book *Catalan's Conjecture* (already quoted) is entirely devoted to this question. Let it be said here only that, with a clever use of BAKER's lower bounds for linear forms in logarithms, TIJDEMAN (1976) showed:

(5.4) There exists an effectively computable number $C > 0$ such that if $a^n + 1 = b^m$ with $a, b \geq 1$, $m \geq 2$, then $a, b, m, n < C$.

LANGEVIN (1976) calculated an upper bound for C:

$$C < e^{e^{e^{e^{e^{730}}}}}$$

which is beyond what imagination can dare.

It would be desirable to lower his bound so that numerical computer calculations may eventually confirm Catalan's conjecture.

Of course, it is easy to show for the special sequence of numbers $2^n - 1$, $2^n + 1$ that they are not higher powers (different from 1). This was done by GÉRONO (1870).

g   Addendum on repunits

A number is called a *repunit* if all its digits in base 10 are equal to 1. Such numbers are of the form

$$\frac{10^n - 1}{10 - 1} = U_n(11, 10).$$

A repunit (different from 1) is not a square, nor a fifth power. This follows from INKERI's result, already quoted. An independent proof was given by BOND (see my book *Catalan's Conjecture* (1994), page 120).

INKERI (1972) showed that a repunit (different from 1) is not a cube. Another proof was given by ROTKIEWICZ (1981) (see *Catalan's Conjecture*, pages 119, 120).

The question of the determination of repunits which are powers has now been completely solved—only the trivial repunit 1 is a power. This result is in a reprint of BUGEAUD (1999). The proof requires bounds in linear forms in two p-adic logarithms plus extensive computations with modular techniques to solve Thue equations.

Sequences	Fibonacci	Lucas	$U_n(2,-1)$	$V_n(2,-1)$	$U_n(3,2)$	$V_n(3,2)$	$\dfrac{a^n-1}{a-1}$ $(a>2)$	a^n+1 $(a>2)$
□	**!** Cohn Wyler	**!** Cohn	**!** Ljungren	**!** Ljungren	**!** trivial	**!** Frénicle de Bessy	**!** Nagell Ljungren	**!** Ko
2□	**!** Cohn	**!** Cohn	**?**	**?**	**!** trivial	**!** trivial	**?**	**?**
Square classes	**!** Cohn Ribenboim	**!** Cohn Ribenboim	**?**	**?**	**!** trivial	**!** trivial	**!** Ribenboim	**!?** Ribenboim
Cubes	**!** London and Finkelstein	**!** London and Finkelstein	**!** Pethö	**?**	**!** Gérono	**!** Gérono	**!?** Nagell Ljungren	**!** Nagell Ljungren
Higher Powers	**!?** Shorey and Stewart or Pethö				**!** Gérono	**!** Gérono	**!?** Nagell	**!?** Tijdeman

h Recapitulation

It is perhaps a good idea to assemble in a table the various results about special Lucas sequences discussed about.

The sign (!) indicates that the problems has been solved; (?) means the problem is completely open, or that I could not find it treated in the literature; the sign (!?) means that only partial results are known, cases remaining still unsettled.

C Uniform explicit determination of multiples, squares, and square-classes for certain families of Lucas sequences

It is an interesting and somewhat unexpected feature in the determination of squares, double-squares, and square-classes, that certain infinite families of Lucas sequences can be treated at the same time, providing uniform results.

In a series of papers, COHN (1966, 1967, 1968, 1972) has linked this problem to the solution of certain quartic equations where he obtained results for all (non-degenerate) sequences with parameters $(P, \pm 1)$, where $P \geq 1$ is odd.

Some results are also valid for a certain infinite, but thin, set of even parameter P, as will be soon indicated.

McDANIEL and I have devised a new method, involving the computation of Jacobi symbols, applicable to parameters (P, Q) with P, Q odd, $P \geq 1$, $\gcd(P, Q) = 1$, and $D > 0$.

These results were announced in (1992), and detailed proofs will soon appear.

a   Squares and double squares

The next results are by McDANIEL and RIBENBOIM .

It is assumed that $P \geq 1$, P, Q are odd, $\gcd(P, Q) = 1$, and $D = P^2 - 4Q > 0$.

(5.5)   1. If $U_n = \square$, then $n = 1$, 2, 3, 6, or 12.
   2. $U_2 = \square$ if and only if $P = \square$.
   3. $U_3 = \square$ if and only if $P^2 - Q = \square$.
   4. $U_6 = \square$ if and only if $P = 3\square$, $P^2 - Q = 2\square$, $P^2 - 3Q = 6\square$.
   5. $U_{12} = \square$ if and only if $P = \square$, $P^2 - Q = 2\square$, $P^2 - 2Q = 3\square$, $P^2 - 3Q = \square$, and $(P^2 - 2Q)^2 - 3Q^2 = 6\square$.

The determination of all allowable (P, Q) for which $U_3(P, Q) = \square$ is obvious, and clearly there are infinitely many such pairs (P, Q).

(5.6) The set of allowable parameters (P, Q) for which $U_6(P, Q) = \square$ is parameterized by the set $\{(s, t) \mid \gcd(s, t) = 1,\ s$ even, t odd, $st \equiv 1 \pmod{3}\}$ by putting

$$P = \frac{(s^2 - t^2)^2}{3}, \qquad Q = (a^2 - b^2)^2 - \frac{8(a^2 + b^2 + ab)^2}{q}$$

with

$$a = \frac{2(s^2 + t^2 + st)}{3}, \qquad b = \frac{s^2 + t^2 + st}{3},$$

and three other similar forms for P, Q (not listed here for brevity). In particular, there are infinitely many (P, Q) for which $U_6(P, Q) = \square$.

$(P, Q) = (1, -1)$ is the only known pair such that $U_{12}(P, Q) = \square$. It is not known if the system of equations given in **(5.5)** part (5) admits other nontrivial solution.

(5.7)   1. If $U_n = 2\square$, then $n = 3$ or 6.
   2. $U_3 = 2\square$ if and only if $P^2 - Q = 2\square$.
   3. $U_6 = 2\square$ if and only if $P = \square$, $P^2 - Q = 2\square$, and $P^2 - 3Q = \square$.

The set of allowable parameters (P, Q) for which $U_3(P, Q) = 2\square$ is clearly infinite and easily parameterized.

The set of allowable (P, Q) for which $U_6(P, Q) = 2\square$ is not completely known. However, the subset of all $(1, Q)$ for which $U_6(1, Q) = 2\square$ may be parameterized and shown to be infinite.

Concerning the sequence V, the results are the following:

(5.8)   1. If $V_n = \square$, then $n = 1, 3,$ or 5.
   2. $V_3 = \square$ if and only if $P = \square$.
   3. $V_3 = \square$ if and only if both P and $P^2 - 3Q$ are squares, or both P and $P^2 - 3Q$ are $3\square$.
   4. $V_5 = \square$ if and only if $P = 5\square$ and $P^4 - 5P^2Q + 5Q^2 = 5\square$.

(5.9) The set of all allowable (P, Q) for which $V_3(P, Q) = \square$ is infinite and parameterized as follows:

First type: $P = s^2$, $Q = \frac{s^4 - t^2}{3}$ where s is odd, t even, 3 does not divide st, $\gcd(s, t) = 1$, and $s^2 < 2t$;

Second type: $P = 3s^2$, $Q = 3s^4 - t^2$, where s is odd, t is even, 3 divides s, $\gcd(s, t) = 1$, and $\sqrt{3}s^2 < 2t$.

(5.10) The set of all allowable (P, Q) for which $V_5(P, Q) = \square$ is infinite and parameterized as follows:

First type: $P = 5s^2t^2$, $Q = -\frac{s^8 - 50s^4t^4 + 125t^8}{4}$ where s, t are odd, 5 does not divide s, $\gcd(s, t) = 1$, and $|s| > \left[\frac{25 + 5\sqrt{5}}{2}\right]^{\frac{1}{4}} t$.

Second type: $P = s^2t^2$, $Q = -\frac{5(s^8 - 10s^4t^4 + 5t^8)}{4}$ where s, t are odd, 5 does not divide s, $\gcd(s, t) = 1$, and $|s| > \left[\frac{49 + \sqrt{1901}}{10}\right]^{\frac{1}{4}} t$.

(5.11)   1. If $V_n = 2\square$, then $n = 3$ or 6.
   2. $V_3 = 2\square$ if and only if either $P = \square$, $P^2 - 3Q = 2\square$, or $P = 3\square$, $P^2 - 3Q = 6\square$.

3. $V_6 = 2\square$ if and only if $P^2 - 2Q = 3\square$, and $(P^2 - 2Q)^2 - 3Q^2 = 6\square$.

(5.12) The set of all allowable (P, Q) for which $V_6(P, Q) = 2\square$ is infinite and parameterized as follows: $P = s^2$, $Q = 3s^4 - 2t^2$ where s is odd, $\gcd(s, t) = 1$, 3 does not divide s, and $\sqrt{6}s^2 < 4t$.

At my request, J. TOP determined the pairs (P, Q) for which $V_6(P, Q) = 2\square$ (see the paper of MCDANIEL and RIBENBOIM already quoted):

(5.13) The allowable (P, Q) for which $V_6(P, Q) = 2\square$ correspond to the rational points of a certain elliptic curve with group of rational points isomorphic to $(\mathbb{Z}/2) \times \mathbb{Z}$. These points give rise to infinitely many pairs of allowable parameters. $(P, Q) = (1, -1)$ corresponds to the points of order 2; $(5, -1)$ corresponds to the generator of the subgroup of infinite order.

Other solutions may be calculated from the group law, that is, with the classical chord and tangent method. Thus

$$(P, Q) = (29, -4801), (4009, 3593279), (58585, -529351744321), \ldots$$

are also possible parameters.

It is much more difficult to deal with the case where P or Q is even. The first known results are due to COHN (1972).

(5.14) Let $Q = -1$ and $P = V_m(A, -1)$, where A is odd, $m \equiv 3 \pmod 6$.
1. If $U_n(P, -1) = \square$, then $n = 1$ or $n = 2$, and $P = 4$ or 36.
2. If $U_n(P, -1) = 2\square$, then $n = 4$, $P = 4$.
3. If $V_n(P, -1) = \square$, then $n = 1$, $P = 4$ or 36.
4. If $U_n(P, -1) = 2\square$, then $n = 2$, and $P = 4$ or 140.

(5.15) Let $Q = 1$ and $P = V_m(A, 1)$ where A is odd and 3 divides m.
1. If $U_n(P, 1) = \square$, then $n = 1$.
2. If $U_n(P, 1) = 2\square$, then $n = 2$, and $P = 18$ or 19602.
3. $V_n(P, 1) = \square$ is impossible.
4. If $V_n(P, 1) = 2\square$, then $n = 1$, and $P = 18$ or 19602.

Note that there are infinitely many even $P = V_m(A, -1)$ with A odd, $m \equiv 3 \pmod 6$, but this set is thin.

For example, for $P < 6000$ the only possibilities are 4, 36, 76, 140, 364, 756, 1364, 2236, 3420, 4964. A similar remark applies to the numbers $P = V_n(A, 1)$, where A is odd and 3 divides m.

In 1983, ROTKIEWICZ published the following partial, but remarkable result:

(5.16) If P is even, $Q \equiv 1 \pmod 4$, $\gcd(P, Q) = 1$, and if $U_n(P, Q) = \square$, then either n is an odd square or n is an even integer, not a power of 2, whose largest prime factor divides the discriminant D.

McDANIEL and RIBENBOIM (1998b) used the result of ROTKIEWICZ to show:

(5.17) Let P be positive and even, let $Q \equiv 1 \pmod 4$ with $D = P^2 - 4Q > 0$, $\gcd(P, Q) = 1$ and let $U_n(P, Q) = \square$. Then n is a square, or twice an odd square; all prime factors of n divide D; if $p^t > 2$ is a prime power dividing n, then for $1 \le u < t$, $U_{p^u} = p\square$ when u is even, and $U_{p^u} = p\square$ when u is odd. If n is even and $U_n = \square$, then, in addition, $p = \square$ or $p = 2\square$.

b Square-classes

In (1992), together with McDANIEL, I proved the following result was proved:

(5.18) Let $(P, Q) \in \mathcal{S}$. Then for every $n > 0$ there exists an effectively computable integer $C_n > 0$, depending on P, Q, n, such that if $n < m$ and $U_n(P, Q)U_m(P, Q) = \square$, or $V_n(P, Q)V_m(P, Q) = \square$, then $M < C_n$.

In particular, all square-classes in sequences U, V are finite.

For parameters $(P, 1)$, $(P, -1)$, with P odd, COHN (1972) used his results on certain quartic equations of type $X^4 - DY^2 = \pm 4, \pm 1$ or $X^2 - DY^4 = \pm 4, \pm - 1$, to obtain results on square-classes:

(5.19) Let $P \ge 1$ be odd.

1. If $1 \le n < m$ and $U_n(P, -1)U_m(P, -1) = \square$, then

$$
\begin{array}{lll}
n = 1, & m = 2, & P = \square, \quad \text{or} \\
n = 1, & m = 12, & P = 1, \quad \text{or} \\
n = 3, & m = 6, & P = 1, \quad \text{or} \\
n = 3, & m = 6, & P = 3.
\end{array}
$$

2. If $P \geq 3$, $1 \leq n \leq m$, and $U_n(P, 1)U_m(P, 1) = \square$, then
$$n = 1, \quad m = 6, \quad P = 3, \quad \text{or}$$
$$n = 1, \quad m = 2, \quad P = \square.$$

(5.20) Let $P \geq 1$ be odd.

1. If $0 \leq n < m$ and $V_n(P, 1)V_m(P, 1) = \square$, then
$$n = 0, \quad m = 6, \quad P = 1, \quad \text{or}$$
$$n = 1, \quad m = 3, \quad P = 1, \quad \text{or}$$
$$n = 0, \quad m = 6, \quad P = 5.$$

2. If $P \geq 3$, $0 \leq n < m$, and $V_n(P, 1)V_m(P, 1) = \square$, then
$n = 0$, $m = 3$, $P = 3$ or 27.

A very special case, but with a more direct proof, was given later by ANDRÉ-JEANNIN (1992).

The following theorem was proved by McDANIEL (1998a):

(5.21) Let $P > 0$, $Q \neq 0$, $\gcd(P, Q) = 1$, $D = P^2 - 4Q > 0$. Assume that P, Q are odd.

1. (a) If $1 < m < n$ and $U_m U_n = \square$, then $(m, n) \in \{(2, 3), (2, 12), (3, 6), (5, 10)\}$ or $n = 3m$,

   (b) If $1 < m$, $U_m U_{3m} = \square$, then m is odd, $3 \nmid m$, $Q \equiv 1 \pmod 4$, $\left(\frac{-Q}{P}\right) = +1$, and $P < |Q + 1|$.

   (c) If P, $m > 1$ are given, there exists an effectively computable constant $C > 0$ such that if Q is as in the hypotheses, and if $U_m U_{3m} = \square$, then $|Q| < C$.

   (d) If P, Q are given as above, there exists an effectively computable $C > 0$ such that if $m > 1$ and $U_m U_{3m} = \square$, then $m < C$.

2. (a) If $1 < m < n$ and $V_m V_n = \square$, then $n = 3m$.

   (b) If $1 < m$ and $V_m V_{3m} = \square$, then m is odd, $3 \nmid m$, $Q \equiv 3 \pmod 4$, $3 \nmid P$, $\left(\frac{-3Q}{P}\right) = +1$ and, $P < |\frac{Q}{k} + k|$, where $k = \sqrt[5]{0.6} \approx 0.9$.

   (c) If $m > 1$ and P are given, there exists an effectively computable $C > 0$ such that if $Q \neq 0$ is as above and $V_m V_{3m} = \square$, then $|Q| < C$.

   (d) If P, Q are given as above, there exists an effectively computable $C > 0$ such that if $1 < m$ and $V_m V_{3m} = \square$, then $m < C$.

c Multiples of squares

There are only a few systematic results, mainly due to COHN (1972).

Let $k \geq 3$ be an odd square-free integer, let $P \geq 1$, with P odd. COHN studied the equations $U_n(P, -1) = k\square$, $U_n(P, -1) = 2k\square$, but could not obtain complete results.

Clearly, there exists a smallest index $r > 0$ such that k divides $U_r(P, -1)$. Since the square-classes have at most two numbers, as indicated before in this case, there exist at most two indices n such that $U_n(P, -1) = k\square$, respectively $2k\square$.

(5.22) With the above hypotheses and notations:

1. If $r \not\equiv 0 \pmod 3$ and $U_n = k\square$, then $n = r$, while $U_n = 2k\square$ is impossible.
2. If $r \equiv 3 \pmod 6$, $U_n(P, -1) = k\square$ is impossible, however, no solution was obtained for $U_n(P, -1) = 2k\square$ in this case.
3. If $n \equiv 0 \pmod 6$, and if the 2-adic value $v_2(r)$ is even, then $U_n(P, -1) = 2k\square$ is impossible; if $v_2(r)$ is odd, then $U_n(P, -1) = k\square$ is impossible except if $P = 5$, $n = 12$, $k = 455$. The other cases are left open.

COHN also stated that for $P \geq 3$, the equations $U_n(P, 1) = k\square$, respectively $2k\square$, can be treated similarly, with partial results.

D   Powerful numbers in Lucas sequences

Let $(P, Q) \in \mathcal{S}$, and let U, respectively V, be the Lucas sequences with parameters (P, Q). If U_n is a powerful number, and if p is a primitive factor of U_n, then p^2 divides U_n. This suggests that the set of indices n such that U_n is powerful should be finite. A similar remark applies to the sequence V.

A proof of this fact, based on MASSER's conjecture, is known for Fibonacci numbers and Lucas numbers.

MASSER's conjecture (1985), also called the (ABC) conjecture, is the following statement (see also OESTERLÉ (1988)):

Given $\epsilon > 0$, there exists a positive number $C(\epsilon)$ such that if a, b, c are positive integers with $\gcd(a, b) = 1$, $a + b = c$, if $g = \prod_{p \mid abc} p$, then $c < C(\epsilon)g^{1+\epsilon}$. It is a great challenge for mathematicians to prove the (ABC) conjecture. A much weaker form of the tantalizing (ABC) conjecture was proved by STEWART (1986). ELKIES (1991) showed that the (ABC) conjecture implies the famous theorem of

FALTINGS (establishing MORDELL's conjecture). It is also known that the (ABC) conjecture implies that there exist at most finitely many integers $n \geq 3$, x, y, $z \neq 0$, such that $x^n + y^n = z^n$. This is just short of proving Fermat's Last Theorem.

I learned the following from G. WALSH:

(5.23) If MASSER's conjecture is true, and if $k \geq 1$ is a given square-free integer, there exist only finitely many indices n such that the Fibonacci number U_n, or the Lucas number V_n, is of the form kt, where t is a powerful number.

The proof is short and simple.

For any integer $N = \prod_{i=1}^{r} p_i^{e_i}$ (where $p_1, \ldots, p_r$ are distinct primes and $e_1, \ldots, e_r \geq 1$), the *powerful part* of N is by definition

$$w(N) = \prod_{e_i > 1} p_i^{e_i}.$$

So, N is powerful exactly when $N = w(N)$.

In 1999, RIBENBOIM and WALSH proved, assuming the (ABC) conjecture to be true,

(5.24) Let U, V be Lucas sequences with positive discriminant. For every $\epsilon > 0$, the sets $\{n \mid w(U_n) > U_n^\epsilon\}$ and $\{n \mid w(V_n) > V_n^\epsilon\}$ are finite. In particular, each of the sequences U, V has only finitely many terms which are powerful.

Noteworthy special cases arise taking $P = 1$, $Q = -1$ (Fibonacci and Lucas numbers), $P = 2$, $Q = -1$ (Pell numbers), $P = 3$, $Q = 2$ and more generally $P = a + 1$ $Q = a$ (while $a > 1$). In particular, the (ABC) conjecture implies that there exist only finitely many Mersenne numbers M_q and Fermat numbers F_m which are powerful.

References

1202 Leonardo Pisano (Fibonacci). *Liber Abbaci* (²*1228*). Tipografia delle Scienze Matematiche e Fisiche, Rome, 1857 edition. B. Boncompagni, editor.

1657 Frénicle de Bessy. Solutio duorum problematum circa numeros cubos et quadratos. Bibliothèque Nationale de Paris.

1843 J. P. M. Binet. Mémoire sur l'intrégation des équations linéaires aux différences finies, d'un ordre quelconque, á coefficients variables. *C. R. Acad. Sci. Paris*, 17:559–567.

1844 E. Catalan. Note extraite d'une lettre addressée á l'éditeur. *J. reine u. angew. Math.*, 27:192.

1870 G. C. Gérono. Note sur la résolution en nombres entiers et positifs de l'équation $x^m = y^n + 1$. *Nouv. Ann. de Math. (2)*, 9:469–471, and 10:204–206 (1871).

1878 E. Lucas. Théorie des fonctions numériques simplement périodiques. *Amer. J. of Math.*, 1:184–240 and 289–321.

1886 A. S. Bang. Taltheoretiske Untersogelser. *Tidskrift Math.*, Ser. 5, 4:70–80 and 130–137.

1892 K. Zsigmondy. Zur Theorie der Potenzreste. *Monatsh. f. Math.*, 3:265–284.

1904 G. D. Birkhoff and H. S. Vandiver. On the integral divisors of $a^n - b^n$. *Ann. Math. (2)*, 5:173–180.

1909 A. Wieferich. Zum letzten Fermatschen Theorem. *J. reine u. angew. Math.*, 136:293–302.

1913 R. D. Carmichael. On the numerical factors of arithmetic forms $\alpha^n \pm \beta^n$. *Ann. of Math. (2)*, 15:30–70.

1920 T. Nagell. Note sur l'équation indéterminée $\frac{x^n-1}{x-1} = y^q$. *Norsk Mat. Tidsskr.*, 2:75–78.

1921a T. Nagell. Des équations indéterminées $x^2 + x + 1 = y^n$ et $x^2 + x + 1 = 3y^n$. *Norsk Mat. Forenings Skrifter, Ser. I*, 1921, No. 2, 14 pages.

1921b T. Nagell. Sur l'équation indéterminée $\frac{x^n-1}{x-1} = y^2$. *Norsk Mat. Forenings Skrifter, Ser. I*, 1921, No. 3, 17 pages.

1930 D. H Lehmer. An extended theory of Lucas' functions. *Ann. of Math.*, 31:419–448.

1935 K. Mahler. Eine arithmetische Eigenschaft der Taylor-Koeffizienten rationaler Funktionen. *Nederl. Akad. Wetensch. Amsterdam Proc.*, 38:50–60.

1938 G. H. Hardy and E. M. Wright. *An Introduction to the Theory of Numbers*. Clarendon Press, Oxford, 5th (1979) edition.

1942a W. Ljunggren. Einige Bemerkungen über die Darstellung ganzer Zahlen durch binäre kubische Formen mit positiver Diskriminante. *Acta Math.*, 75:1–21.

1942b W. Ljunggren. Über die Gleichung $x^4 - Dy^2 = 1$. *Arch. Math. Naturvid.*, 45(5):61–70.

1942c W. Ljunggren. Zur Theorie der Gleichung $x^2 + 1 = Dy^4$. *Avh. Norsk Vid. Akad. Oslo.*, 1(5):1–27.

1943 W. Ljunggren. New propositions about the indeterminate equation $\frac{x^n - 1}{x - 1} = y^q$. *Norsk Mat. Tidsskr.*, 25:17–20.

1950 H.-J. Kanold. Sätze über Kreisteilungspolynome und ihre Anwendungen auf einige zahlentheoretische Probleme. *J. reine u. angew. Math.*, 187:355–366.

1953 C. G. Lekkerkerker. Prime factors of elements of certain sequences of integers. *Nederl. Akad. Wetensch. Proc. (A)*, 56:265–280.

1954 M. Ward. Prime divisors of second order recurring sequences. *Duke Math. J.*, 21:607–614.

1955 E. Artin. The order of the linear group. *Comm. Pure Appl. Math.*, 8:335–365.

1955 M. Ward. The intrinsic divisors of Lehmer numbers. *Ann. of Math. (2)*, 62:230–236.

1958 D. Jarden. *Recurring Sequences*. Riveon Lematematike, Jerusalem. [3]1973, revised and enlarged by J. Brillhart, Fibonacci Assoc., San Jose, CA.

1960 A. A. Brauer. Note on a number theoretical paper of Sierpiński. *Proc. Amer. Math. Soc.*, 11:406–409.

1960 Chao Ko. On the Diophantine equation $x^2 = y^n + 1$. *Acta Sci. Natur. Univ. Szechuan*, 2:57–64.

1961 L. K. Durst. Exceptional real Lucas sequences. *Pacific J. Math.*, 11:489–494.

1961 M. Ward. The prime divisors of Fibonacci numbers. *Pacific J. Math.*, 11:379–389.

1962 A. Rotkiewicz. On Lucas numbers with two intrinsic prime divisors. *Bull. Acad. Polon. Sci. Sér. Sci. Math. Astron. Phys.*, 10:229–232.

1962a A. Schinzel. The intrinsic divisions of Lehmer numbers in the case of negative discriminant. *Ark. Math.*, 4:413–416.

1962b A. Schinzel. On primitive prime factors of $a^n - b^n$. *Proc. Cambridge Phil. Soc.*, 58:555–562.

1963a A. Schinzel. On primitive prime factors of Lehmer numbers, I. *Acta Arith.*, 8:213–223.

1963b A. Schinzel. On primitive prime factors of Lehmer numbers, II. *Acta Arith.*, 8:251–257.

1963 N. N. Vorob'ev. *The Fibonacci Numbers.* D. C. Heath, Boston.

1964a J. H. E. Cohn. On square Fibonacci numbers. *J. London Math.Soc.*, 39:537–540.

1964b J. H. E. Cohn. Square Fibonacci numbers etc. *Fibonacci Q.*, 2:109–113.

1964 Chao Ko. On the Diophantine equation $x^2 = y^n + 1$. *Scientia Sinica (Notes)*, 14:457–460.

1964 O. Wyler. Squares in the Fibonacci series. *Amer. Math. Monthly*, 7:220–222.

1965 J. H. E. Cohn. Lucas and Fibonacci numbers and some Diophantine equations. *Proc. Glasgow Math. Assoc.*, 7:24–28.

1965 P. Erdös. Some recent advances and current problems in number theory. In *Lectures on Modern Mathematics, Vol. III*, edited by T. L. Saaty, 169–244. Wiley, New York.

1965 H. Hasse. Über die Dichte der Primzahlen p, für die eine vorgegebene ganzrationale Zahl $a \neq 0$ von durch eine vorgegebene Primzahl $l \neq 2$ teilbarer bzw. unteilbarer Ordnung mod p ist. *Math. Annalen*, 162:74–76.

1966 J. H. E. Cohn. Eight Diophantine equations. *Proc. London Math. Soc. (3)*, 16:153–166, and 17:381.

1966 H. Hasse. Über die Dichte der Primzahlen p, für die eine vorgegebene ganzrationale Zahl $a \neq 0$ von gerader bzw. ungerader Ordnung mod p ist. *Math. Annalen*, 168:19–23.

1966 K. Mahler. A remark on recursive sequences. *J. Math. Sci.*, 1:12–17.

1966 E. Selmer. *Linear Recurrences over Finite Fields.* Lectures Notes, Department of Mathematics, University of Bergen.

1967 J. H. E. Cohn. Five Diophantine equations. *Math. Scand.*, 21:61–70.

1967 C. Hooley. On Artin's conjecture. *J. reine u. angew. Math.*, 225:209–220.

1968 J. H. E. Cohn. Some quartic Diophantine equations. *Pacific J. Math.*, 26:233–243.

1968 L. P. Postnikova and A. Schinzel. Primitive divisors of the expression $a^n - b^n$. *Math. USSR-Sb.*, 4:153–159.

1968 A. Schinzel. On primitive prime factors of Lehmer numbers, III. *Acta Arith.*, 15:49–70.

1969 V. E. Hoggatt. *Fibonacci and Lucas Numbers.* Houghton-Mifflin, Boston.

1969 R. R. Laxton. On groups of linear recurrences, I. *Duke Math. J.*, 36:721–736.

1969 H. London and R. Finkelstein (alias R. Steiner). On Fibonacci and Lucas numbers which are perfect powers. *Fibonacci Q.*, 7:476-481 and 487.

1972 J. H. E. Cohn. Squares in some recurrence sequences. *Pacific J. Math.*, 41:631–646.

1972 K. Inkeri. On the Diophantic equation $a\frac{x^n-1}{x-1} = y^m$. *Acta Arith.*, 21:299–311.

1973 A. Baker. A sharpening for the bounds of linear forms in logarithms, II. *Acta Arith.*, 24:33–36.

1973 H. London and R. Finkelstein (alias R. Steiner). *Mordell's Equation $y^2 - k = x^3$.* Bowling Green State University Press, Bowling Green, OH.

1974 A. Schinzel. Primitive divisions of the expression $A^n - B^n$ in algebraic number fields. *J. reine u. angew. Math.*, 268/269: 27–33.

1975 A. Baker. *Transcendental Number Theory.* Cambridge Univ. Press, Cambridge.

1975 C. L. Stewart. The greatest prime factor of $a^n - b^n$. *Acta Arith.*, 26:427–433.

1976 E. Z. Chein. A note on the equation $x^2 = y^n + 1$. *Proc. Amer. Math. Soc.*, 56:83–84.

1976 S. V. Kotov. Über die maximale Norm der Idealteiler des Polynoms $\alpha x^m + \beta y^n$ mit den algebraischen Koeffizenten. *Acta Arith.*, 31:210–230.

1976 M. Langevin. Quelques applications des nouveaux résultats de van der Poorten. *Sém. Delange-Pisot-Poitou,* 17e *année,* 1976, No. G12, 1–11.

1976 P. J. Stephens. Prime divisors of second order linear recurrences, I. and II. *J. Nb. Th.*, 8:313–332 and 333–345.

1976 R. Tijdeman. On the equation of Catalan. *Acta Arith.*, 29: 197–209.

1977 A. Baker. The theory of linear forms in logarithms. In *Transcendence Theory: Advances and Applications (Proceedings of a conference held in Cambridge 1976)*, edited by A. Baker and D. W. Masser, 1–27. Academic Press, New York.

1977 T. N. Shorey, A. J. van der Porten, R. Tijdeman, and A. Schinzel. Applications of the Gel'fond-Baker method to Diophantine equations. In *Transcendence theory: Advances and Applications*, edited by A. Baker and D. W. Masser, 59–77. Academic Press, New York.

1977a C. L. Stewart. On divisors of Fermat, Fibonacci, Lucas and Lehmer numbers. *Proc. London Math. Soc.*, 35:425–447.

1977b C. L. Stewart. Primitive divisors of Lucas and Lehmer numbers. In *Trancendence Theory: Advances and Applications*, edited by A. Baker and D. W. Masser, 79–92. Academic Press, New York.

1977 A. J. van der Poorten. Linear forms in logarithms in p-adic case. In *Transcendence Theory: Advances and Applications*, edited by A. Baker and D. W. Masser, 29–57. Academic Press, New York.

1978 P. Kiss and B. M. Phong. On a function concerning second order recurrences. *Ann. Univ. Sci. Budapest. Eötvös Sect Math.*, 21:119–122.

1978 N. Robbins. On Fibonacci numbers which are powers. *Fibonacci Q.*, 16:515–517.

1980 R. Steiner. On Fibonacci numbers of the form $v^2 + 1$. In *A Collection of Manuscripts Related to the Fibonacci Sequence*, edited by W. E. Hogatt and M. Bicknell-Johnson, 208–210. The Fibonacci Association, Santa Clara, CA.

1980 C. L. Stewart. On some Diophantine equations and related recurrence sequences. In *Séminaire de Théorie des Nombres Paris 1980/81 (Séminare Delange-Pisot-Poitou), Progress in Math.*, 22:317–321 (1982). Birkhäuser, Boston.

1980 M. Waldschmidt. A lower bound for linear forms in logarithms. *Acta Arith.*, 37:257–283.

1981 K. Györy, P. Kiss, and A. Schinzel. On Lucas and Lehmer sequences and their applications to Diophantine equations. *Colloq. Math.*, 45:75–80.

1981 J. C. Lagarias and D. P. Weissel. Fibonacci and Lucas cubes. *Fibonacci Q.*, 19:39–43.

1981 H. Lüneburg. Ein einfacher Beweis für den Satz von Zsigmondy über primitive Primteiler von $A^n - B^n$. In *Ge-*

ometries and Groups, Lect. Notes in Math., 893:219–222, edited by M. Aigner and D. Jungnickel. Springer-Verlag, New York.

1981 T. N. Shorey and C. L. Stewart. On divisors of Fermat, Fibonacci, Lucas and Lehmer numbers, II. *J. London Math. Soc.*, 23:17–23.

1982 K. Györy. On some arithmetical properties of Lucas and Lehmer numbers. *Acta Arith.*, 40:369–373.

1982 A. Pethö. Perfect powers in second order linear recurrences. *J. Nb. Th.*, 15:5–13.

1982 C. L. Stewart. On divisors of terms of linear recurrence sequences. *J. reine u. angew. Math.*, 333:12–31.

1983 A. Pethö. Full cubes in the Fibonacci sequence. *Publ. Math. Debrecen*, 30:117–127.

1983a N. Robbins. On Fibonacci numbers of the form px^2, where p is a prime. *Fibonacci Q.*, 21:266–271.

1983b N. Robbins. On Fibonacci numbers which are powers, II. *Fibonacci Q.*, 21:215–218.

1983 A. Rotkiewicz. Applications of Jacobi symbol to Lehmer's numbers. *Acta Arith.*, 42:163–187.

1983 T. N. Shorey and C. L. Stewart. On the Diophantine equation $ax^{2t} + bx^ty + cy^2 = 1$ and pure powers in recurrence sequences. *Math. Scand.*, 52:24–36.

1984 N. Robbins. On Pell numbers of the form px^2, where p is prime. *Fibonacci Q. (4)*, 22:340–348.

1985 J. C. Lagarias. The set of primes dividing the Lucas numbers has density 2/3. *Pacific J. Math.*, 118:19–23.

1985 D. W. Masser. Open problems. In *Proceedings Symposium Analytic Number Theory*, edited by W. W. L. Chen, London. Imperial College.

1985 C. L. Stewart. On the greatest prime factor of terms of a linear recurrence sequence. *Rocky Mountain J. Math.*, 15: 599–608.

1986 T. N. Shorey and R. Tijdeman. *Exponential Diophantine Equations*. Cambridge University Press, Cambridge.

1986 C. L. Stewart and R. Tijdeman. On the Oesterlé-Masser conjecture. *Monatshefte Math.*, 102:251–257.

1987 A. Rotkiewicz. Note on the Diophantine equation $1 + x + x^2 + \ldots + x^m = y^m$. *Elem. of Math.*, 42:76.

1988 J. Brillhart, P. L. Montgomery, and R. D. Silverman. Tables of Fibonacci and Lucas factorizations. *Math. of Comp.*, 50: 251–260.

1988 M. Goldman. Lucas numbers of the form px^2, where $p = 3$, 7, 47 or 2207. *C. R. Math. Rep. Acad. Sci. Canada*, 10: 139–141.

1988 J. Oesterlé. Nouvelles approches du "théorème" de Fermat. Séminaire Bourbaki, 40ème anée, 1987/8, No. 694, *Astérisque*, 161–162, 165–186.

1989a P. Ribenboim. Square-classes of Fibonacci numbers and Lucas numbers. *Portug. Math.*, 46:159–175.

1989b P. Ribenboim. Square-classes of $\frac{a^n-1}{a-1}$ and a^n+1. *J. Sichuan Univ. Nat. Sci. Ed.*, 26:196–199. Spec. Issue.

1989 N. Tzanakis and B. M. M. de Weger. On the practical solution of the Thue equation. *J. Nb. Th.*, 31:99–132.

1991 W. D. Elkies. *ABC* implies Mordell. *Internat. Math. Res. Notices (Duke Math. J.)*, 7:99–109.

1991 A. Pethö. The Pell sequence contains only trivial perfect powers. In *Colloquia on Sets, Graphs and Numbers, Soc. Math., János Bolyai*, 561–568. North-Holland, Amsterdam.

1991a P. Ribenboim. *The Little Book of Big Primes*. Springer-Verlag, New York.

1991b P. Ribenboim and W. L McDaniel. Square-classes of Lucas sequences. *Portug. Math.*, 48:469–473.

1992 R. André-Jeannin. On the equations $U_n = U_q x^2$, where q is odd and $V_n = V_q x^2$, where q is even. *Fibonacci Q.*, 30: 133–135.

1992 W. L. McDaniel and P. Ribenboim. Squares and double squares in Lucas sequences. *C. R. Math. Rep. Acad. Sci. Canada*, 14:104–108.

1994 P. Ribenboim. *Catalan's Conjecture*. Academic Press, Boston.

1995 P. M. Voutier. Primitive divisors of Lucas and Lehmer sequences. *Math. of Comp.*, 64:869–888.

1998a W. L. McDaniel and P. Ribenboim. Square classes in Lucas sequences having odd parameters. *J. Nb. Th.*, 73:14–23.

1998b W. L. McDaniel and P. Ribenboim. Squares in Lucas sequences having one even parameter. *Colloq. Math.*, 78: 29–34.

1999 Y. Bugeaud and M. Mignotte. On integers with identical digits. Preprint.

1999 H. Dubner and W. Keller. New Fibonacci and Lucas primes. *Math. of Comp.*, 68:417–427.

1999a P. Ribenboim. Números primos, Mistérios e Récordes. Instituto de Matématica Puru e Aplicado, Rio de Janeiro.

1999b P. Ribenboim and P. G. Walsh. The *ABC* conjecture and the powerful part of terms in binary recurring sequences. *J. Nb. Th.*, 74:134–147.

2

Representation of Real Numbers by Means of Fibonacci Numbers

Our aim is to derive a new representation of positive real numbers as sums of series involving Fibonacci numbers. This will be an easy application of an old result of KAKEYA (1941). The paper concludes with a result of LANDAU (1899), relating the sum $\sum_{n=1}^{\infty} \frac{1}{F_n}$ with values of theta series. We believe it worthwhile to unearth LANDAU's result which is now rather inaccessible.

1. Let $(s_i)_{i \geq 1}$ be a sequence of positive real numbers such that $s_1 > s_2 > s_3 > \cdots$, and $\lim_{i \to \infty} s_i = 0$. Let $S = \sum_{i=1}^{\infty} s_i \leq \infty$.

We say that $x > 0$ is representable by the sequence $(s_i)_{i \geq 1}$ if $x = \sum_{j=1}^{\infty} s_{i_j}$ (with $i_1 < i_2 < i_3 < \cdots$). Then, necessarily, $x \leq S$.

The first result is due to KAKEYA; for the sake of completeness, we give a proof:

Proposition 1. *The following conditions are equivalent:*

1. *Every x, $0 < x \leq S$, is representable by the sequence $(s_i)_{i \geq 1}$, $x = \sum_{j=1}^{\infty} s_{i_j}$, where i_1 is the smallest index such that $s_{i_1} < x$.*
2. *Every x, $0 < x \leq S$, is representable by the sequence $(s_i)_{i \geq 1}$.*
3. *For every $n \geq 1$, $s_n \leq \sum_{i=n+1}^{\infty} s_i$.*

PROOF. $(1) \Rightarrow (2)$. This is trivial.

$(2) \Rightarrow (3)$. If there exists $n \geq 1$ such that $s_n > \sum_{i=n+1}^{\infty} s_i$, let x be such that $s_n > x > \sum_{i=n+1}^{\infty} s_i$. By hypothesis, $x = \sum_{j=1}^{\infty} s_{i_j}$ with

$i_1 < i_2 < \ldots$. Since $s_n > x > s_{i_1}$, then $n < i_1$, hence $x = \sum_{j=1}^{\infty} s_{i_j} \leq \sum_{k=n+1}^{\infty} s_k$, which is absurd.

(3) $\Rightarrow$ (1). Because $\lim_{i \to \infty} s_i = 0$, there exists a smallest index i_1 such that $s_{i_1} < x$. Similarly, there exists a smallest index i_2 such that $i_1 < i_2$ and $s_{i_2} < x - s_{i_1}$.

More generally, for every $n \geq 1$ we define i_n to be the smallest index such that $i_{n-1} < i_n$ and $s_{i_n} < x - \sum_{j=1}^{n-1} s_{i_j}$. Thus, $x \geq \sum_{j=1}^{\infty} s_{i_j}$.

Suppose that $x > \sum_{j=1}^{\infty} s_{i_j}$. We note that there exists N such that if $m \geq N$, then $s_{i_m} < x - \sum_{j=1}^{m} s_{i_j}$. Otherwise, there exist infinitely many indices $n_1 < n_2 < n_3 < \cdots$ such that $s_{i_{n_k}} \geq x - \sum_{j=1}^{n_k} s_{i_j}$. At the limit, we have

$$0 = \lim_{k \to \infty} s_{i_{n_k}} \geq x - \sum_{j=1}^{\infty} s_{i_j} > 0,$$

and this is a contradiction.

We choose N minimal with the above property.

We show: for every $m \geq N$, $i_m + 1 = i_{m+1}$. In fact,

$$s_{i_m+1} < s_{i_m} < x - \sum_{j=1}^{m} s_{i_j},$$

so by definition of the sequence of indices, $i_m + 1 = i_{m+1}$. Therefore,

$$\{i_N, i_N + 1, i_N + 2, \ldots\} = \{i_N, i_{N+1}, i_{N+2}, \ldots\}.$$

Next, we show that $i_N = 1$. If $i_N > 1$ we consider the index $i_N - 1$, and by hypothesis (3),

$$s_{i_N-1} \leq \sum_{k=i_N}^{\infty} s_k = \sum_{j=N}^{\infty} s_{i_j} < x - \sum_{j=1}^{N-1} s_{i_j}.$$

We have $i_{N-1} \leq i_N - 1 < i_N$. If $i_{N-1} < i_N - 1$ this is impossible because i_N was defined to be the smallest index such that $i_{N-1} < i_N$ and $s_{i_N} < x - \sum_{j=1}^{N-1} s_{i_j}$. Thus, $i_{N-1} = i_N - 1$, that is, $s_{i_{N-1}} < x - \sum_{j=1}^{N-1} s_{i_j}$ and this contradicts the choice of N as minimal with the property indicated.

Thus $i_N = 1$ and $x > \sum_{j=1}^{\infty} s_{i_j} = \sum_{i=1}^{\infty} s_i = S$, contradicting the hypothesis. $\qquad\square$

We remark now that if the above conditions are satisfied for the sequence $(s_i)_{i\geq 1}$, and if $m \geq 0$, then for every x such that $0 < x < S = \sum_{i=m+1}^{\infty} s_i$, x is representable by the sequence $(s_i)_{i\geq m+1}$, with i_1 the smallest index such that $m + 1 \leq i_1$ and $s_{i_1} < x$.

Indeed, condition (3) holds for $(s_i)_{i\geq 1}$, hence also for $(s_i)_{i\geq m+1}$. Since $0 < x < S$, the remark follows from the proposition.

Proposition 1 has been generalized (see, for example, FRIDY (1966)). Now we consider the question of unique representation (this was generalized by BROWN (1971)).

Proposition 2. *With the above notations, the following conditions are equivalent:*

(2') *Every x, $0 < x < S$, has a unique representation $x = \sum_{j=1}^{\infty} s_{i_j}$.*
(3') *For every $n \geq 1$, $s_n = \sum_{i=n+1}^{\infty} s_i$.*
(4') *For every $n \geq 1$, $s_n = \frac{1}{2^{n-1}} s_1$ (hence $S = 2s_1$).*

PROOF. (2') $\Rightarrow$ (3'). Suppose there exists $n \geq 1$ such that $s_n \neq \sum_{i=n+1}^{\infty} s_i$. Since (2') implies (2), hence also (3), then $s_n < \sum_{i=n+1}^{\infty} s_i$. Let x be such that $s_n < x < \sum_{i=n+1}^{\infty} s_i$. By the above remark, x is representable by the sequence $\{s_i\}_{i\geq n+1}$, that is, $x = \sum_{j=1, k_j \geq n+1}^{\infty} s_{k_j}$. On the other hand, (2') implies (2), hence also (1), and x has a representation $x = \sum_{j=1}^{\infty} s_{i_j}$, where i_1 is the smallest index such that $s_{i_1} < x$. From $s_n < x$ it follows that $i_1 \leq n$, and so x would have two distinct representations, contrary to the hypothesis.

(3') $\Rightarrow$ (4'). We have $s_n = s_{n+1} + \sum_{i=n+2}^{\infty} s_i = 2s_{n+1}$ for every $n \geq 1$, hence $s_n = \frac{1}{2^{n-1}} s_1$ for every $n \geq 1$.

(4') $\Rightarrow$ (2'). Suppose that there exists an x such that $0 < x < S$ and has two distinct representations

$$x = \sum_{j=1}^{\infty} s_{i_j} = \sum_{j=1}^{\infty} s_{k_j}.$$

Let j_0 be the smallest index such that $i_{j_0} \neq k_{j_0}$, say $i_{j_0} < k_{j_0}$. Then

$$\sum_{j=j_0}^{\infty} s_{i_j} = \sum_{j=j_0}^{\infty} s_{k_j} \leq \sum_{n=i_{j_0}+1}^{\infty} s_n.$$

By hypothesis, after dividing by s_1 we have

$$\sum_{n=i_{j_0}}^{\infty} \frac{1}{2^n} \geq \sum_{j=j_0}^{\infty} 2^{1-k_j} = \sum_{j=j_0}^{\infty} 2^{1-i_j} = 2^{1-i_{j_0}} + \sum_{j=j_0+1}^{\infty} 2^{1-i_j}$$

$$= \sum_{n=i_{j_0}}^{\infty} 2^{-n} + \sum_{j=j_0+1}^{\infty} 2^{1-i_j},$$

hence $\sum_{j=j_0+1}^{\infty} 2^{1-i_j} \leq 0$, which is impossible.   □

For practical applications, we note: If $s_n \leq 2s_{n+1}$ for every $n \geq 1$, then condition (3) is satisfied.

Indeed,

$$\sum_{i=n+1}^{\infty} s_i \leq 2 \sum_{i=n+1}^{\infty} s_{i+1} = 2 \sum_{i=n+2}^{\infty} s_i,$$

hence

$$s_{n+1} \leq \sum_{i=n+2}^{\infty} s_i$$

and $s_n \leq 2s_{n+1} \leq \sum_{i=n+1}^{\infty} s_i$.

2. Now we give various ways of representing real numbers.

First, the dyadic representation, which may of course be easily obtained directly:

Corollary 3. *Every real number* x, $0 < x < 1$, *may be written uniquely in the form* $x = \sum_{j=1}^{\infty} \frac{1}{2^{n_j}}$ *(with* $1 \leq n_1 < n_2 < n_3 < \cdots$*).*

PROOF.   This has been shown in Proposition 2, taking $s_1 = \frac{1}{2}$.   □

Corollary 4. *Every positive real number* x *may be written in the form* $x = \sum_{j=1}^{\infty} \frac{1}{n_j}$ *(with* $n_1 < n_2 < n_3 < \cdots$*).*

PROOF.   We consider the sequence $(1/n)_{n \geq 1}$, which is decreasing with limit equal to zero, and we note that $\sum_{n=1}^{\infty} \frac{1}{n} = \infty$ and $\frac{1}{n} \leq \frac{2}{n+1}$ for every $n \geq 1$. Thus, by KAKEYA's theorem and the above remark, every $x > 0$ is representable as indicated.   □

Corollary 5. *Every positive real number* x *may be written as* $x = \sum_{j=1}^{\infty} \frac{1}{p_{i_j}}$ *(where* $p_1 < p_2 < p_3 < \cdots$ *is the increasing sequence of prime numbers).*

PROOF.   We consider the sequence $(1/p_i)_{i \geq 1}$, which is decreasing with limit equal to zero. As EULER proved, $\sum_{i=1}^{\infty} \frac{1}{p_i} = \infty$. By Chebyshev's theorem (proof of Bertrand's "postulate") there is a prime in each interval $(n, 2n)$; thus $p_{i+1} < 2p_i$ and $\frac{1}{p_i} < \frac{2}{p_{i+1}}$ for every $i \geq 1$. By Kakeya's theorem and the above remark, every $x > 0$ is representable as indicated.   □

3. Now we shall represent real numbers by means of Fibonacci numbers and we begin giving some properties of these numbers.

The Fibonacci numbers are: $F_1 = F_2 = 1$, and for every $n \geq 3$, F_n is defined by the recurrence relation $F_n = F_{n-1} + F_{n-2}$.

Thus the sequence of Fibonacci numbers is

$$1, 1, 2, 3, 5, 8, 13, 21, 34, 55, 89, 144, \ldots.$$

In the following proposition, we give a closed form expression for the Fibonacci numbers; this is due to BINET (1843).

Let $\alpha = \frac{\sqrt{5}+1}{2}$ (the golden number) and $\beta = -\frac{\sqrt{5}-1}{2}$, so $\alpha + \beta = 1$, $\alpha\beta = -1$, and hence α, β are the roots of $X^2 - X - 1 = 0$ and $-1 < \beta < 0 < 1 < \alpha$.

We have

Lemma 6. *For every* $n \geq 1$, $F_n = \frac{\alpha^n - \beta^n}{\sqrt{5}}$ *and* $\frac{\alpha^{n-1}}{\sqrt{5}} < F_n < \frac{\alpha^{n+1}}{\sqrt{5}}$.

PROOF. We consider the sequence of numbers $G_n = \frac{\alpha^n - \beta^n}{\sqrt{5}}$ for $n \geq 1$. Then $G_1 = G_2 = 1$; moreover,

$$
\begin{aligned}
G_{n-1} + G_{n-2} &= \frac{\alpha^{n-1} - \beta^{n-1}}{\sqrt{5}} + \frac{\alpha^{n-2} - \beta^{n-2}}{\sqrt{5}} \\
&= \frac{\alpha^{n-2}(\alpha + 1) - \beta^{n-2}(\beta + 1)}{\sqrt{5}} = \frac{\alpha^n - \beta^n}{\sqrt{5}} \\
&= G_n
\end{aligned}
$$

because $\alpha^2 = \alpha + 1$ and $\beta^2 = \beta + 1$. Therefore, the sequence $(G_n)_{n \geq 1}$ coincides with the Fibonacci sequence.

Now we establish the estimates.

If $n \geq 1$, then

$$(-\beta)^n = \frac{1}{\alpha^n} < \alpha^{n-1} = -\alpha^n\beta = \alpha^n(\alpha - 1) = \alpha^{n+1} - \alpha^n,$$

so

$$F_n = \frac{\alpha^n - \beta^n}{\sqrt{5}} \leq \frac{\alpha^n + (-\beta)^n}{\sqrt{5}} < \frac{\alpha^{n+1}}{\sqrt{5}}.$$

Similarly, if $n \geq 2$, then

$$(-\beta)^n = \frac{1}{\alpha^n} < \alpha^{n-2} = -\alpha^{n-1}\beta = n - 1(\alpha - 1) = \alpha^n - \alpha^{n-1},$$

so

$$F_n = \frac{\alpha^n - \beta^n}{\sqrt{5}} \geq \frac{\alpha^n - (-\beta)^n}{\sqrt{5}} > \frac{\alpha^{n-1}}{\sqrt{5}};$$

this is also true when $n = 1$. □

For every $m \geq 1$ let $I_m = \sum_{n=1}^{\infty} \frac{1}{F_n^{1/m}}$. We have:

Lemma 7. *For every $m \geq 1$, $I_m < \infty$, $I_1 < I_2 < I_3 < \cdots$, and $\lim_{m \to \infty} I_m = \infty$.*

PROOF.   We have

$$I_m < \sum_{n=1}^{\infty} \left(\frac{\sqrt{5}}{\alpha^{n-1}} \right)^{1/m} = (\sqrt{5})^{1/m} \sum_{n=1}^{\infty} \left(\frac{1}{\alpha^{1/m}} \right)^{n-1} = \frac{(\sqrt{5})^{1/m} \alpha^{1/m}}{\alpha^{1/m} - 1},$$

noting that $\frac{1}{\alpha^{1/m}} < 1$.
    Next, we have

$$I_{m-1} = \sum_{n=1}^{\infty} \frac{1}{F_n^{1/m-1}} < \sum_{n=1}^{\infty} \frac{1}{F_n^{1/m}} = I_m.$$

Finally,

$$I_m = \sum_{n=1}^{\infty} \frac{1}{F_n^{1/m}} > \sum_{n=1}^{\infty} \left(\frac{\sqrt{5}}{\alpha^{n+1}} \right)^{1/m} = \frac{(\sqrt{5})^{1/m}}{\alpha^{1/m}} \times \frac{1}{\alpha^{1/m} - 1};$$

thus $\lim_{m \to \infty} I_m = \infty$. □

Proposition 8. *For every positive real number x there exists a unique $m \geq 1$ such that $x = \sum_{j=1}^{\infty} \frac{1}{F_{i_j}^{1/m}}$, but x is not of the form $\sum_{j=1}^{\infty} \frac{1}{F_{i_j}^{1/(m-1)}}$.*

PROOF.   First, we note that each of the sequences $(1/F_n^{1/m})_{n \geq 1}$ is decreasing with limit equal to zero. By the above proposition, there exists $m \geq 1$ such that $I_{m-1} < x \leq I_m$ (with $I_0 = 0$).
    We observe that

$$\frac{1}{F_n} \leq \frac{2}{F_{n+1}} \leq \frac{2^m}{F_{n+1}} \qquad \text{for } m \geq 1$$

because $F_{n+1} = F_n + F_{n-1} < 2F_n$. By Proposition 1 and a previous remark, x is representable as indicated, while the last assertion follows from $x > I_{m-1} = \sum_{i=1}^{\infty} \frac{1}{F_i^{1/m-1}}$. □

The number $I_1 = \sum_{n=1}^{\infty} \frac{1}{F_n}$ appears to be quite mysterious. As we have seen, $\sqrt{5} < I_1 < \sqrt{5}\frac{\alpha}{\alpha-1}$.

4. In 1899, LANDAU gave an expression of I_1 in terms of Lambert series and Jacobi theta series. The Lambert series is $L(x) = \sum_{n=1}^{\infty} \frac{x^n}{1-x^n}$; it is convergent for $0 < x < 1$, as is easily verified by the ratio test.

Jacobi theta series, which are of crucial importance (for example, in the theory of elliptic functions), are defined as follows, for $0 < |q| < 1$ and $z \in C$:

$$\theta_1(z, q) = i \sum_{n=-\infty}^{\infty} (-1)^n q^{\left(n-\frac{1}{2}\right)^2} e^{(2n-1)\pi i z}$$
$$= 2q^{1/4} \sin \pi z - 2q^{9/4} \sin 3\pi z + 2q^{25/4} \sin 5\pi z - \cdots$$
$$\theta_2(z, q) = \sum_{n=-\infty}^{\infty} q^{\left(n+\frac{1}{2}\right)^2} e^{(2n-1)\pi i z}$$
$$= 2q^{1/4} \cos \pi i z + 2q^{9/4} \cos 3\pi z + 2q^{25/4} \cos 5\pi z + \cdots$$
$$\theta_3(z, q) = \sum_{n=-\infty}^{\infty} q^{n^2} e^{2n\pi i z}$$
$$= 1 + 2q \cos 2\pi z + 2q^4 \cos 4\pi z + 2q^9 \cos 6\pi z + \cdots$$
$$\theta_4(z, q) = \sum_{n=-\infty}^{\infty} (-1)^n q^{n^2} e^{2n\pi i z}$$
$$= 1 - 2q \cos 2\pi z + 2q^4 \cos 4\pi z - 2q^9 \cos 6\pi z + \cdots$$

In particular, we have

$$\theta_1(0, q) = 0$$
$$\theta_2(0, q) = 2q^{1/4} + 2q^{9/4} + 2q^{25/4} + \cdots$$
$$\theta_3(0, q) = 1 + 2q + 2q^4 + 2q^9 + \cdots$$
$$\theta_4(0, q) = 1 - 2q + 2q^4 - 2q^9 + \cdots$$

Now we prove LANDAU's result:

Proposition 9. *We have:*

$$\sum_{n=1}^{\infty} \frac{1}{F_{2n}} = \sqrt{5} \left[L \left(\frac{3 - \sqrt{5}}{2} \right) - L \left(\frac{7 - 3\sqrt{5}}{2} \right) \right]. \tag{1}$$

$$\sum_{n=0}^{\infty} \frac{1}{F_{2n-1}} = -\sqrt{5}(1 + 2\beta^4 + 2\beta^{16} + 2\beta^{36} + \cdots)(\beta + \beta^9 + \beta^{25} + \cdots)$$

$$= -\frac{\sqrt{5}}{2} [\theta_3(0, \beta) - \theta_2(0, \beta^4)] \theta_2(0, \beta^4). \tag{2}$$

PROOF.   (1) We have

$$\frac{1}{F_n} = \frac{\sqrt{5}}{\alpha^n - \beta^n} = \frac{\sqrt{5}}{\frac{(-1)^n}{\beta^n} - \beta^n} = \frac{\sqrt{5}\beta^n}{(-1)^n - \beta^{2n}},$$

so

$$\frac{1}{\sqrt{5}} \sum_{n=1}^{\infty} \frac{1}{F_{2n}} = \sum_{n=1}^{\infty} \frac{\beta^{2n}}{1 - \beta^{4n}} = \sum_{n=1}^{\infty} \sum_{k=0}^{\infty} \beta^{(4k+2)n} = \sum_{k=0}^{\infty} \sum_{n=1}^{\infty} \beta^{(4k+2)n}$$

$$= \sum_{k=0}^{\infty} \frac{\beta^{4k+2}}{1 - \beta^{4k+2}} = \frac{\beta^2}{1 - \beta^2} + \frac{\beta^6}{1 - \beta^6} + \frac{\beta^{10}}{1 - \beta^{10}} + \cdots.$$

Because $|\beta| < 1$, it follows that

$$\sum_{n=1}^{\infty} \frac{1}{F_{2n}} = \sqrt{5}\left[L(\beta^2) - L(\beta^4)\right]$$

$$= \sqrt{5}\left[L\left(\frac{3 - \sqrt{5}}{2}\right) - L\left(\frac{7 - 3\sqrt{5}}{2}\right)\right].$$

(2) Now we have

$$\frac{1}{\sqrt{5}} \sum_{n=0}^{\infty} \frac{1}{F_{2n-1}} = -\sum_{n=0}^{\infty} \frac{\beta^{2n+1}}{1 + \beta^{4n+2}}$$

$$= -\sum_{n=0}^{\infty} \beta^{2n+1}\left(1 - \beta^{4n+2} + \beta^{8n+4} + \cdots\right)$$

$$= (-\beta + \beta^3 - \beta^5 + \beta^7 - \beta^9 + \cdots)$$
$$+ (-\beta^3 + \beta^9 - \beta^{15} + \beta^{21} - \cdots)$$
$$+ (-\beta^5 + \beta^{15} - \beta^{25} + \beta^{35} - \cdots)$$
$$+ (-\beta^7 + \beta^{21} - \beta^{35} + \beta^{49} - \cdots) + \cdots.$$

Now we need to determine the coefficient of β^m (for m odd), remarking that since the series is absolutely convergent its terms may be rearranged.

If m is odd and d divides m, then β^m appears in the horizontal line beginning with $-\beta^{m/d}$ with the sign

$$\begin{cases} + & \text{when } d \equiv 3 \pmod 4, \\ - & \text{when } d \equiv 1 \pmod 4. \end{cases}$$

Thus the coefficient ϵ_m of β^m is $\epsilon_m = \delta_3(m) - \delta_1(m)$ where

$$\delta_1(m) = \#\{d \mid 1 \le d \le m,\ d \mid m \text{ and } d \equiv 1 \pmod 4\},$$
$$\delta_3(m) = \#\{d \mid 1 \le d \le m,\ d \mid m \text{ and } d \equiv 3 \pmod 4\}.$$

A well-known result of JACOBI (see HARDY and WRIGHT's book, page 241) relates the difference $\delta_1(m) - \delta_3(m)$ with the number $r(m) = r_2(m)$ of representations of m as sums of two squares. Precisely, let $r(m)$ denote the number of pairs (s, t) of integers (including the zero and negative integers) such that $m = s^2 + t^2$. JACOBI showed that

$$r(m) = 4[\delta_1(m) - \delta_3(m)].$$

It follows that the number $r'(m)$ of pairs (s, t) of integers with $s > t \ge 0$ and $m = s^2 + t^2$ is

$$r'(m) = \begin{cases} \dfrac{r(m)}{8} & \text{when } m \text{ is not a square,} \\[2mm] \dfrac{r(m) - 4}{8} + 1 = \dfrac{r(m) + 4}{8} & \text{when } m \text{ is a square;} \end{cases}$$

(the first summand above corresponds to the representation of m as a sum of two non-zero squares).

Therefore,

$$\epsilon_m = -\frac{r(m)}{4} = \begin{cases} -2r'(m) & \text{when } m \text{ is not a square,} \\[1mm] -(2r'(m) - 1) & \text{when } m \text{ is a square.} \end{cases}$$

Because m is odd, it follows that $s \not\equiv t \pmod 2$, and therefore

$$\frac{1}{\sqrt{5}} \sum_{n=0}^{\infty} \frac{1}{F_{2n+1}} = \sum_{\substack{m=1 \\ m \text{ odd}}}^{\infty} \epsilon_m \beta^m$$
$$= -2(1 + \beta^4 + \beta^{16} + \beta^{36} + \cdots)(\beta + \beta^9 + \beta^{25} + \cdots)$$
$$+ (\beta + \beta^9 + \beta^{25} + \cdots)$$
$$= -(1 + 2\beta^4 + 2\beta^{16} + 2\beta^{36} + \cdots)(\beta + \beta^9 + \beta^{25} + \cdots).$$

So,

$$\sum_{n=0}^{\infty} \frac{1}{F_{2n+1}} = -\sqrt{5}(1 + 2\beta^4 + 2\beta^{16} + 2\beta^{36} + \cdots)(\beta + \beta^9 + \beta^{25} + \cdots).$$

We may now express this formula in terms of Jacobi series. Namely,

$$1 + 2\beta^4 + 2\beta^{16} + 2\beta^{36} + \cdots = (1 + 2\beta + 2\beta^4 + 2\beta^9 + 2\beta^{16} + \cdots$$
$$-(2\beta + 2\beta^9 + 2\beta^{25} + \cdots) = \theta_3(0, \beta) - \theta_2(0, \beta^4),$$

so,

$$\sum_{n=0}^{\infty} \frac{1}{F_{2n+1}} = -\frac{\sqrt{5}}{2} \left[\theta_3(0, \beta) - \theta_2(0, \beta^4) \right] \theta_2(0, \beta^4). \qquad \square$$

A formula of ALMQVIST (1983) kindly communicated to me gives another expression of I_1 only in terms of Jacobi theta series:

$$I_1 = \frac{\sqrt{5}}{4} \left\{ \left[\theta_2 \left(0, -\frac{1}{\beta^2} \right) \right]^2 + \frac{1}{\pi} \int_0^1 \left(\frac{d}{dx} \log \theta_4 \left(x, -\frac{1}{\beta^2} \right) \right) \cot \pi x \, dx \right\}.$$

The following question remained unanswered for a long time: is $I_1 = \sum_{n=1}^{\infty} \frac{1}{F_n}$ an irrational number? Yes—this was proved by ANDRÉ-JEANNIN in 1989, with a method reminiscent of the one of APÉRY for the proof of the irrationality of $\zeta(3) = \sum_{n=1}^{\infty} \frac{1}{n^3}$.

CARLITZ considered also in 1971 the following numbers:

$$S_k = \sum_{n=1}^{\infty} \frac{1}{F_n F_{n+1} \cdots F_{n+k}},$$

so, $S_0 = \sum_{n=1}^{\infty} \frac{1}{F_n} = I_1$.

Clearly, all the above series are convergent. CARLITZ showed that $S_3, S_7, S_{11}, \ldots \in \mathbb{Q}(\sqrt{5})$, while $S_{4k} = r_k + r_k' S_0$ for $k \geq 1$ and $r_k, r_k' \in \mathbb{Q}$.

One may ask: are the numbers S_0, S_1, S_2 algebraically independent?

References

1899 E. Landau.   Sur la série des inverses de nombres de Fibonacci. *Bull. Soc. Math. France* 27, 298–300.

1941 S. Kakeya. On the partial sums of an infinite series. *Science Reports Tôhoku Imp. Univ. (1)*, 3:159–163.

1960 G. Sansone and J. Gerretsen. *Lectures on the Theory of Functions of a Complex Variable, Vol. I.* P. Noordhoff, Groningen.

1966 J. A. Fridy.   Generalized bases for the real numbers. *Fibonacci Q.*, 4:193–201.

1971 J. L. Brown.   On generalized bases for real numbers. *Fibonacci Q.*, 9:477–496.

1971 L. Carlitz. Reduction formulas for Fibonacci summations. *Fibonacci Q.*, 9:449–466, and 510.

1987 J. M. Borwein and P. B. Borwein. *Pi and the AGM.* John Wiley & Sons, New York.

1989 R. André-Jeannin. Irrationalité de la somme des inverses de certaines suites récurrentes. *C. R. Acad. Sci. Paris, Sér. I*, 308:539–541.

3

Prime Number Records

The theory of prime numbers can be roughly divided into four main inquiries: How many prime numbers are there? How can one produce them? How can one recognize them? How are the primes distributed among the natural numbers? In answering these questions, calculations arise that can be carried out only for numbers up to a certain size. This chapter records the biggest sizes reached so far—the prime number records.

All the world loves records. They fascinate us and set our imaginations soaring. The famous *Guinness Book of Records*, which has appeared in surprisingly many editions, contains many noteworthy and interesting occurrences and facts. Did you know, for example, that the longest uninterrupted bicycle trip was made by Carlos Vieira of Leiria, Portugal? During the period June 8–16, 1983, he pedalled for 191 hours nonstop, covering a distance of 2407 km. Or did you know that the largest stone ever removed from a human being weighed 6.29 kg? The patient was an 80-year-old woman in London, in 1952. And nearer our usual lines of interest: Hideaki Tomoyoki, born in Yokohama in 1932, quoted 40 000 digits of π from memory, a heroic exploit that required 17 hours and 20 minutes, with pauses totalling 4 hours. Leafing through the *Guinness Book*, one finds very few scientific records, however, and even fewer records about numbers.

Not long ago I wrote *The Book of Prime Number Records* (RIBEN-BOIM (1996)), in which I discuss the feats of mathematicians in this domain so neglected by Guinness. How this book originated is a story worth telling. Approached by my university to give a colloquium lecture for undergraduate students, I sought a topic that would be not only understandable but interesting. I came up with the idea of speaking about *prime number records*, since the theme of records is already popular with students in connection with sports. The interest of the students so exceeded my expectations that I resolved to write a monograph based on this lecture. In the process, I learned of so many new facts and records that the brief text I had planned kept on expanding. Thanks to colleagues who supplied me with many helpful references, I was at last able to complete this work.

I must confess that when preparing the lecture I did not know a lot (indeed I knew very little!) about the theorems for primes and prime number records. For me, all these facts, although quite interesting, were not tied together. They seemed to be just isolated theorems about prime numbers, and it was not clear how they could be woven into a connected theory. But when one wishes to write a book, the first task is to shape the subject matter into a coherent whole.

The scientific method may be considered as a two-step process: first, observation and experiment—*analysis*; then formulation of the rules, theorems, and orderly relationships of the facts—*synthesis*. Stated in these terms, my task was thus to present a synthesis of the known observations about prime numbers, with an emphasis on the records achieved. Any originality of my work undoubtedly lies in the systematic investigation of the interplay between theory and calculation. This undertaking needs no justification if one keeps in mind what role the prime numbers have in the theory of numbers. After all, the fundamental theorem of elementary number theory says that every natural number $N > 1$ can be expressed in a unique way (except for the order of the factors) as a product of primes. Prime numbers are thus the foundation stones on which the structure of arithmetic is raised.

Now, how did I go about organizing the theory of prime numbers? I began by posing four direct, unambiguous questions:

1. How many prime numbers are there?
2. How can one generate primes?
3. How can one know if a given number is prime?

4. Where are the primes located?

As I shall show, out of these four questions the theory of prime numbers naturally unfolds.

How Many Primes Are There?

As is well known, EUCLID in his *Elements* proved that there are infinitely many primes, proceeding as follows: Assume that there are only finitely many primes. Let p be the largest prime number and P be the product of all primes less than or equal to p; then consider the number P plus 1:

$$P + 1 = (\prod_{q \leq p} q) + 1.$$

Two cases are possible: either (a) $P + 1$ is prime, or (b) $P + 1$ is not prime. But if (a) is true, $P + 1$ would be a prime number larger than p. And if (b) holds, none of the primes $q \leq p$ is a prime factor of $P + 1$, so the prime factors of $P + 1$ are all larger than p. In both cases the assumption that there is a largest prime p leads to a contradiction. This shows that there must be an infinite number of primes.

From this indirect proof one cannot deduce a method for generating prime numbers, but it prompts a question: Are there infinitely many primes p such that the corresponding number $P + 1$ is also prime? Many mathematicians have devoted calculations to this question.

Record. $p = 24029$ *is the largest known prime for which $P + 1$ is also prime; here, $P + 1$ has 10387 decimal digits. This was found by* C. CALDWELL *in 1995.*

There are many other proofs (but not quite infinitely many) of the existence of infinitely many primes; each reveals another interesting aspect of the set of all prime numbers. EULER showed that the sum of the reciprocals of the prime numbers is divergent:

$$\sum \frac{1}{p} = \infty.$$

From this we again see that there cannot be only finitely many primes. EULER's proof can be found in many elementary books on number theory or real analysis, such as HARDY and WRIGHT (1979), and permits an interesting deduction. For any $\epsilon > 0$, no matter how small, we know

$$\sum_{n=1}^{\infty} \frac{1}{n^{1-\epsilon}} < \infty.$$

Hence the prime numbers are closer together, or are less sparsely scattered along the number line, than are numbers of the form $n^{1-\epsilon}$. For example, the primes lie closer together than the squares n^2, for which EULER showed

$$\sum_{n=1}^{\infty} \frac{1}{n^2} = \frac{\pi^2}{6}.$$

Another simple and elegant proof that infinitely many primes exist was given by GOLDBACH. It clearly suffices to find an infinite sequence $F_0, F_1, F_2, F_3, \ldots$ of pairwise relatively prime natural numbers (i.e., no two having a common divisor greater than 1); since each F_n has at least one prime factor, then there are infinitely many primes. It is easy to prove that the sequence of Fermat numbers $F_n = 2^{2^n} + 1$ has this property. Clearly, neither F_n nor F_{n+k} $(k > 0)$ is divisible by 2; and if p is an odd prime factor of F_n, then $2^{2^n} \equiv -1 \pmod{p}$, so that $2^{2^{n-k}} = (2^{2^n})^{2^k} \equiv 1 \pmod{p}$. Thus $F_{n+k} \equiv 2 \pmod{p}$, and since $p > 2$, it follows that p does not divide F_{n+k}. I will devote further attention to the Fermat numbers after the next section.

Generating Prime Numbers

The problem is to find a "good" function $f : \mathbb{N} \to \{prime\ numbers\}$. This function should be as easy to calculate as possible and, above all, should be representable by previously well-known functions. One may place additional conditions on this function, for example:

Condition (a). $f(n)$ equals the nth prime number (in the natural order); this amounts to a "formula" for the nth prime number.

Condition (b). For $m \neq n$, $f(m) \neq f(n)$; this amounts to a function that generates distinct primes, but not necessarily all the primes.

One can also seek a function f defined on $\mathbb{N}$ with integer values (but not necessarily positive values) that fulfills

Condition (c).   The set of prime numbers coincides with the set of positive values of the function. This is a far looser requirement and one that can be fulfilled in unexpected ways, as I shall later show.

To begin, let's discuss formulas for prime numbers. There are plenty of them! In fact many of us in younger days sought—often with success—a formula for the nth prime number. Unfortunately, all these formulas have one thing in common: They express the nth prime number through functions of the preceding primes that are difficult to compute. Consequently, these formulas are useless for deriving properties of the prime numbers. Nevertheless, I will give as an illustration one such formula found in 1971. I do so in honor of its discoverer, J. M. GANDHI, a mathematician who also worked on Fermat's Last Theorem.

To simplify the statement of the formula, I will introduce the Möbius function $\mu : \mathbb{N} \rightarrow \mathbb{Z}$, given by

$$\mu(n) = \begin{cases} 1 & \text{if } n = 1, \\ (-1)^r & \text{if } n \text{ is square-free and a product of} \\ & \quad r \text{ distinct prime factors,} \\ 0 & \text{otherwise.} \end{cases}$$

Now if $p_1, p_2, p_3, \ldots$ is the sequence of prime numbers in increasing order, set $P_{n-1} = p_1 p_2 \ldots p_{n-1}$; then Gandhi's formula is

$$p_n = \left[1 - \log_2 \left(-\frac{1}{2} + \sum_{d, P_{n-1}} \frac{\mu(d)}{2^d - 1} \right) \right].$$

Here, $\log_2$ indicates the logarithm in base 2, and $[x]$ denotes, as usual, the largest integer less than or equal to the real number x. One can see how difficult it is to calculate p_n using Gandhi's formula!

Now I sketch the construction of a function that generates prime numbers. E. M. WRIGHT and G. H. HARDY in their famous book (HARDY and WRIGHT (1979)) showed that if $\omega = 1.9287800\ldots$, and if

$$f(n) = \left[2^{2^{\cdot^{\cdot^{2^\omega}}}} \right] \quad \text{(with } n \text{ twos)},$$

then $f(n)$ is prime for all $n \geq 1$. Thus $f(1) = 3$, $f(2) = 13$, and $f(3) = 16381$, but $f(4)$ is rather hard to calculate and has almost 5000 decimal places. However, as the exact value of ω depends on knowledge of the prime numbers, this formula is ultimately uninteresting.

Do any truly simple functions generate prime numbers? There are no such polynomial functions because of the following negative result:

For every $f \in \mathbb{Z}[X_1, \ldots, X_m]$ there are infinitely many m-tuples of integers $(n_1, \ldots, n_m)$ for which $|f(n_1, \ldots, n_m)|$ is a composite number.

Other similar negative results are plentiful.

Well, then, are there polynomials in just *one* indeterminate for which many consecutive values are primes? More precisely: Let q be a prime number. Find a polynomial of degree 1, in fact, a polynomial of the form $f_q(X) = dX + q$ whose values at the numbers $0, 1, \ldots, q-1$ are all prime. Then f_q generates a sequence of q prime numbers in arithmetic progression with difference d and initial value q.

For small values of q finding f_q is easy:

q	d	Values at $0, 1, \ldots, q-1$				
2	1	2	3			
3	2	3	5	7		
5	6	5	11	17	23	29
7	150	7	157	307	... ...	907

However, it is not known how to prove that this is possible for every prime number q.

Records. *In* 1986, *G.* LÖH *gave the smallest values of d for two primes:*

$$\text{For } q = 11, \quad d = 1\,536\,160\,080.$$
$$\text{For } q = 13, \quad d = 9\,918\,821\,194\,590.$$

One can also examine the related problem: to search for the longest sequences of primes in arithmetic progression.

Record. *The longest known sequence of primes in arithmetic progression consists of* 22 *terms in the sequence with first term* $a = 11\,410\,337\,850\,553$ *and difference* $d = 4\,609\,098\,694\,200$ *(work coordinated by* A. MORAN, P. PRITCHARD, *and* A. THYSSEN, 1993*).*

EULER discovered quadratic polynomials for which many values are primes. He observed that if q is the prime 2, 3, 5, 11, 17, or 41, then the values $f_q(0), f_q(1), \ldots, f_q(q-2)$ of the polynomial $f_q(X) = X^2 + X + q$ are prime. (Evidently $f_q(q-1) = q^2$ is not prime, so this sequence of consecutive prime values is the best one can hope for.) For $q = 41$ this gives 40 prime numbers: $41, 43, 47, 53, \ldots, 1447, 1523, 1601$.

The next question is obvious: Can one find primes $q > 41$ for which the first $q-1$ values of EULER's quadratic are all prime? If infinitely many such primes q exist, I could generate arbitrarily long sequences of primes! However, the following theorems say this is not to be:

Theorem. *Let q be a prime number. The integers $f_q(0)$, $f_q(1)$, $\ldots$, $f_q(q-2)$ are all primes if and only if the imaginary quadratic field $\mathbb{Q}(\sqrt{1-4q})$ has class number 1 (G. RABINOVITCH, 1912).*

(A quadratic field K has class number 1 if every algebraic integer in K can be expressed as a product of primes in K, and if any two such representations differ only by a unit, i.e., an algebraic integer that is a divisor in 1 in K.)

Theorem. *Let q be a prime number. An imaginary quadratic field $\mathbb{Q}(\sqrt{1-4q})$ has class number 1 if and only if $4q-1 = 7, 11, 19, 43, 67$, or 163, that is, $q = 2, 3, 5, 11, 17$, or 41.*

The imaginary quadratic fields of class number 1 were determined in 1966 by A. BAKER and H. M. STARK, independently and free of the doubt that clung to HEEGNER's earlier work in 1952.

Thus the following unbeatable record has been attained:

Record. *$q = 41$ is the largest prime number for which the values $f_q(0), f_q(1), \ldots, f_q(q-2)$ of the polynomial $f_q(X) = X^2 + X + q$ are all primes.*

It is worth mentioning that in the solution of this quite harmless-looking problem a rather sophisticated theory was required. Details are given in Chapter 5.

I now turn to some polynomials whose positive values coincide with the set of prime numbers. The astonishing fact that such polynomials exist was discovered in 1971 by YU. V. MATIJASEVIČ in connection with the tenth Hilbert problem. Here are the records, which depend on the number of unknowns n and the degree d of the polynomial:

Records.

n	d	Year	
21	21	1971	YU. V. MATIJASEVIČ (not explicit)
26	25	1976	J. P. JONES, D. SALO, H. WADA, D. WIENS
42	5	1976	JONES ET AL. (not explicit): Lowest d
10	$\sim 1.6 \times 10^{48}$	1978	YU. V. MATIJASEVIČ (not explicit): Lowest n

It is not known whether the minimum values for n and d are 10 and 5, respectively.

Recognizing Prime Numbers

Given a natural number N, is it possible to determine with a finite number of calculations whether N is a prime? Yes! It suffices to divide N by every prime number d for which $d^2 < N$. If the remainder is nonzero every time, then N is prime. The trouble with this method is that a large N requires a large number of calculations. The problem, therefore, is to find an algorithm A where the number of computations is bounded by a function f_A of the number of digits of N, so $f_A(N)$ does not grow too fast with N. For example, $f_A(N)$ should be a polynomial function of the number of binary digits of N, which is $1 + [\log_2(N)]$. Essentially, this number is proportional to the natural logarithm $\log N$, since $\log_2(N) = \log N / \log 2$.

This problem remains open—it is not known whether such a polynomial algorithm exists. On the one hand, I cannot prove the impossibility of its existence; on the other hand, no such algorithm has yet been found. Efforts in this direction have produced several primality-testing algorithms. According to the point of view, they may be classified as follows:

$$\begin{cases} \text{Algorithms for arbitrary numbers} \\ \text{Algorithms for numbers of special form} \end{cases}$$
$$\begin{cases} \text{Algorithms that are fully justified by theorems} \\ \text{Algorithms that are based on conjectures} \end{cases}$$
$$\begin{cases} \text{Deterministic algorithms} \\ \text{Probabilistic algorithms} \end{cases}$$

To clarify these notions I offer some examples.

One algorithm applicable to arbitrary numbers is that of G. L. MILLER (1976), the complexity of which can be estimated only with the help of the generalized Riemann conjecture. Assuming this conjecture, for Miller's algorithm the estimate $f_A(N) \leq C(\log N)^5$ is valid, where C is a positive constant. Thus this is an algorithm whose polynomial growth rate remains uncertain. By contrast, the algorithm of L. M. ADLEMAN, C. POMERANCE, and R. S. RUMELY (1983) possesses a completely assured complexity estimate, and the number of computation operations as a function of the number of binary digits of N is bounded by $(\log N)^{C \log \log \log N}$ where C is a constant. The complexity is therefore in practice not far from polynomial, and this algorithm can be applied to an arbitrary integer N.

Both of these algorithms are deterministic, unlike those I shall now describe. First, I must introduce the so-called pseudoprime numbers. Let $a > 1$ be an integer. For every prime p that does not divide a, Fermat's Little Theorem says $a^{p-1} \equiv 1 \pmod{p}$. But it is quite possible for a number $N > 1$ with $a^{N-1} \equiv 1 \pmod{N}$ to be composite—in which case we say N is *pseudoprime for the base a*. For example, 341 is the smallest pseudoprime for the base 2. Every base a has infinitely many pseudoprimes. Some among them satisfy an additional congruence condition and are called *strong pseudoprimes for the base a*; they, too, are infinite in number.

An algorithm is called a *probabilistic prime number test* if its application to a number N leads either to the conclusion that N is composite or to the conclusion that with high probability N is a prime number. Tests of this type include those of R. BAILLIE and S. S. WAGSTAFF (1980), and M. O. RABIN (1980). In these tests one examines certain "witnesses." Let $k > 1$ (for example, $k = 30$) and let $a_1 = 2, a_2 = 3, \ldots, a_k$ be primes that will serve as witnesses. Should a witness fail to satisfy the condition $a_j^{N-1} \equiv 1 \pmod{N}$, then N is surely composite. If for every witness a_j the preceding congruence holds (that is, if N is pseudoprime for the base a_j, for $j = 1, 2, \ldots, k$) then N is with high probability a prime number. Rabin's test is similar, using more restrictive congruences which lead to better probabilities. This test leads to the conclusion that N either is certainly composite or with probability $1 - (1/4^k)$ is prime. For $k = 30$, then, the test gives a false result only once out of every 10^{18} values of N. These probabilistic tests are clearly very easy to apply.

Now I turn to prime number tests applicable to numbers of the form $N \pm 1$, where many if not all of the prime factors of N are known. The tests for $N + 1$ depend on a weak converse, due to PEPIN, of Fermat's Little Theorem, while those for $N - 1$ use the Lucas sequence.

In 1877, PEPIN showed that the Fermat numbers $F_n = 2^{2^n} + 1$ are prime if and only if $3^{(F_n-1)/2} \equiv -1 \pmod{F_n}$. The search for primes among the Fermat numbers F_n has produced several records.

Record. *The largest Fermat number known to be prime is $F_4 = 65537$.*

Record. *F_{11} is the largest Fermat number all of whose prime factors have been determined (R. P. BRENT and F. MORAIN, 1988)*

Record. *F_{303088} is the largest Fermat number known to be composite; it has the factor $3 \times 2^{303093} + 1$ (J. YOUNG, 1998).*

Record. *F_{24} is the smallest Fermat number not yet proven prime or composite.*

For the Mersenne numbers, $M_q = 2^q - 1$, with q a prime, one applies the Lucas test (1878): Let $S_0 = 4, S_{k+1} = S_k^2 - 2$, for $k \geq 0$. Then M_q is prime if and only if M_q is a divisor of S_{q-2}. This test makes it possible to discover very large primes.

Record. *To date, 37 Mersenne primes are known. The largest Mersenne prime known, which is also the largest known prime today, is M_q where $q = 3\,021\,377$ (you may easily compute that it has more than 2 million digits). It was discovered in January 1998 by R. CLARKSON, G. WOLTMAN, S. KUNOWSKI, et. al. The next smaller Mersenne primes are M_q, with $q = 2\,976\,221$, and $1\,398\,269$.*

The search for new primes has intensified with the creation by G. WOLTMAN of a club (one could almost say) or, better, a true co-operative research program entitled "The Great Internet Mersenne Primes Search" (GIMPS). It mobilizes thousands of computers throughout the world, and has led to the discovery of the three largest known Mersenne primes. The algorithm implemented is a modification of the one due to R. E. CRANDALL and has been of basic importance in this search.

Record. *The largest known composite Mersenne number is M_q for $q = 72021 \times 2^{23630} - 1$* (Y. GALLOT, 1998).

From 1876, when E. LUCAS proved M_{127} prime, until 1989, the title "largest prime number" was always held by a Mersenne prime. That became true again in 1992, but in the three intervening years another champion reigned: $391581 \times 2^{216193} - 1$.

Record. *The largest prime known today that is not a Mersenne prime is $302627325 \times 2^{530101} + 1$ with 159 585 digits, discovered in 1999 by* R. BURROWES, P. JOBLING, *and* Y. GALLOT.

The Distribution of the Prime Numbers

At this point we know the following:

1. There are infinitely many prime numbers.
2. There is no reasonably simple formula for the prime numbers.
3. One can determine whether a given number is prime if it is not too large.

What can one say about the way the primes are distributed among the natural numbers? Earlier I gave a hint in connection with EU-LER's proof of the existence of infinitely many primes: The primes are closer together than are, for example, the squares. A quite simple way to discuss the distribution of the primes is to count the number of primes less than a given number. For every real $x > 0$, set $\pi(x) = \#\{\text{prime numbers } p \mid p \leq x\}$. Thus, π is the function that counts the prime numbers. To have a good idea of the behavior of π we can compare it with simpler functions. This approach leads to results of an asymptotic nature.

When only 15 years old, C. F. GAUSS conjectured from his studies of prime number tables that

$$\pi(x) \sim \frac{x}{\log x}.$$

That is, the limit of the quotient

$$\frac{\pi(x)}{x/\log x},$$

as $x \to \infty$, exists and equals 1. An equivalent formulation is

$$\pi(x) \sim \int_1^x \frac{dt}{\log t}.$$

The function on the right is called the logarithmic integral and is denoted Li. GAUSS's assertion was proved in 1896 by J. HADAMARD and C. DE LA VALLÉE POUSSIN ; previously, P. L. CHEBYSHEV had shown that the limiting value, if it exists, must be 1.

This theorem belongs among the most significant results in the theory of prime numbers, for which reason it is customarily referred to as the *Prime Number Theorem*. However, this theorem obviously says nothing about the exact value of $\pi(x)$. For that purpose we have the famous formula that D. F. E. MEISSEL found in 1871 expressing the exact value of $\pi(x)$ in terms of $\pi(y)$ for all $y \le x^{2/3}$ and prime numbers $p \le x^{1/2}$.

Record. *The largest value $\pi(N)$ which has been exactly computed is $\pi(10^{20}) = 2\,220\,819\,602\,560\,918\,840$ done by M. DELEGLISE in 1997. He also showed that $\pi(4\,185\,296\,581\,467\,695\,669) = 10^{17}$.*

The differences

$$\left|\pi(x) - \frac{x}{\log x}\right| \qquad \text{and} \qquad |\pi(x) - Li(x)|$$

do not remain bounded as $x \to \infty$. Evaluating these error terms as exactly as possible is enormously important in applications of the Prime Number Theorem. On the basis of tables it was first conjectured, and then proved (J. B. ROSSER and L. SCHOENFELD, 1962), that for all $x \ge 17$, $x/\log x \le \pi(x)$. This is interesting because, by contrast, the difference $Li(x) - \pi(x)$ changes sign infinitely many times, as J. E. LITTLEWOOD (1914) showed. In 1933, S. SKEWES showed that the difference $Li(x) - \pi(x)$ is negative for some x_0 with $x_0 \le e^{e^{e^{e^{7.7}}}}$. As a matter of fact, this change in sign occurs much earlier:

Record. *The smallest x_0 for which $Li(x) < \pi(x)$ must be less than 6.69×10^{370}* (H. J. J. TE RIELE, 1987).

The most important function for studying the distribution of primes is the Riemann *zeta function*: For every complex number s

with $\text{Re}(s) > 1$, the series $\sum_{n=1}^{\infty} 1/n^s$ is absolutely convergent; it is also uniformly convergent in every half-plane $\{s \mid \text{Re}(s) > 1 + \epsilon\}$ for any $\epsilon > 0$. The function ζ thus defined can be extended by analytic continuation to a meromorphic function defined in the entire complex plane, with only one pole. The pole is at the point $s = 1$, has order 1, and the residue there is 1. It was the study of the properties of this function that ultimately made the proof of the Prime Number Theorem possible. The function ζ has zeros at $-2, -4, -6, \ldots$, as one can easily show with the help of the functional equation satisfied by ζ. All other zeros of ζ are complex numbers $\sigma + it$ (t real) with $0 < \sigma < 1$.

The so far unproved *Riemann hypothesis* says: The nontrivial zeros of the Riemann zeta function are located on the *critical line* $\frac{1}{2} + it$ (t real). Without going into the details, I will just observe that many theorems about the distribution of primes can be proved with the assumption of the Riemann hypothesis. It is therefore of fundamental importance to determine the nontrivial zeros of ζ. By symmetry considerations, it suffices to determine the zeros with $t > 0$, which can be listed in a sequence $\sigma_n + it_n$, where $t_n \leq t_{n+1}$, and in case $t_n = t_{n+1}$ we require that $\sigma_n < \sigma_{n+1}$. (It must first be shown that there are at most a finite number of zeros of ζ for each value of t.)

Record. *For $n \leq 1\,500\,000\,001$ all the zeros $\sigma_n + it_n$ of the Riemann zeta function are located on the critical line; that is, $\sigma_n = \frac{1}{2}$. These calculations were carried out in 1986 by* J. VAN DE LUNE, H. J. J. TE RIELE, *and* D. T. WINTER.

Record. *In 1974,* N. LEVINSON *showed that at least one third of the zeros of the Riemann zeta function are on the critical line, and in 1989,* J. B. CONREY *improved this result, replacing 1/3 by 2/5.*

The foregoing considerations are based on the asymptotic behavior of the function π and on the function ζ, which is very useful for estimating the error terms. One can say that they deal with the estimation of π "at infinity." Next I turn to the *local* behavior of π—estimating the gaps between the prime numbers. Here the fundamental question is: Knowing the nth prime p_n, where will one find the following prime p_{n+1}? Thus, one is concerned with the sequence of differences $d_n = p_{n+1} - p_n$. It is easy to see that $\limsup d_n = \infty$, that is, arbitrarily long blocks of consecutive composite numbers exist. Here is one: For any N, the N consecutive numbers

$$(N+1)! + 2, (N+1)! + 3, \ldots, (N+1)! + (N+1)$$

are composite. It has amused some mathematicians to find the largest blocks of consecutive composite numbers between fairly small primes—the widest gaps between such primes.

Record. *The largest gap between consecutive prime numbers that has been effectively computed consists of the* 1131 *composite numbers which follow the prime* $p = 1\,693\,182\,318\,746\,371$. *This was discovered by* B. NYMAN *in* 1999.

The question about wide gaps between not too large primes can be made more precise. Let us look at the sequence d_n/p_n of relative gaps. As early as 1845, J. BERTRAND postulated from a study of tables that a prime always lies between p_n and $2p_n$, for every $n \geq 1$. It was CHEBYSHEV who first proved this result, which can be written in the form $p_{n+1} < 2p_n$ or, better, $d_n/p_n < 1$. This result, while amusing, is much weaker than what can be deduced by using the Prime Number Theorem:

$$\lim_{n \to \infty} \frac{d_n}{p_n} = 0.$$

The investigation of gaps between prime numbers has led to the following conjecture: For every $\epsilon > 0$ the inequality $p_{n+1} < p_n + p_n^{1/2+\epsilon}$ holds for all sufficiently large n.

Record. *The latest entry in a long line, the current record is in the work of* S. LOU *and* Q. YAO *in* 1993: $p_{n+1} < p_n + p_n^{6/11}$.

What about the limit inferior of the difference sequence d_n? Two prime numbers p and p' $(p < p')$ are said to be *twin primes* if $p' - p = 2$. It is still not known if there are infinitely many twin primes, i.e., if $\liminf d_n = 2$. The question is delicate. In 1919, V. BRUN showed that the sum over all pairs of twin primes

$$\sum \left(\frac{1}{p} + \frac{1}{p+2} \right) = B < \infty.$$

It follows that if there are infinitely many twin primes, which one expects to be the case, then they are thinly dispersed. In 1976, Brun's constant was calculated by R. P. BRENT: $B = 1.90216054$.

Record. *The largest known pair of twin primes is* $361700055 \times 2^{39020} \pm 1$ *with* 11755 *digits, discovered by* H. LIFCHITZ *in* 1999.

Conclusion

Lest this presentation grow to long, I have had to pass over many fascinating questions, such as the behavior of primes in arithmetic progression, to say nothing of the Goldbach conjecture. Fortunately, these and many other facts have been both recorded and amply explained in a book (RIBENBOIM (1991)) that is just waiting to be read! I will close with two curiosities to work into your repertoire.

A *repunit* is an integer of the form $R_n = 111 \ldots 1$, with n decimal digits equal to 1. It is not known if there are infinitely many prime repunits, but there is the following record.

Record. H. C. WILLIAMS *and* H. DUBNER *showed in* 1986 *that* R_{1031} *is a prime number.*

Only four other repunits that are primes are known: R_2, R_{19}, R_{23}, and R_{317}.

I offer one final noteworthy record—but if you want to know why and how it was found, you must ask H. DUBNER.

Record. *The largest known prime number whose digits are also all prime is*

$$723\,232\,523\,232\,272\,325\,252 \times \frac{10^{3120} - 1}{10^{20} - 1} + 1.$$

It has 3120 *digits and was discovered by* H. DUBNER *in* 1992.

The observation and study of the prime numbers is a fruitful as well as diverting activity. Mathematicians derive much enjoyment from it, and that alone is worth the labor. In time, one comes to consider the prime numbers as friends—friends who bring us problems!

References

1979 G. H. Hardy and E. M. Wright. *An Introduction to the Theory of Numbers.* Clarendon Press, Oxford, 5th edition.

1988 P. Ribenboim. Euler's famous prime generating polynomial and the class number of imaginary quadratic fields. **34**: 23–42. See also this volume, Chapter 5.

1991 P. Ribenboim. *The Little Book of Big Primes.*
1996 P. Ribenboim. *The New Book of Prime Number Records.* New York.

4

Selling Primes

I am a big shot in a factory that produces primes.

And I will tell you an interesting dialogue with a buyer, coming from an exotic country.

The Dialogue

—**Buyer**: I wish to buy some primes.

—**I** (generously): I can give to you, free of charge, many primes: 2, 3, 5, 7, 11, 13, 17, 19, ...

—**Buyer** (interrupting my generous offer): Thank you, sir; but I want primes with 100 digits. Do you have these for sale?

—**I**: In this factory we can produce primes as large as you wish. There is in fact an old method of Euclid, that you may have heard about. If I have any number n of primes, say $p_1, p_2, \ldots, p_n$, we multiply them and add 1, to get the number $N = p_1 p_2 \ldots p_n + 1$. Either N is a prime or, if it is not a prime, we pick any prime dividing N. In this way, it is easy to see that we get a prime, which is different from the ones we mixed. Call it p_{n+1}. If we now mix $p_1, p_2, \ldots, p_n, p_{n+1}$ as I already said, we get still another prime p_{n+2}. Repeating this procedure we get as many primes as we wish and so, we are bound to get primes as large as we wish, for sure with at least 100 digits.

—**Buyer**: You are very nice to explain your procedure. Even in my distant country, I have heard about it. It yields primes that may be arbitrarily large. However, I want to buy primes that have exactly 100 digits, no more, no less. Do you have them?

—**I**: Yes. Long ago—at the beginning of last century, BERTRAND observed that between any number $N > 1$ and its double $2N$, there exists at least one prime number. This experimental observation was confirmed by a rigorous proof by CHEBYSHEV. So I can find the primes p_1, p_2, p_3 where

$$10^{99} < p_1 < 2 \times 10^{99}$$
$$2 \times 10^{99} < p_2 < 4 \times 10^{99}$$
$$4 \times 10^{99} < p_3 < 8 \times 10^{99}$$

—**Buyer**: This means that you have guaranteed 3 primes with 100 digits, and perhaps a few more. But I want to buy many primes with 100 digits. How many can you produce?

—**I**: I have never counted how many primes of 100 digits could eventually be produced. I have been told that my colleagues in other factories have counted the total number of primes up to 10^{20}. We usually write $\pi(N)$ to denote the number of primes up to the number N. Thus, the count I mentioned has given:

$$\pi(10^8) = 5\ 761\ 455$$
$$\pi(10^9) = 50\ 847\ 534$$
$$\pi(10^{12}) = 37\ 607\ 912\ 018$$
$$\pi(10^{17}) = 2\ 625\ 557\ 157\ 654\ 233$$
$$\pi(10^{18}) = 24\ 739\ 954\ 287\ 740\ 860$$
$$\pi(10^{20}) = 2\ 220\ 819\ 602\ 560\ 918\ 840.$$

Even though all primes up to 10^{20} have not yet been produced by any factory, the count of $\pi(10^{20})$ is exact.

—**Buyer** (a bit astonished): If you cannot—as I understand—know how many primes of each large size there are in stock, how can you operate your factory and guarantee delivery of the merchandise?

—**I**: Your country sells oil, does it not? You can estimate the amount of oil at shallow depths quite accurately, but you cannot measure exactly the entire amount underground. It is just the same with us.

GAUSS, one of the foremost scientists, discovered that

$$\pi(N) \sim \frac{N}{\log N}$$

for large values of N. This was confirmed, just over a century ago, with proofs given by HADAMARD and DE LA VALLÉE POUSSIN.

—**Buyer**: Do you mean that $\pi(N)$ is approximately equal to $N/\log N$, with a small error?

—**I**: Yes. To be more precise, the relative error, namely the absolute value of the difference $|\pi(N) - N/\log N|$, divided by $\pi(N)$, tends to 0, as N increases indefinitely.

—**Buyer**: Then, because of the error, you cannot be very specific in your estimate. Unless you estimate the error.

—**I**: Correct (the buyer is not stupid ...). CHEBYSHEV showed, even before the Prime Number Theorem was proved, that if N is large, then

$$0.9 \frac{N}{\log N} < \pi(N) < 1.1 \frac{N}{\log N}.$$

To count primes with 100 digits:

$$0.9 \frac{10^{99}}{99 \log 10} < \pi(10^{99}) < 1.1 \frac{10^{99}}{990. \log 10}$$

$$0.9 \frac{10^{100}}{100 \log 10} < \pi(10^{100}) < 1.1 \frac{10^{100}}{100 \log 10}.$$

It is easy to estimate the difference $\pi(10^{100}) - \pi(10^{99})$, which gives the number of primes with exactly 100 digits:

$$3.42 \times 10^{97} < \pi(10^{100}) - \pi(10^{99}) < 4.38 \times 10^{97}.$$

—**Buyer**: You are rich! I think you have more primes than we have oil. But I wonder how your factory produces the primes with 100 digits. I have an idea but I'm not sure how efficient my method would be.

1. Write all the numbers with 100 digits.
2. Cross out, in succession, all the multiples of 2, of 3, of 5, ..., of each prime p less than 10^{99}. For this purpose, spot the first multiple of p, then cross out every pth number.

What remains are the primes between 10^{99} and 10^{100}, that is, the primes with exactly 100 digits.

—**I**: This procedure is correct and was already discovered by ERATOSTHENES (in the 3rd century B.C.). In fact, you may stop when you have crossed out the multiples of all the primes less than 10^{50}.

However, this method of production is too slow. This explains why the archeologists never found a factory of primes amongst the Greek ruins, but just temples to Apollo, statues of Aphrodite (known as Venus, since the time of Romans) and other ugly remains that bear witness to a high degree of decadence.

Even with computers this process is too slow to be practical. Think of a computer that writes 10^6 digits per second.

- There are $10^{100} - 10^{99} = 10^{99} \times 9$ numbers with 100 digits.
- These numbers have a total of $10^{101} \times 9$ digits.
- One needs $10^{95} \times 9$ seconds to write these numbers, that is about 1.5×10^{94} minutes, that is about 25×10^{92} hours, so more than 10^{91} days, that is of the order of 3×10^{88} years, that is 3×10^{86} centuries!

And after writing the numbers (if there is still an After ...) there is much more to be done!

Before the buyer complained, I added:

—**I**: There are short cuts, but even then the method would still be too slow. So, instead of trying to list the primes with 100 digits, our factory uses fast algorithms to produce enough primes to cover our orders.

—**Buyer**: I am amazed. I never thought how important it is to have a fast method. Can you tell me the procedure used in your factory? I am really curious. [Yes, this buyer was being too nosy. Now I became convinced that he was a spy.]

—**I**: When you buy a Mercedes, you don't ask how it was built. You choose your favorite color, pink, purple, or green with orange dots, you drive it and you are happy because everyone else is envious of you.

Our factory will deliver the primes you ordered and we do better than Mercedes. We support our product with a lifetime guarantee. Goodbye, Sir.

[He may have understood: Good buy, Sir ...]

After the Dialogue

I hope that after the dialogue with the spy-buyer, you became curious to know about our fast procedure to produce large primes. I shall tell you some of our most cherished secrets. In our factory there are two main divisions.

1) Production of primes
2) Control of quality.

Production of Primes

One of the bases of our production methods was discovered long ago by POCKLINGTON (1914/16). I will state and prove his theorem in the particular situation adapted to our production requirements. Then, I shall discuss how it may be used to obtain, in a surprisingly short time, primes with the required number of digits.

CRITERION OF POCKLINGTON. *Let p be an odd prime, let k be a natural number such that p does not divide k and $1 \leq k < 2(p+1)$; and let $N = 2kp + 1$. Then the following conditions are equivalent:*

1) *N is a prime.*
2) *There exists a natural number a, $2 \leq a < N$, such that*

$$a^{kp} \equiv -1 \pmod{N}$$

and

$$\gcd(a^k + 1, N) = 1.$$

PROOF.   (1) $\Rightarrow$ (2). Assume that N is a prime. As is known, there is some integer a, $1 < a < N$, such that $a^{N-1} \equiv 1 \pmod{N}$, but $a^m \not\equiv 1 \pmod{N}$ if $1 < m < N - 1$; such a number a is called a *primitive root modulo N*. Thus $a^{2kp} \equiv 1 \pmod{N}$, but $a^{kp} \not\equiv 1 \pmod{N}$; then $a^{kp} \equiv -1 \pmod{N}$. Also, $a^k \not\equiv -1 \pmod{N}$, otherwise $a^{2k} \equiv 1 \pmod{N}$, which is not true; so $\gcd(a^k + 1, N) = 1$.

(2) $\Rightarrow$ (1). In order to show that N is a prime, we shall prove: If q is any prime dividing N, then $\sqrt{N} < q$. It follows that N cannot have two (equal or distinct) prime factors, so N is a prime.

So, let q be any prime factor of N. Then $a^{kp} \equiv -1 \pmod{q}$ and $a^{2kp} \equiv 1 \pmod{q}$. Hence, $\gcd(a, q) = 1$. Let e be the order of a

modulo q, hence e divides $q - 1$, by Fermat's Little Theorem. Similarly, e divides $2kp = N - 1$, because $a^{2kp} \equiv 1 \pmod{q}$. Note that $a^k \not\equiv 1 \pmod{q}$, otherwise $a^{kp} \equiv 1 \pmod{q}$; from $a^{kp} \equiv -1 \pmod{q}$, it follows that $q = 2$ and N would be even, which is false.

From $\gcd(a^k + 1, N) = 1$, it follows that $a^k \not\equiv -1 \pmod{q}$. Hence, $a^{2k} \not\equiv 1 \pmod{q}$, thus $e \nmid 2k = (N-1)/p$. But $e \mid N-1$, so $(N-1)/e$ is an integer, hence $p \nmid (N-1)/e$. Since $N - 1 = e((N-1)/e)$ and $p \mid N - 1$, then $p \mid e$, thus $p \mid q - 1$. Also $2 \mid q - 1$, hence $2p \mid q - 1$, so $2p \leq q - 1$ and $2p + 1 \leq q$. It follows that $N = 2kp + 1 < 2 \times 2(p+1)p + 1 = 4p^2 + 4p + 1 = (2p+1)^2 \leq q^2$, therefore $\sqrt{N} < q$. This concludes the proof. $\qquad\square$

The criterion of POCKLINGTON is applied as follows to obtain primes of a required size, say with 100 digits.

First step: Choose, for example, a prime p_1 with $d_1 = 5$ digits. Find $k_1 < 2(p_1 + 1)$ such that $p_2 = 2k_1p_1 + 1$ has $d_2 = 2d_1 = 10$ digits or $d_2 = 2d_1 - 1 = 9$ digits and there exists $a_1 < p_2$ satisfying the conditions $a_1^{k_1 p_1} \equiv -1 \pmod{p_2}$ and $\gcd(a_1^{k_1} + 1, p_2) = 1$. By Pocklington's criterion, p_2 is a prime.

Subsequent steps: Repeat the same procedure starting with the prime p_2 to obtain the prime p_3, etc. . . . In order to produce a prime with 100 digits, the process must be iterated five times. In the last step, k_5 should be chosen so that $2k_5p_5 + 1$ has 100 digits.

Feasibility of the Algorithm

Given p and k, with $1 \leq k < 2(p + 1)$, k not a multiple of p, if $N = 2kp+1$ is a prime, then it has a primitive root. It would be much too technical to explain in detail the following results, some known to experts, others still unpublished. It follows from a generalized form of the Riemann hypothesis that if x is a large positive real number and the positive integer a is not a square, then the ratio

$$\frac{\#\{\text{primes } q \leq x \text{ such that } a \text{ is a primitive root modulo } q\}}{\#\{\text{primes } q \leq x\}}$$

converges; if a is a prime, the limit is at least equal to Artin's constant

$$\prod_{q \text{ prime}} \left(1 - \frac{1}{q(q-1)}\right) \approx 0.37.$$

Better, given positive integers, a, b, not squares, and a large prime q, the probability that a or b is a primitive root modulo q is much larger. Taking $a = 2$, $b = 3$, it is at least 58%. The corresponding probability increases substantially when taking three positive integers a, b, c which are not squares.

This suggests that we proceed as follows. Given the prime p, choose k, not a multiple of p, $1 \leq k < 2(p + 1)$. If $N = 2kp + 1$ is a prime, then very likely 2, 3 or 5 is a primitive root modulo N. If this is not the case, it is more practical to choose another integer k', like k, and investigate whether $N' = 2k'p + 1$ is a prime.

The question arises: what are the chances of finding k, such that N is a prime? I now discuss this point.

1. According to a special case of DIRICHLET's famous theorem, (see RIBENBOIM (1996), RIBENBOIM (1991)), given p there exist infinitely many integers $k \geq 1$ such that $2kp + 1$ is a prime. This may be proved in an elementary way.

2. How small may k be, so that $2kp + 1$ is a prime? A special case of a deep theorem of LINNIK asserts:
   For every sufficiently large p, in the arithmetic progression with first term 1 and difference $2p$, there exists a prime $p_1 = 2kp + 1$ satisfying $p_1 \leq (2p)^L$; here L is a positive constant (that is, L is independent of p) (see RIBENBOIM (1996)).

3. Recently, HEATH-BROWN has shown that $L \leq 5.5$.

4. In POCKLINGTON's criterion, it is required to find $k < 2(p + 1)$ such that $p_1 = 2kp + 1$ is a prime. This implies that $p_1 < (2p+1)^2$. No known theorem guarantees that such small values of k lead to a prime.

5. Recent work of BOMBIERI, FRIEDLANDER, and IWANIEC deals with primes p for which there are small primes $p_1 = 2kp + 1$. Their results, which concern averages, point to the existence of a sizable proportion of primes p with small prime $p_1 = 2kp + 1$.

The problems considered above are of great difficulty. In practice, we may ignore these considerations and find, with a few trials, the appropriate value of k.

Estimated time to produce primes with 100 digits

The time required to perform an algorithm depends on the speed of the computer and on the number of bit operations (i.e., operations with digits) that are necessary.

As a basis for this discussion, we may assume that the computer performs 10^6 bit operations per second. If we estimate an upper bound for the number of bit operations, dividing by 10^6 gives an upper bound for the number of seconds required.

A closer look at the procedure shows that it consists of a succession of the following operations on natural numbers: multiplication ab modulo n, power a^b modulo n, calculation of greatest common divisor.

It is well-known (see LeVeque (1975), Mignotte (1991)) and not difficult to show that for each of the above operations there exist $C > 0$ and an integer $e \geq 1$ such that the number of bit operations required to perform the calculation is at most Cd^e, where d is the maximum of the number of digits of the numbers involved. Combining these estimates gives an upper bound of the same form Cd^e for the method ($C > 0, e \geq 1$, and d is the maximum of the number of digits of all integers involved in the calculation).

It is not my purpose to give explicit values for C and e, when p, k, a are given. Let me just say that C, e are rather small, so the algorithm runs very fast. I stress that in this estimate the time required in the search of k, a is not taken into account.

The above discussion makes clear that much more remains to be understood in the production of primes and the feasibility of the algorithm. This task is delegated to our company's division of research and development, and I admire our colleagues in the research subdivision who face the deep mysteries of prime numbers.

Before I rapidly tour our division of quality control, I would like to make a few brief comments about our preceding considerations. They concern the complexity of an algorithm.

An algorithm A, performed on natural numbers, is said to run in *polynomial time* if there exist positive integers C, e (depending on the algorithm) such that the number of bit operations (or equivalently, the time) required to perform the algorithm on natural numbers with at most d digits is at most Cd^e.

An algorithm which does not run in polynomial time is definitely too costly to implement and is rejected by our factory. It is one of the main subjects of research to design algorithms that run in polynomial time. The algorithm to produce primes of a given size, for all practical purposes, runs in polynomial time, even though this has not yet been supported by a proof.

Quality Control

The division of quality control in our factory watches that the primes we sell are indeed primes. When POCKLINGTON's method is used we need only worry that no silly calculation error was made, because it leads automatically to prime numbers. If other methods are used, as I shall soon invoke, there must be a control. The division of quality control also engages in consulting work. A large number N is presented with the question: Is N a prime number?

Thus, our division of quality control also deals with tests of primality. Since this is a cash rewarding activity, there are now many available tests of primality. I may briefly classify them from the following four points of view:

1. Tests for generic numbers.
   Tests for numbers of special forms, like $F_n = 2^{2^n} + 1$ (Fermat numbers), $M_p = 2^p - 1$, (p prime, Mersenne numbers), etc....
2. Tests fully justified by theorems.
   Tests based on justification that depends on forms of Riemann's hypothesis of the zeros of the zeta function, or on heuristic arguments.
3. Deterministic tests.
4. Probabilistic or Monte Carlo tests.

A deterministic test applied to a number N will certify that N is a prime or that N is a composite number. A Monte Carlo test applied to N will certify either that N is composite, or that, with large probability, N is a prime.

Before I proceed, let me state that the main problem tempting the researchers is the following: Will it be possible to find a fully justified and deterministic test of primality for generic numbers, that runs in polynomial time? Or will it be proven that there cannot exist a deterministic, fully justified test of primality which runs in polynomial time, when applied to any natural number?

This is a tantalizing and deep problem.

It would be long-winded and complex even to try to describe for you all the methods and algorithms used in primality testing. So, I shall concentrate only on the strong pseudoprime test, which is of Monte Carlo type.

Pseudoprimes

Let N be a prime, let a be such that $1 < a < N$. By Fermat's Little Theorem, $a^{N-1} \equiv 1 \pmod{N}$.

However, the converse is not true. The smallest example is $N = 341 = 11 \times 31$, with $a = 2$, $2^{340} \equiv 1 \pmod{341}$.

The number N is called a *pseudoprime in base a*, where $\gcd(a, N) = 1$, if N is composite and $a^{N-1} \equiv 1 \pmod{N}$. For each $a \geq 2$, there are infinitely many pseudoprimes in base a. Now observe that every odd prime N satisfies the following property:

> For any a, $2 \leq a < N$, with $\gcd(a, N) = 1$, writing $N - 1$ in the form $N - 1 = 2^s d$ (where $1 \leq s$, d is odd), either $a^d \equiv 1 \pmod{N}$ or there exists r, $0 \leq r < s$, such that $a^{2^r d} \equiv -1 \pmod{N}$. $\qquad (*)$

Again, the converse is not true, as illustrated by $N = 2047 = 23 \times 89$, with $a = 2$.

The number N is called a *strong pseudoprime in base a*, where $\gcd(a, N) = 1$, if N is composite and the condition $(*)$ is satisfied.

It has been shown by POMERANCE, SELFRIDGE, and WAGSTAFF that for every $a \geq 2$ there exist infinitely many strong pseudoprimes in base a.

The strong pseudoprime test

The main steps in the strong pseudoprime test for a number N are the following:

1. Choose $k > 1$ numbers a, $2 \leq a < N$, such that $\gcd(a, N) = 1$. This is easily done by trial division and does not require the knowledge of the prime factors of N. If $\gcd(a, N) > 1$ for some a, $1 < a < N$, then N is composite.
2. For each chosen base a, check if the condition $(*)$ is satisfied.

If there is a such that $(*)$ is not satisfied, then N is composite. Thus, if N is a prime, then $(*)$ is satisfied for each base a. The events that condition (∗) is satisfied for different bases may be legitimately considered as independent if the bases are randomly chosen.

Now, RABIN proved (see RIBENBOIM (1996)): Let N be composite. Then the number of bases a for which N is a strong pseudoprime in base a is less than $\frac{1}{4}(N-1)$. Thus, if N is composite, the probability that (∗) is satisfied for k bases is at most $1/4^k$. Hence, certification that N is a prime when (∗) is satisfied for k distinct bases is incorrect in only one out of 4^k numbers; for example if $k = 30$, the certification is incorrect only once in every 10^{18} numbers.

The strong pseudoprime test runs in polynomial time and it is applicable to any number.

If a generalized form of Riemann's hypothesis is assumed to be true, MILLER showed (see RIBENBOIM (1996)): If N is composite, there exists a base a, with $\gcd(a, N) = 1$, such that $a < (\log N)^{2+\varepsilon}$, for which (∗) is not satisfied.

A New Production Method

We may use Rabin's test to produce numbers with 100 digits which may be certified to be prime numbers, with only a very small probability of error.

1. Pick a number N with 100 digits. Before doing any hard work, it is very easy, with trial division, to find out if this number does, or does not have, any prime factor less than, say, 1000. In the latter case, keep this number; in the first case, discard N, pick another number N', and proceed similarly.

2. Use $k = 30$ small numbers a, prime to N, as bases to verify if condition (∗) is satisfied. Discard N if, for some base a, the condition (∗) is not satisfied and repeat the process with some other number N'. If (∗) is satisfied for all a then, according to Rabin's calculation, we may certify that N is prime; in doing so, we are incorrect only in at most one out of 10^{18} numbers. How unlucky can we be and choose, in succession, numbers which are composite? According to the inequalities of CHEBYSHEV already indicated, the proportion of numbers with 100 digits that are prime is not less than $\frac{3.42}{9 \times 10^2} \approx \frac{1}{260}$ and not more than $\frac{4.38}{9 \times 10^9} \approx \frac{1}{205}$. Unintelligent employees who would pick even numbers, or numbers divisible by $3, 5, \ldots$ or small primes (say up to 1000) are

sure to be fired. Thus the luck of picking a prime increases and it becomes quite reasonable to use Rabin's method of production.

We may sell the number N as if it were a prime, even with a "money back guarantee" because the probability that we are selling a composite number is only 1 in every 1 000 000 000 000 000 000 sales! This is a better guarantee than anyone can get in any deal. We are sure that our company will not be bankrupt and will continue to support generously my trips to advertise our products—all complemented by lavish dinners and the finest wines, to help convince our customers that primes are the way of life.

Appendix

Here is an example of a prime with 100 digits which was calculated by L. ROBERTS, using the method of POCKLINGTON.

$p_1 = 2333$ $k_1 = 2001$ $a_1 = 2$
$p_2 = 9336667$ $k_2 = 9336705$ $a_2 = 3$
$p_3 = 174347410924471$ $k_3 = 174347410924479$ $a_3 = 2$
$p_4 = 6079403939213548914830805121 9$
            $k_4 = 6079403939213548914830805125 6$ $a_4 = 3$
$p_5 = 739183045122504318980574950295193587260702667372985056$
       2129
$k_5 = 500000000000000000000000000000000000000137$ $a_5 = 2$
$p_6 = 739183045122504318980574950295193587260905203527348622$
       396300677536380883042909432530860197905402334 7

References

1914/16 H. C. Pocklington. The determination of the prime or composite nature of large numbers by Fermat's theorem. *Proc. Cambridge Phil. Soc.*, 18:29–30.

1975 W. J. LeVeque, editor. *Studies in Number Theory.* Math. Assoc. of America, Washington.

1979 D. A. Plaisted. Fast verification, testing and generation of large primes. *Theor. Comp. Sci.*, 9:1–16.

1991 M. Mignotte. *Mathematics for Computer Algebra.* Springer-Verlag, New York.

1991 P. Ribenboim. *The Little Book of Big Primes.* Springer-Verlag, New York.

1996 P. Ribenboim. *The New Book of Prime Number Records.* Springer-Verlag, New York.

5

Euler's Famous Prime Generating Polynomial and the Class Number of Imaginary Quadratic Fields*

Introduction

Can a non-constant polynomial with integral coefficients assume only prime values?

No! because of the following.

Theorem. *If $f(X) \in \mathbb{Z}[X]$, $\deg(f) > 0$, there exist infinitely many natural numbers n such that $f(n)$ is composite.*

PROOF. It is true if $f(n)$ is composite for every $n \geq 1$. Assume that there exists $n_0 \geq 1$ such that $f(n_0) = p$ is a prime. Since $\lim_{n \to \infty} |f(n)| = \infty$, there exists $n_0 \geq 1$ such that if $n \geq n_1$ then $|f(n)| > p$. Take any h such that $n_0 + ph \geq n_1$. Then $|f(n_0 + ph)| > p$, but $f(n_0 + ph) = f(n_0) + (\text{multiple of } p) = \text{multiple of } p$, so $|f(n_0 + ph)|$ is composite. $\square$

On the other hand, must a non-constant polynomial $f(X) \in \mathbb{Z}[X]$ always assume at least one prime value?

---

*This is the text of a lecture at the University of Rome, on May 8, 1986. The original notes disappeared when my luggage was stolen in Toronto (!); however, I had given a copy to my friend Paolo Maroscia, who did not have his luggage stolen in Rome (!) and was very kind to let me consult his copy. It is good to have friends.

The question is interesting if $f(X)$ is irreducible, primitive (that is, the greatest common divisor of its coefficients is equal to 1), and even more, there is no prime p dividing all values $f(n)$ (for arbitrary integers n).

BOUNIAKOWSKY, and later SCHINZEL and SIERPIŃSKI (1958), conjectured that any polynomial $f(X) \in \mathbb{Z}[X]$ satisfying the above conditions assumes a prime value. This has never been proved for arbitrary polynomials. For the specific polynomials $f(X) = aX + b$, with $\gcd(a, b) = 1$, it is true—this is nothing else than the famous theorem of DIRICHLET: every arithmetic progression

$$\{a + kb \mid k = 0, 1, 2, \ldots\} \quad \text{with} \quad \gcd(a, b) = 1,$$

contains infinitely many primes.

In my book entitled *The New Book of Prime Number Records* (1995), I indicated many astonishing consequences of the hypothesis of BOUNIAKOWSKY derived by SCHINZEL and SIERPIŃSKI. But this is not the subject of the present chapter.

Despite the theorem and what I have just said, for many polynomials it is easy to verify that they assume prime values, and it is even conceivable that they assume prime values at many consecutive integers. For example, EULER's famous polynomial $f(X) = X^2 + X + 41$ has the property that $f(n)$ is a prime for $n = 0, 1, \ldots, 39$ (40 successive prime values):

41, 43, 47, 53, 61, 71, 83, 97, 113, 131, 151, 173, 197, 223, 251, 281, 313, 347, 383, 421, 461, 503, 547, 593, 641, 691, 743, 797, 853, 911, 971, 1033, 1097, 1163, 1231, 1301, 1373, 1447, 1523, 1601.

However, $f(40) = 40^2 + 40 + 41 = 40 \times 41 + 41 = 41^2$.

Note that if $n > 0$, then $(-n)^2 + (-n) + 41 = (n-1)^2 + (n-1) + 41$, so $X^2 + X + 41$ assumes also prime values for all integers

$$n = -40, -39, \ldots, -2, -1.$$

Which other polynomials are like the above?

Some of these polynomials may be easily obtained from $X^2 + X + c$ by just changing X into $X - a$, for some $a \geq 1$. For example, $(X - a)^2 + (X - a) + 41 = X^2 - (2a - 1)X + (a^2 - a + 41)$; taking $a = 1$ gives $X^2 - X + 41$, which assumes primes values for every integer n,

$-39 \leq n \leq 40$, while taking $a = 40$, gives $X^2 - 79X + 1601$, which assumes primes values for every integer n, $0 \leq n \leq 79$, but these are the same values assumed by $X^2 + X + 41$, taken twice. In summary, it is interesting to concentrate the attention on polynomials of the form $X^2 + X + c$ and their values at consecutive integers $n = 0, 1, \ldots$. If the value at 0 is a prime q, then $c = q$. Since $(q-1)^2 + (q-1) + q = q^2$, then at best $X^2 + X + q$ assumes prime values for $0, 1, 2, \ldots, q - 2$ (such as $q = 41$). For example, if $f(X) = X^2 + X + q$ and $q = 2, 3, 5, 11, 17, 41$, then $f(n)$ is a prime for $n = 0, 1, \ldots, q-2$. However, if $q = 7, 13, 19, 23, 29, 31, 37$ this is not true, as may be easily verified.

Can one find $q > 41$ such that $X^2 + X + q$ has prime value for $n = 0, 1, \ldots, q - 2$? Are there infinitely many, or only finitely many such primes q? If so, what is the largest possible q?

The same problem should be asked for polynomials of first degree $f(X) = aX + b$, with $a, b \geq 1$. If $f(0)$ is a prime q, then $b = q$. Then $f(q) = aq + q = (a + 1)q$ is composite. So, at best, $aX + q$ assumes prime values for X equal to $0, 1, \ldots, q - 1$.

Can one find such polynomials? Equivalently, can one find arithmetic progressions of q prime numbers, of which the first number is equal to q?

For small values of q this is not difficult.

If $q = 3$, take: 3, 5, 7, so $f(X) = 2X + 3$.

If $q = 5$, take: 5, 11, 17, 23, 29, so $f(X) = 6X + 5$.

If $q = 7$, take: 7, 157, 307, 457, 607, 757, 907, so $f(X) = 150X + 7$.

Quite recently, KELLER communicated to me that for $q = 11, 13$ the smallest such arithmetic progressions are given by polynomials $f(X) = d_{11}X + 11$, respectively $f(X) = d_{13}X + 13$, with

$$d_{11} = 1536160080 = 2 \times 3 \times 5 \times 7 \times 7315048,$$
$$d_{13} = 9918821194590 = 2 \times 3 \times 5 \times 7 \times 11 \times 4293861989;$$

this determination required a considerable amount of computation, done by KELLER and LÖH.

It is not known whether for every prime q there exists an arithmetic progression of q primes of which the first number is q. Even the problem of finding arbitrarily large arithmetic progressions consisting only of prime numbers (with no restriction on the initial term or the difference) is still open. The largest known such arithmetic progression consists of 22 primes and was found by PRITCHARD, MORAN, and THYSSEN (1995).

The determination of all polynomials $f(X) = X^2 + X + q$ such that $f(n)$ is a prime for $n = 0, 1, \ldots, q - 2$ is intimately related with the theory of imaginary quadratic fields. In order to understand this relationship, I shall indicate now the main results which will be required.

1   Quadratic extensions

Let d be an integer which is not a square, and let $K = \mathbb{Q}(\sqrt{d})$ be the field of all elements $\alpha = a + b\sqrt{d}$, where $a, b \in \mathbb{Q}$. There is no loss of generality to assume that d is square-free, hence $d \not\equiv 0 \pmod 4$. The field extension $K|\mathbb{Q}$ is quadratic, that is, K is a vector space of dimension 2 over $\mathbb{Q}$.

Conversely, if K is a quadratic extension field of $\mathbb{Q}$, then it is necessarily of the form $K = \mathbb{Q}(\sqrt{d})$, where d is a square-free integer.

If $d > 0$, then K is a subfield of the field $\mathbb{R}$ of real numbers: it is called a *real quadratic field*.

If $d < 0$, then K is not a subfield of $\mathbb{R}$, and it is called an *imaginary quadratic field*.

If $\alpha = a + b\sqrt{d} \in K$, with $a, b \in \mathbb{Q}$, its conjugate is $\alpha' = a - b\sqrt{d}$. Clearly, $\alpha = \alpha'$ exactly when $\alpha \in \mathbb{Q}$.

The norm of α is $N(\alpha) = \alpha\alpha' = a^2 - db^2 \in \mathbb{Q}$. It is obvious that $N(\alpha) \neq 0$ exactly when $\alpha \neq 0$. If $\alpha, \beta \in K$, then $N(\alpha\beta) = N(\alpha)N(\beta)$; in particular, if $\alpha \in \mathbb{Q}$, then $N(\alpha) = \alpha^2$.

The trace of α is $\mathrm{Tr}(\alpha) = \alpha + \alpha' = 2a \in \mathbb{Q}$. If $\alpha, \beta \in K$, then $\mathrm{Tr}(\alpha + \beta) = \mathrm{Tr}(\alpha) + \mathrm{Tr}(\beta)$; in particular, if $\alpha \in \mathbb{Q}$, then $\mathrm{Tr}(\alpha) = 2\alpha$.

It is clear that α, α' are the roots of the quadratic equation $X^2 - \mathrm{Tr}(\alpha)X + N(\alpha) = 0$.

2   Rings of integers

Let $K = \mathbb{Q}(\sqrt{d})$, where d is a square-free integer.

The element $\alpha \in K$ is an algebraic integer when there exist integers $m, n \in \mathbb{Z}$ such that $\alpha^2 + m\alpha + n = 0$.

Let A be the set of all algebraic integers of K. The set A is a subring of K, which is the field of fractions of A, and $A \cap \mathbb{Q} = \mathbb{Z}$. If $\alpha \in A$, then the conjugate α' is an element of A. Clearly, $\alpha \in A$ if and only if both $N(\alpha)$ and $\mathrm{Tr}(\alpha)$ are in $\mathbb{Z}$.

Here is a criterion for the element $\alpha = a + b\sqrt{d}(a, b \in \mathbb{Q})$ to be an algebraic integer: $\alpha \in A$ if and only if

$$\begin{cases} 2a = u \in \mathbb{Z}, \quad 2b = v \in \mathbb{Z} \\ u^2 - dv^2 \equiv 0 \pmod 4. \end{cases}$$

Using this criterion, it may be shown:

If $d \equiv 2$ or $3 \pmod 4$, then $A = \{a + b\sqrt{d} \mid a, b \in \mathbb{Z}\}$.

If $d \equiv 1 \pmod 4$, then $A = \{\frac{1}{2}(a + b\sqrt{d}) \mid a, b \in \mathbb{Z}, a \equiv b \pmod 2\}$.

If $\alpha_1, \alpha_2 \in A$ are such that every element $\alpha \in A$ is uniquely of the form $\alpha = m_1\alpha_1 + m_2\alpha_2$, with $m_1, m_2 \in \mathbb{Z}$, then $\{\alpha_1, \alpha_2\}$ is called an *integral basis* of A. In other words, $A = \mathbb{Z}\alpha_1 \oplus \mathbb{Z}\alpha_2$.

If $d \equiv 2$ or $3 \pmod 4$, then $\{1, \sqrt{d}\}$ is an integral basis of A.

If $d \equiv 1 \pmod 4$, then $\left\{1, \frac{1+\sqrt{d}}{2}\right\}$ is an integral basis of A.

3 Discriminant

Let $\{\alpha_1, \alpha_2\}$ be an integral basis. Then

$$D = D_k = \det \begin{pmatrix} \mathrm{Tr}(\alpha_1^2) & \mathrm{Tr}(\alpha_1\alpha_2) \\ \mathrm{Tr}(\alpha_1\alpha_2) & \mathrm{Tr}(\alpha_2^2) \end{pmatrix}$$

is independent of the choice of the integral basis. It is called the *discriminant* of K. It is a non-zero integer.

If $d \equiv 2$ or $3 \pmod 4$, then

$$D = \det \begin{pmatrix} \mathrm{Tr}(2) & \mathrm{Tr}(\sqrt{d}) \\ \mathrm{Tr}(\sqrt{d}) & \mathrm{Tr}(d) \end{pmatrix} = \det \begin{pmatrix} 2 & 0 \\ 0 & 2d \end{pmatrix},$$

so, $D = 4d$.

If $d \equiv 1 \pmod 4$, then

$$D = \det \begin{pmatrix} \mathrm{Tr}(1) & \mathrm{Tr}\left(\dfrac{1+\sqrt{d}}{2}\right) \\ \mathrm{Tr}\left(\dfrac{1+\sqrt{d}}{2}\right) & \mathrm{Tr}\left(\dfrac{1+\sqrt{d}}{2}\right)^2 \end{pmatrix} = \det \begin{pmatrix} 2 & 1 \\ 0 & \dfrac{1+d}{2} \end{pmatrix},$$

so $D = d$.

Every discriminant D is congruent to 0 or $1 \pmod 4$.

In terms of the discriminant,

$$A = \{\tfrac{1}{2}(a + b\sqrt{D}) \mid a, b \in \mathbb{Z}, a^2 \equiv Db^2 \pmod 4\}.$$

4   Decomposition of primes

Let $K = \mathbb{Q}(\sqrt{d})$, where d is a square-free integer, and let A be the ring of integers of K.

The ideal $P \neq 0$ of A is a prime ideal if the residue ring A/P has no zero-divisor.

If P is a prime ideal there exists a unique prime number p such that $P \cap \mathbb{Z} = \mathbb{Z}p$, or equivalently, $P \supseteq Ap$.

If I, J are non-zero ideals of A, it is said that I divides J when there exists an ideal I_1 of A such that $I \cdot I_1 = J$.

The prime ideal P containing the prime number p divides the ideal Ap.

If I is a non-zero ideal of A, then the residue ring A/I is finite. The norm of I is $N(I) = \#(A/I)$.

A   Properties of the norm

If I, J are non-zero ideals, then $N(I, J) = N(I)N(J)$. If I divides J, then $N(I)$ divides $N(J)$.

If $\alpha \in A$, $\alpha \neq 0$, then $N(A\alpha) = |N(\alpha)|$ (absolute value of the norm of α). In particular, if $a \in \mathbb{Z}$ then $N(Aa) = a^2$.

If the prime ideal P divides Ap then $N(P)$ is equal to p or to p^2.

Every non-zero ideal I is, in a unique way, the product of powers of prime ideals:

$$I = \prod_{i=1}^{n} P_i^{e_i}.$$

If I, J are non-zero ideals, and if $I \supseteq J$, then I divides J.

Every ideal $I \neq 0$ may be generated by two elements, of which one may be chosen in $\mathbb{Z}$; if $I \cap \mathbb{Z} = \mathbb{Z}n$, then $I = An + A\alpha$ for some $\alpha \in A$. In this case, the notation $I = (n, \alpha)$ is used.

Consider now the special case where p is a prime number. Then Ap is of one of the following types:

$$
\begin{cases}
Ap = P^2, & \text{where } P \text{ is a prime ideal: } p \text{ is ramified in } K. \\
Ap = P, & \text{where } P \text{ is a prime ideal: } p \text{ is inert in } K. \\
Ap = P_1 P_2, & \text{where } P_1, P_2 \text{ are distinct prime ideals: } p \text{ is} \\
& \text{decomposed or splits in } K.
\end{cases}
$$

Note also that if $Ap = I \cdot J$, where I and J are any ideals (different from A), not necessarily distinct, then I and J must in fact be prime ideals.

I shall now indicate when a prime number p is ramified, inert, or decomposed, and also give generators of the prime ideals of A. There are two cases: $p \neq 2$ and $p = 2$.

Denote by $\left(\frac{d}{p}\right)$ the Legendre symbol, so

$$
\begin{cases}
\left(\frac{d}{p}\right) = 0 & \text{when } p \text{ divides } d, \\
\left(\frac{d}{p}\right) = +1 & \text{when } d \text{ is a square modulo } p, \\
\left(\frac{d}{p}\right) = -1 & \text{when } d \text{ is not a square modulo } p.
\end{cases}
$$

Let $p \neq 2$.

1) If p divides d, then $Ap = (p, \sqrt{d})^2$.

2) If p does not divide d and there does not exist $a \in \mathbb{Z}$ such that $d \equiv a^2 \pmod{p}$, then Ap is a prime ideal.

3) If p does not divide d and there exists $a \in \mathbb{Z}$ such that $d \equiv a^2 \pmod{p}$, then $Ap = (p, a + \sqrt{d})(p, a - \sqrt{d})$.

Hence,

1) p is ramified if and only if $\left(\frac{d}{p}\right) = 0$.

2) p is inert if and only if $\left(\frac{d}{p}\right) = -1$.

3) p is decomposed if and only if $\left(\frac{d}{p}\right) = +1$.

PROOF. The proof is divided into several parts. (a) If $\left(\frac{d}{p}\right) = -1$, then Ap is a prime ideal.

Otherwise, $Ap = P \cdot P'$ or P^2, with $P \cap \mathbb{Z} = \mathbb{Z}p$. Let $\alpha \in A$ be such that $P = (p, \alpha) \supseteq A\alpha$, so $P \mid A\alpha$, hence p divides $N(P)$, which in turn divides $N(A\alpha) = |N(\alpha)|$. If $p \mid \alpha$, then $\frac{\alpha}{p} \in A$ and $P = Ap \cdot \left(1, \frac{\alpha}{p}\right) = Ap$, which is absurd. So $p \nmid \alpha$. Then,

$$
\begin{cases}
d \equiv 2 \text{ or } 3 \pmod 4 \\
d \equiv 1 \pmod 4
\end{cases}
\Rightarrow
\begin{cases}
\alpha = a + b\sqrt{d}, & \text{with } a, b \in \mathbb{Z} \\
\alpha = \frac{a + b\sqrt{d}}{2}, & \text{with } a, b \in \mathbb{Z}, \, a \equiv b \pmod 2
\end{cases}
$$

$$
\Rightarrow
\begin{cases}
N(\alpha) = a - db^2 \\
N(\alpha) = \frac{a^2 - db^2}{4}
\end{cases}
$$

$$
\Rightarrow p \text{ divides } a^2 - db^2,
$$

hence $a^2 \equiv db^2 \pmod{p}$, and so $p \nmid b$ (otherwise $p \mid a$, hence $p \mid \alpha$, which is absurd).

Let b' be such that $bb' \equiv 1 \pmod{p}$, so $(ab')^2 \equiv d \pmod{p}$, therefore either $p \mid d$ or $\left(\frac{d}{p}\right) = +1$, which is a contradiction.

(b) If $\left(\frac{d}{p}\right) = 0$, then $Ap = (p, \sqrt{d})^2$.

Indeed, let $P = (p, \sqrt{d})$, so

$$P^2 = (p^2, p\sqrt{d}, d) = Ap\left(p, \sqrt{d}, \frac{d}{p}\right)$$

since $\frac{d}{p} \in \mathbb{Z}$. But d is square-free, so $\gcd\left(p, \frac{d}{p}\right) = 1$, hence $P^2 = Ap$, and this implies that P is a prime ideal.

(c) If $\left(\frac{d}{p}\right) = -1$, then $Ap = (p, a + \sqrt{d})(p, a - \sqrt{d})$, where $1 \leq a \leq p - 1$ and $a^2 \equiv d \pmod{p}$.

Indeed,

$$(p, a + \sqrt{d})(p, a - \sqrt{d}) = (p^2, pa + p\sqrt{d}, pa - p\sqrt{d}, a^2 - d)$$
$$= Ap\left(p, a + \sqrt{d}, a - \sqrt{d}, \frac{a^2 - d}{p}\right)$$
$$= Ap\left(p, a + \sqrt{d}, a - \sqrt{d}, 2a, \frac{a^2 - d}{p}\right)$$
$$= Ap,$$

because $\gcd(p, 2a) = 1$. If one of the ideals $(p, a + \sqrt{d})$, $(p, a - \sqrt{d})$ is equal to A, so is the other, which is not possible.

So $(p, a + \sqrt{d})$, $(p, a - \sqrt{d})$ are prime ideals. They are distinct: if $(p, a + \sqrt{d}) = (p, a - \sqrt{d})$, then they are equal to their sum

$$(p, a + \sqrt{d}, a - \sqrt{d}) = (p, a + \sqrt{d}, a - \sqrt{d}, 2a) = A,$$

which is absurd.

Finally, these three cases are exclusive and exhaustive, so the converse assertions are also true.   □

Note. If $d \equiv 1 \pmod{4}$ and $d \equiv a^2 \pmod{p}$, then

$$(p, a + \sqrt{d}) = (p, l(a - 1) + \omega),$$

where $\omega = \frac{1 + \sqrt{d}}{2}$ and $2l \equiv 1 \pmod{p}$. Hence, if $\left(\frac{d}{p}\right) \neq -1$ there exists $b \in \mathbb{Z}$, $0 \leq b \leq p - 1$, such that p divides $N(b + \omega)$ and, moreover, if $b = p - 1$, then $d \equiv 1 \pmod{p}$.

Indeed, $a + \sqrt{d} = a - 1 + 2\omega$. If $2l \equiv 1 \pmod{p}$, then

$$(p, a + \sqrt{d}) = (p, (a-1) + 2\omega) = (p, l(a-1) + \omega).$$

If $\left(\frac{d}{p}\right) \neq -1$, then there exists a prime ideal P dividing Ap, where

$$P = (p, a + \sqrt{d}), \qquad 0 \leq a \leq p - 1.$$

So, $P = (p, b + \omega)$ with $0 \leq b \leq p - 1$ and $b \equiv l(a-1) \pmod{p}$.

Since $P \supseteq A(b+\omega)$, it follows that p divides $N(P)$, which divides $N(b + w)$. Finally, if p divides $N(p - 1 + w) = N\left(\frac{2p-1+\sqrt{d}}{2}\right) = \frac{(2p-1)^2-d}{4}$, then p divides $\frac{1-d}{4}$, so $d \equiv 1 \pmod{p}$.

Let $p = 2$.

If $d \equiv 2 \pmod 4$, then $A2 = (2, \sqrt{d})^2$.

If $d \equiv 3 \pmod 4$, then $A2 = (2, 1 + \sqrt{d})^2$.

If $d \equiv 1 \pmod 8$, then $A2 = (2, \omega)(2, \omega')$.

If $d \equiv 5 \pmod 8$, then $A2$ is a prime ideal.

Hence,

(1) 2 is ramified if and only if $d \equiv 2$ or $3 \pmod 4$.

(2) 2 is inert if and only if $d \equiv 5 \pmod 8$.

(3) 2 is decomposed if and only if $d \equiv 1 \pmod 8$.

PROOF.   The proof is divided into several parts.

(a) If $d \equiv 5 \pmod 8$ then $A2$ is a prime ideal.

Otherwise, $A2 = P \cdot P'$ or P^2, with $P \cap \mathbb{Z} = \mathbb{Z}2$. Then there exists $\alpha \in A$ such that $P = (2, \alpha) \supseteq A\alpha$, so P divides $A\alpha$ and 2 divides $N(P)$, which divides $N(\alpha)$.

If $2 \mid a$, then $P = A2\left(l, \frac{\alpha}{2}\right) = A2$, which is absurd. Thus

$$2 \nmid \alpha = \frac{a + b\sqrt{d}}{2}, \qquad \text{with } a \equiv b \pmod 2,$$

so $N(\alpha) = \frac{a^2 - db^2}{4}$. From $2 \mid N(\alpha)$, then 8 divides $a^2 - db^2 \equiv a^2 - 5b^2 \equiv a^2 + 3b^2 \pmod 8$.

If a, b are odd, then $a^2 \equiv b^2 \equiv 1 \pmod 8$, so $a^2 + 3b^2 = 4 \pmod 8$, which is absurd. So a, b are even, $a = 2a'$, $b = 2b'$, and $\alpha = a' + b'\sqrt{d}$. 2 divides $N(\alpha) = (a')^2 - d(b')^2$.

Since d is odd, then a', b' are both even or both odd. If a', b' are even, then 2 divides α, which is absurd. If a', b' are odd, then $\alpha = a' + b'\sqrt{d} = (\text{multiple of } 2) + 1 + \sqrt{d} = (\text{multiple of } 2) + 2\omega = (\text{multiple of } 2)$, which is absurd.

(b) If $d \equiv 1 \pmod 8$, then $A2 = (2, \omega)(2, \omega')$.
Indeed,

$$(2, \omega)(2, \omega') = \left(4, 2\omega, 2\omega', \frac{1-d}{4}\right) = A2\left(2, \omega, \omega', \frac{1-d}{8}\right) = A2,$$

because $\omega + \omega' = 1$.

Also, $(2, \omega) \neq (2, \omega')$, otherwise these ideals are equal to their sum $(2, \omega, \omega') = A$, because $\omega + \omega' = 1$.

(c) If $d \equiv 2$ or $3 \pmod 4$, then $A2 = (2, \sqrt{d})^2$, respectively $(2, 1 + \sqrt{d})^2$.

First, let $d = 4e + 2$; then

$$(2, \sqrt{d})^2 = (4, 2\sqrt{d}, d) = A2(2, \sqrt{d}, 2e + 1) = A2,$$

so $(2, \sqrt{d})$ is a prime ideal.

Now, let $d = 4e + 3$; then

$$\begin{aligned}
(2, 1 + \sqrt{d})^2 &= (4, 2 + 2\sqrt{d}, 1 + d + 2\sqrt{d}) \\
&= (4, 2 + 2\sqrt{d}, 4(e + 1) + 2\sqrt{d}) \\
&= A2(2, 1 + \sqrt{d}, 2(e + 1) + \sqrt{d}) \\
&= A2(2, 2e + 1, 1 + \sqrt{d}, 2(e + 1) + \sqrt{d}) \\
&= A2,
\end{aligned}$$

and so $(2, 1 + \sqrt{d})$ is a prime ideal.

Finally, these three cases are exclusive and exhaustive, so the converse assertions also hold. □

5 Units

The element $\alpha \in A$ is a unit if there exists $\beta \in A$ such that $\alpha\beta = 1$. The set U of units is a group under multiplication. Here is a description of the group of units in the various cases. First, let $d < 0$.

Let $d \neq -1, -3$; then $U = \{\pm 1\}$.

Let $d = -1$; then $U = \{\pm 1, \pm i\}$, with $i = \sqrt{-1}$.

Let $d = -3$; then $U = \{\pm 1, \pm\rho, \pm\rho^2\}$, with $\rho^3 = 1$, $\rho \neq 1$, i.e. $\rho = \frac{-1+\sqrt{-3}}{2}$.

Let $d > 0$. Then the group of units is the product $U = \{\pm 1\} \times C$, where C is a multiplicative cyclic group. Thus $C = \{\epsilon^n \mid n \in \mathbb{Z}\}$, where ϵ is the smallest unit such that $\epsilon > 1$. The element ϵ is called the *fundamental unit*.

6   The class number

The theory of quadratic number fields originated with the study of binary quadratic forms $aX^2 + bXY + cY^2$ (where a, b, c are integers, and $ac \neq 0$). The discriminant of the form is, by definition, $D = b^2 - 4ac$. Note that $D \equiv 0$ or $1 \pmod 4$; let $d = D/4$ or $d = D$, respectively.

An integer m is said to be represented by the form if there exist integers x, y such that $m = ax^2 + bxy + cy^2$.

If a form $a'(X')^2 + b'X'Y + c'(Y')^2$ is obtained from the above form by a linear change of variables

$$\begin{cases} X = hX' + kY' \\ Y = mX' + nY' \end{cases}$$

where h, k, m, n are integers and the determinant is $hn - km = 1$, then the two forms represent the same integers. In this sense, it is reasonable to consider such forms as being equivalent. Clearly, equivalent forms have the same discriminant.

In *Disquisitiones Arithmeticae*, GAUSS classified the binary quadratic forms with a given discriminant D. GAUSS defined an operation of composition between equivalence classes of forms of a given discriminant. The classes constitute a group under this operation. GAUSS showed that, for any given discriminant D, there exist only finitely many equivalence classes of binary quadratic forms.

The theory was later reinterpreted, associating to each form $aX^2 + bXY + cY^2$ of discriminant D, the ideal I of $\mathbb{Q}(\sqrt{d}) = \mathbb{Q}(\sqrt{D})$ generated by a and $\frac{-b+\sqrt{D}}{2}$. Define two non-zero ideals I, I' to be equivalent when there exists a non-zero element $\alpha \in \mathbb{Q}(\sqrt{d})$ such that $I = A\alpha \cdot I'$. Then, equivalent binary quadratic forms correspond to equivalent ideals, and the composition of classes of forms corresponds to the multiplication of equivalence classes of deals. Thus, $\mathbb{Q}(\sqrt{d})$ has finitely many classes of ideals. Denote by $h = h(d)$ the number of classes of ideals, or class number of the field $\mathbb{Q}(\sqrt{d})$.

The class number $h(d) = 1$ exactly when every ideal of $\mathbb{Q}(\sqrt{d})$ is a principal ideal.

GAUSS conjectured that for every $h \geq 1$ there exist only finitely many imaginary quadratic fields $\mathbb{Q}(\sqrt{d})$ (with $d < 0$) such that the class number is equal to h. Soon, I shall say more about this conjecture.

I shall now indicate how to calculate the class number of the quadratic field $\mathbb{Q}(\sqrt{D})$. Define the real number θ as follows:

$$
\theta = \begin{cases} \frac{1}{2}\sqrt{D} & \text{if } D > 0, \\ \frac{2}{\pi}\sqrt{-D} & \text{if } D < 0. \end{cases}
$$

A non-zero ideal I of A is said to be *normalized* if $N(I) \leq [\theta]$ (the largest integer less than or equal to θ). The ideal I is said to be *primitive*if there does not exist any prime number p such that Ap divides I.

Let $\mathcal{N}$ denote the set of normalized primitive ideals of A.

If $I \in \mathcal{N}$, and if p is a ramified prime, then $p^2 \nmid N(I)$, and if p is an inert prime, then $p \nmid N(I)$. So,

$$
N(I) = \prod_{r \text{ ramified}} r \times \prod_{p \text{ decomposed}} p^{e(p)}.
$$

It may be shown that every class of ideals contains a primitive normalized ideal. Since for every $m \geq 1$ there exist at most finitely many ideals I of A such that $N(I) = m$, this implies, once more, that the number of classes of ideals is finite.

Note that if $\mathcal{N}$ consists only of the unit ideal $A = A \cdot 1$, then $h = 1$. Thus, if every prime p such that $p \leq [\theta]$ is inert, then $h = 1$. Indeed, if $I \in \mathcal{N}$, then $N(I) = 1$, so I is the unit ideal, hence $h = 1$.

Denote by $N(\mathcal{N})$ the set of integers $N(I)$, where $I \in \mathcal{N}$.

In order to decide if the ideals I, $J \in \mathcal{N}$ are equivalent, it will be necessary to decide which integers $m \in N(\mathcal{N})$ are of the form $m = N(A\alpha)$.

Let $m \geq 1$, and let

$$
\alpha = \begin{cases} u + v\sqrt{d} & \text{when } d \equiv 2 \text{ or } 3 \ (\text{mod } 4), \text{ with } u, v \in \mathbb{Z}, \\ \frac{u+v\sqrt{d}}{2} & \text{when } d \equiv 1 \ (\text{mod } 4), \text{ with } u, v \in \mathbb{Z}, u \equiv v \ (\text{mod } 2). \end{cases}
$$

It now follows that $A\alpha$ is a primitive ideal with $N(A\alpha) = m$ if and only if

$$
\begin{cases} m = |u^2 - dv^2| & \gcd(u, v) = 1 & \text{if } d \equiv 2 \text{ or } 3 \ (\text{mod } 4), \\ m = \frac{|u^2 - dv^2|}{4} & \gcd\left(\frac{u-v}{2}, v\right) = 1 & \text{if } d \equiv 1 \ (\text{mod } 4). \end{cases}
$$

(this is called the *primitive representation* of m).

PROOF.   Let $d \equiv 2$ or 3 $(\mathrm{mod}\,4)$, $m = N(A\alpha) = |u^2 - dv^2|$, also $\gcd(u, v) = 1$, because $A\alpha$ is primitive.

Let $d \equiv 1$ $(\mathrm{mod}\,4)$, $m = N(A\alpha) = \frac{|u^2 - dv^2|}{4}$, also if p divides $\frac{u-v}{2}$ and p divides v then p divides $\alpha = \frac{u-v}{2} - v\left(\frac{1+\sqrt{d}}{2}\right)$, contrary to the hypothesis.

Conversely, let $d \equiv 2$ or 3 $(\mathrm{mod}\,4)$, so $N(A\alpha) = m$: if p divides $A\alpha$, since $\{1, \sqrt{d}\}$ is an integral basis then $p \mid u$ and $p \mid v$, which is absurd.

Let $d \equiv 1$ $(\mathrm{mod}\,4)$, so $N(A\alpha) = m$; if p divides $A\alpha$, because

$$\alpha = \frac{u-v}{2} + v\left(\frac{1+\sqrt{d}}{2}\right) \qquad \text{and} \qquad \left\{1, \frac{1+\sqrt{d}}{2}\right\}$$

is an integral basis, then p divides $\frac{u-v}{2}$ and v, which is absurd.   □

A   Calculation of the class number

Let $d > 0$, so $\theta = \frac{1}{2}\sqrt{D}$.

$[\theta] = 1$.

Since $1 \leq \frac{1}{2}\sqrt{D} < 2$, it follows that $4 \leq D < 16$, with $D \equiv 0$ or 1 $(\mathrm{mod}\,4)$, hence $D \in \{4, 5, 8, 9, 12, 13\}$, and therefore $d \in \{5, 2, 3, 13\}$.

Now $N(\mathcal{N}) = \{1\}$, hence $\mathcal{N}$ consists only of the unit ideal, and therefore $h = 1$.

$[\theta] = 2$.

Since $2 \leq \frac{1}{2}\sqrt{D} < 3$, it follows that $16 \leq D < 36$, with $D \equiv 0$ or 1 $(\mathrm{mod}\,4)$, hence $D \in \{16, 17, 20, 21, 24, 25, 28, 29, 32, 33\}$ and therefore $d \in \{17, 21, 6, 7, 29, 33\}$.

Now, $N(\mathcal{N}) = \{1, 2\}$.

Take, for example, $d = 17$. Since $17 \equiv 1$ $(\mathrm{mod}\,8)$, then it follows that $A^2 = P \cdot P'$, $N(P) = N(P') = 2$, $2 = \frac{1}{4}|3^2 - 17 \times 1^2|$, and $\gcd\left(\frac{3-17}{2}, 17\right) = 1$, hence

$$P = A\alpha, \qquad \alpha = \frac{3 + \sqrt{17}}{2},$$

$$P' = A\alpha, \qquad \alpha' = \frac{3 - \sqrt{17}}{2}.$$

Therefore, the class number is $h = 1$.

Let $d = 21$. Since $21 \equiv 5 \pmod 8$, then $A2$ is a prime ideal, 2 is inert, hence $h = 1$.

Let $d = 6$, then 2 divides $24 = D$, so 2 is ramified, $A2 = P^2$, and $2 = |2^2 - 6 \times 1^2|$, $\gcd(2,1) = 1$, hence $P = A\alpha$, with $\alpha = 2 + \sqrt{6}$. Therefore $h = 1$.

$[\theta] = 3$.

Since $3 \le \frac{1}{2}\sqrt{D} < 4$, then $36 \le D < 64$, with $D \equiv 0$ or $1 \pmod 4$, hence

$$D \in \{36, 37, 40, 41, 44, 45, 48, 49, 52, 53, 56, 57, 60, 61\}$$

and, therefore,

$$d \in \{37, 10, 41, 11, 53, 14, 57, 15, 61\}.$$

Now, $N(\mathcal{N}) = \{1, 2, 3\}$.

Take, for example, $d = 10$. Since 2 divides $40 = D$, then 2 is ramified, $A2 = R^2$. Since $\left(\frac{10}{3}\right) = \left(\frac{1}{3}\right) = 1$, then 3 is decomposed, $A3 = P \cdot P'$. The ideals R, P, P' are primitive.

2 has no primitive representation: If $2 = |u^2 - 10v^2|$, then $u^2 = 10v^2 \pm 2 \equiv \pm 2 \pmod{10}$, which is impossible.

3 has no primitive representation: If $3 = |u^2 - 10v^2|$, then $u^2 = 10v^2 \pm 3 \equiv \pm 3 \pmod{10}$, which is impossible.

Thus, R, P, P' are not principal ideals. The ideals RP, RP' are primitive. Also,

$$-2 \times 3 = -6 = 2^2 - 10 \times 1^2, \quad \gcd(2,1) = 1,$$
$$2 \times 3 = N(RP) = N(RP'),$$

hence RP, RP' are principal ideals. In conclusion, $h' = 2$.

Let $d < 0$, so $\theta = \frac{2}{\pi}\sqrt{-D}$.

$[\theta] = 1$.

Since $1 \le \frac{2}{\pi}\sqrt{-D} < 2$, then $\frac{\pi^2}{4} \le |D| < \pi^2$, and $|D| \equiv 0$ or $3 \pmod 4$, hence $|D| \in \{3, 4, 7, 8\}$, therefore $d \in \{-3, -1, -7, -2\}$. Now $N(\mathcal{N}) = 1$, hence $\mathcal{N}$ consists only of the unit ideal, so $h = 1$.

$[\theta] = 2$.

Since $2 \le \frac{2}{\pi}\sqrt{-D} < 3$, then $\pi^2 \le |D| < \frac{9}{4}\pi^2$, and $|D| \equiv 0$ or $3 \pmod 4$, hence $|D| \in \{11, 12, 15, 16, 19, 20\}$, therefore $d \in \{-11, -15, -19, -5\}$.

Take, for example, $d = -11$. Since $-11 \equiv 5 \pmod{8}$, then 2 is inert, and therefore $h = 1$.

Let $d = -5$. Since 2 divides $D = -20$, 2 is ramified, $A2 = P^2$.

2 has no primitive representation: If $2 = |u^2 + 5v^2|$ then $u^2 = -5v^2 + 2 \equiv 2 \pmod{5}$, which is impossible. Also, $-5 \equiv 3 \pmod{4}$. So, P is not principal and $h = 2$.

Let $d = -15$. Since $-15 \equiv 1 \pmod{8}$, then $A2 = P \cdot P'$.

2 has no primitive representation: if

$$2 = \frac{|u^2 + 15v^2|}{4}, \quad \text{with} \quad \gcd\left(\frac{u-v}{2}, v\right) = 1,$$

then $u^2 + 15v^2 = 8$, so $u^2 \equiv 3 \pmod{5}$, which is impossible. Also $-15 \equiv 1 \pmod{4}$. Since P, P' are not principal ideals, then $h = 2$.

Let $d = -19$. Since $-19 \equiv 5 \pmod{8}$, 2 is inert, hence $h = 1$.

$[\theta] = 3$.

Since $3 \leq \frac{2}{\pi}\sqrt{-D} < 4$, then $\frac{9\pi^2}{4} \leq |D| < 4\pi^2$, and $|D| \equiv 0$ or $3 \pmod{4}$, hence

$$|D| \in \{23, 24, 27, 28, 31, 32, 35, 36, 39\},$$

and, therefore,

$$d \in \{-23, -6, -31, -35, -39\}.$$

Take $d = -31$. Since $-31 \equiv 1 \pmod{8}$, then $A2 = P \cdot P'$. Since $-31 \equiv 1 \pmod{8}$, then $A2 = P \cdot P'$. Since $\left(\frac{-31}{3}\right) = \left(\frac{-1}{3}\right)\left(\frac{1}{3}\right) = -1$, it follows that $A3$ is a prime ideal.

2 has no primitive representation: If

$$2 = \frac{|u^2 + 31v^2|}{4}, \quad \text{with} \quad \gcd\left(u - v\frac{u-v}{2}, v\right) = 1,$$

then $8 = u^2 + 31v^2$, which is impossible. Since $-31 \equiv 1 \pmod{4}$, then P, P' are not principal ideals. If P, P' are equivalent, then $P = P'.A\alpha$, so $P^2 = P \cdot P' \cdot A\alpha = A(2\alpha)$, so $4 = N(P^2) = 4N(A\alpha)$, hence $N(A\alpha) = 1$, thus $A\alpha = A$, and $P = P'$, which is absurd. In conclusion, $h = 3$.

These examples are enough to illustrate how to compute the class number, at least for discriminants with small absolute value. There are more sophisticated methods for calculating the class number which are effecient even for large values of $|d|$. These algorithms are desribed in the books of BUELL (1989) and COHEN (1993) which, of course, also deal with real quadratic fields.

B   Determination of all quadratic fields with class number 1

Let $d > 0$.

It is conjectured that there exist infinitely many $d > 0$ such that $\mathbb{Q}(\sqrt{d})$ has class number 1. This question is difficult to settle, but it is expected that the conjecture is true.

For example, there exist 142 fields $\mathbb{Q}(\sqrt{d})$ with $2 \leq d < 500$ having class number 1.

Let $d < 0$.

It was seen that if $\mathcal{N}$ consists only of the unit ideal, then $h = 1$. But conversely:

If $d < 0$ and $h = 1$, then $\mathcal{N} = \{A\}$.

PROOF.   If $|D| \leq 7$, it is true. Let $|D| > 7$, $I \in \mathcal{N}$, and $I \neq A$, so there exists a prime ideal P dividing I. Then $N(P) = p$ or p^2, where p is a prime number. If $N(P) = p^2$, then p is inert and $Ap = P$ divides I, so I would not be primitive, a contradiction. If $N(P) = p$, since P divides I, then $p \leq N(I) \leq [\theta] \leq \frac{2}{\pi}\sqrt{|D|}$. If p has a primitive representation:

If $d \equiv 2$ or $3 \pmod 4$, then $d = \frac{D}{4}$, so $p = u^2 - dv^2$, hence $v \neq 0$, therefore $\frac{2}{\pi}\sqrt{|D|} \geq p \geq |d| = \frac{|D|}{4}$, so $7 \geq \frac{64}{\pi^2} \geq |D|$, which is absurd.

If $d \equiv 1 \pmod 4$, then $d = D$, so $p = \frac{u^2 - dv^2}{4}$, hence $v \neq 0$, therefore $\frac{2}{\pi}\sqrt{|D|} \geq p \geq \frac{|d|}{4} = \frac{|D|}{4}$, and again $7 \geq D$, which is absurd.

Therefore, P is not a principal ideal, and $h \neq 1$, which contradicts the hypothesis. $\square$

GAUSS developed a theory of genera and proved:

If $d < 0$, and if t is the number of distinct prime factors of D, then 2^{t-1} divides the class number of $\mathbb{Q}(\sqrt{d})$.

Hence, if $h = 1$, then $D = -4$, -8, or $-p$, where p is a prime, $p \equiv 3 \pmod 4$, hence $d = -1$, -2, or $-p$.

From this discussion, it follows:

If $D = -3, -4, -7, -8$, then $h = 1$.

If $D \neq -3, -4, -7, -8$, and $D = -p$, $p \equiv 3 \pmod 4$, then $h = 1$ if and only if $\mathcal{N} = \{A\}$, and this is equivalent to the following conditions:

2 is inert in $\mathbb{Q}(\sqrt{-p})$ and if q is any odd prime, $q \leq [\theta]$, then $\left(\frac{-p}{q}\right) = -1$, i.e., q is inert in $\mathbb{Q}(\sqrt{-p})$.

This criterion is used in the determination of all $D < 0$, $|D| \leq 200$, such that $h = 1$.

$[\theta] = 1$.

This gives the discriminants $D = -3, -4, -7, -8$.

$[\theta] = 2$.

Now $-20 \leq D \leq -11$, with $D = p$, $p \equiv 3 \pmod 4$, so $D = -11$ or -19.

Since $-11 \equiv 5 \pmod 8$, then 2 is inert, so if $D = -11$, then $h = 1$.

Similarly, since $-19 \equiv 5 \pmod 8$, 2 is inert, so if $D = -19$, then $h = 1$.

$[\theta] = 3$.

Now $-39 \leq D \leq -23$, with $D = -p$, $p \equiv 3 \pmod 4$, so $D = -23$ or -31. But $-23 \not\equiv 5 \pmod 8$, $-31 \not\equiv 5 \pmod 8$, so the class numbers of $\mathbb{Q}(\sqrt{-23})$ and of $\mathbb{Q}(\sqrt{-31})$ are not 1.

$[\theta] = 4$.

Now $-59 \leq D \leq -40$, $D = -p$, $p \equiv 3 \pmod 4$, so $D = -43, -47, -59$. Since $-43 \equiv 5 \pmod 8$ and $\left(\frac{-43}{3}\right) = -1$, then $\mathbb{Q}(\sqrt{-43})$ has class number 1. Since $-47 \not\equiv 5 \pmod 8$ and $\left(\frac{-59}{3}\right) = 1$, then 3 is not inert. So the class numbers of $\mathbb{Q}(\sqrt{-47})$ and of $\mathbb{Q}(\sqrt{-59})$ are not equal to 1.

The same calculations yield:

$[\theta] = 5$: $D = -67$, with class number 1.
$[\theta] = 6$: no discriminant.
$[\theta] = 7$: no discriminant.
$[\theta] = 8$: $D = -163$, with class number 1.

This process may continued beyond 200, but leads to no other discriminant for which the class number is 1. Of course, this does not allow us to decide whether there exists any other such discriminant, nor to decide whether there are only finitely many imaginary quadratic fields with class number 1.

In a classic paper, HEILBRONN and LINFOOT showed in 1934, with analytical methods, that besides the above examples there exists at most one other value of $d < 0$ for which $\mathbb{Q}(\sqrt{d})$ has class number 1. LEHMER showed that if such a discriminant d exists at all, then $|d| > 5 \times 10^9$. In 1952, HEEGNER proved that no other such d could exist, but his proof contained some steps which were unclear, perhaps even a gap. BAKER reached the same conclusion in 1966 with his method involving effective lower bounds on linear forms of three logarithms; this is also reported in his article of 1971. At about the same time, unaware of HEEGNER's result, but with similar ideas concerning elliptic modular functions, STARK proved that no further possible value for d exists. So were determined all the imaginary quadratic fields with class number 1. It was somewhat of an anticlimax when in 1968 DEURING was able to straighten out HEEGNER's proof. The technical details involved in these proofs are far beyond the scope of the present article.

This is the place to say that GAUSS's conjecture was also solved in the affirmative. Thanks to the work of HECKE, DEURING, MORDELL, and HEILBRONN, it was established that if $d < 0$ and $|d|$ tends to infinity, then so does the class number of $\mathbb{Q}(\sqrt{d})$. Hence, for every integer $h \geq 1$ there exists only finitely many fields $\mathbb{Q}(\sqrt{d})$ with $d < 0$, having class number h.

The determination of all imaginary quadratic fields with class number 2 was achieved by BAKER, STARK, and WEINBERGER.

An explicit estimate of the number of imaginary quadratic fields with a given class number was obtained by the efforts of SIEGEL, GOLDFELD, and GROSS and ZAGIER. For this matter, I suggest reading the paper of GOLDFELD (1985).

7   The main theorem

Theorem. *Let q be a prime, let $f_q(X) = X^2 + X + q$. The following conditions are equivalent:*

(1) $q = 2$, 3, 5, 11, 17, 41.
(2) $f_q(n)$ is a prime for $n = 0, 1, \ldots, q - 2$.
(3) $\mathbb{Q}(\sqrt{1 - 4q})$ has class number 1.

PROOF.   The implication (1) $\Rightarrow$ (2) is a simple verification.

The equivalence of the assertions (2) and (3) was first shown by RABINOVITCH in 1912. In 1936, LEHMER proved once more that (2) $\Rightarrow$ (3), while (3) $\Rightarrow$ (2) was proved again by SZEKERES (1974), and by AYOUB and CHOWLA (1981) who gave the simplest proof. The proof of (3) $\Rightarrow$ (1) follows from the complete determination of all imaginary quadratic fields with class number 1. Since this implication requires deep results, I shall also give the proof of $(3) \Rightarrow (2)$.

$(2) \Rightarrow (3)$. Let $d = 1 - 4q < 0$, so $d \equiv 1 \pmod 4$. If $q = 2$ or 3, then $d = -7$ or -11 and $\mathbb{Q}(\sqrt{d})$ has class number 1, as was already seen. Assume now that $q \geq 5$. It suffices to show that every prime $p \leq \frac{2}{\pi}\sqrt{|d|}$ is inert in $\mathbb{Q}(\sqrt{d})$.

First, let $p = 2$; since $q = 2t - 1$, then $d = 1 - 4q = 1 - 4(2t - 1) = 5 \pmod 8$, so 2 is inert in $\mathbb{Q}(\sqrt{d})$.

Now let $p \neq 2$, $p \leq \frac{2}{\pi}\sqrt{|d|} < \sqrt{|d|}$ and assume that p is not inert. Then $\left(\frac{d}{p}\right) \neq -1$ and, as was noted, there exists $b \in \mathbb{Z}$, $0 \leq b \leq p - 1$, such that p divides $N(b + \omega)$, where $\omega = \frac{1+\sqrt{d}}{2}$, that is, p divides

$$
\begin{aligned}
(b + \omega)(b + \omega') &= b^2 + b(\omega + \omega') + \omega\omega' \\
&= b^2 + b + \frac{1 - d}{4} \\
&= b^2 + b + q = f_q(b).
\end{aligned}
$$

It should be also noted that $b \neq p - 1$, otherwise as was shown, p divides $1 - d = 4q$, hence $p = q < \sqrt{|d|} = \sqrt{|1 - 4q|}$, so $q^2 < 4q - 1$, hence $q = 2$ or 3, against the hypothesis.

By hypothesis, $f_q(b)$ is therefore a prime number, hence $\sqrt{4q - 1} > p = f_q(b) \geq f_q(0) = q$ and, again, $q = 2$ or 3, against the hypothesis.

This shows that every prime p less than $\frac{2}{\pi}\sqrt{|d|}$ is inert, hence $h = 1$.

$(3) \Rightarrow (1)$. If $\mathbb{Q}(\sqrt{1 - 4q})$ has class number 1, then $d = 1 - 4q = -7, -11, -19, -43, -67, -163$, hence $q = 2, 3, 5, 11, 17, 41$.   □

As I have already said, the proof is now complete, but it is still interesting to indicate the proof of $(3) \Rightarrow (2)$.

PROOF.   Assume that $d = 1 - 4q$ and that the class number of $\mathbb{Q}(\sqrt{-d})$ is 1. Then either $d = -1, -2, -3, -7$, or $d < -7$, so $d = -p$ with $p \equiv 3 \pmod 4$ and $q > 2$.

As noted before, 2 is inert in $\mathbb{Q}(\sqrt{-p})$, so $p \equiv 3 \pmod 8$. Next, I show that if l is any odd prime with $l < q$, then $\left(\frac{l}{p}\right) = -1$. Indeed, if $\left(\frac{l}{p}\right) = 1$ then l splits in $\mathbb{Q}(\sqrt{-p})$. But $h = 1$, so there exists an algebraic integer $\alpha = \frac{a+b\sqrt{-p}}{2}$ such that $Al = A\alpha \cdot A\alpha'$. Then

$$
l^2 = N(Al) = N(A\alpha) \cdot N(A\alpha') = N(A\alpha)^2 = N(\alpha)^2,
$$

so $l = N(\alpha) = \frac{a^2+b^2p}{4}$. Hence $p+1 = 4q > 4l = a^2 + b^2p$, thus $1 > a^2 + (b^2 - 1)p$ and necessarily $a^2 = 0$, $b^2 = 1$, hence $4l = p$, which is absurd.

Now assume that there exists m, $0 \leq m \leq q-2$, such that $f_q(m) = m^2 + m + q$ is not a prime. Then there exists a prime l such that $l^2 \leq m^2 + m + q$ and $m^2 + m + q = al$, with $a \geq 1$. Since $m^2 + m + q$ is odd, then $l \neq 2$. Also,

$$4l^2 \leq (2m + 1)^2 + p < \left(\frac{p-1}{2}\right)^2 + p = \left(\frac{p+1}{2}\right)^2,$$

hence $l < (p+1)/4 = q$. As was shown, $\left(\frac{l}{p}\right) = -1$. However,

$$4al = (2m + 1)^2 + 4q - 1 = (2m + 1)^2 + p,$$

hence $-p$ is a square modulo l, so by Gauss' reciprocity law,

$$1 = \left(\frac{-p}{l}\right) = \left(\frac{-1}{l}\right)\left(\frac{p}{l}\right) = (-1)^{\frac{l-1}{2}}\left(\frac{l}{p}\right)(-1)^{\frac{l-1}{2}\times\frac{p-1}{2}} = \left(\frac{l}{p}\right),$$

and this is absurd. $\square$

References

1912 G. Rabinovitch. Eindeutigkeit der Zerlegung in Primzahl-faktoren in quadratischen Zahlkörper. 418–421.

1936 D. H. Lehmer. On the function $x^2 + x + A$. *Sphinx*, 6: 212–214.

1958 A. Schinzel and W. Sierpiński. Sur certaines hypothèses concernant les nombres premiers. Remarques. *Acta Arith.*, 4:185–208 and 5:259 (1959).

1961/62 A. Schinzel. Remarks on the paper "Sur certaines hypothèses concernant les nombres premiers". *Acta Arith.*, 7:1–8.

1962 H. Cohn. *Advanced Number Theory*. Dover, New York.

1966 Z. I. Borevich and I. R. Shafarevich. *Number Theory*. Academic Press, New York.

1972 P. Ribenboim. *Algebraic Numbers*. Wiley-Interscience, New York.

1974 G. Szekeres. On the number of divisors of $x^2 + x + A$. *J. Nb. Th.*, 6:434–442.

1981 R. G. Ayoub and S. Chowla. On Euler's polynomial. *J. Nb. Th.*, 13:443–445.

1985 D. M. Goldfeld. Gauss's class number problem for imaginary quadratic fields. *Bull. Amer. Math. Soc.*, 13: 23–37.

1989 D. A. Buell. *Binary Quadratic Forms*. Springer-Verlag, New York.

1993 H. Cohen. *A Course in Computational Algebraic Number Theory*. Springer-Verlag, Berlin.

1995 P. A. Pritchard, A. Moran, and A. Thyssen. Twenty-two primes in arithmetic progression. *Math. of Comp.*, 64:1337–1339.

1996 P. Ribenboim. *The New Book of Prime Number Records*. Springer-Verlag, New York.

6

Gauss and the Class Number Problem

1 Introduction

The theory of binary quadratic forms, one of the great achievements of GAUSS in number theory.

Some conjectures formulated by GAUSS are still the object of considerable research. This text* contains also a succinct description of the most significant recent results concerning the conjectures of GAUSS on the class number.

2 Highlights of Gauss' life

Carl Friedrich Gauss was born in Braunschweig in 1777 and died in Göttingen in 1855.

He was a precocious child. This is illustrated by the following well-known anecdote.

At age 8, when the pupils in his class angered the teacher, they were given the following task: to add all the numbers from 1 to 100:

$$1 + 2 + 3 + \cdots + 100.$$

---

*This is a much enlarged version of a lecture given at the First Gauss Symposium, in Guarujá, Brazil, July 1989.

The teacher thought he would have long moments of respite. But he was wrong, as the young GAUSS readily handed the solution: 5050. Astonished, the teacher asked how he got such a quick answer. Explained the child:

"I imagined the numbers 1 to 100 written in a row, and then again the same numbers written in another row, but backwards:

$$
\begin{array}{cccccc}
1 & 2 & 3 & \dots & 98 & 99 & 100 \\
100 & 99 & 98 & \dots & 3 & 2 & 1
\end{array}
$$

I noted that the two numbers in each column added to 101. There are 100 columns; this gives a total of 10100, but I have counted each number twice, so the sum asked is one-half of 10100, that is, 5050."

I cannot swear that this had happened as I have just told, but as the Italians say, "*se non è vero, è ben trovato*".

The young GAUSS gave many more indications of his superior intelligence, being excellent in all subjects, especially classical languages and mathematics.

At age 11, GAUSS entered the Gymnasium. His talents were recognized as soon as 1792 when at age 15 he received a stipend from the Duke of Braunschweig which would allow him to continue his studies without financial worries.

GAUSS calculated with gusto. For example, he computed tables of prime numbers (when still very young) and also quadratic residues, primitive roots modulo primes, inverses of prime numbers written with many decimal digits, etc . . . .

The results of his calculations served as a basis for conjectures and statements, which, throughout his career, he would try to prove, often with great success.

GAUSS entered the university at Göttingen, where he could benefit from a rich library. There he studied BERNOULLI's *Ars Conjectandi*, NEWTON's *Principia*, as well as the works of EULER, LAGRANGE, and LEGENDRE.

GAUSS had very wide interests in Mathematics, and also in Astronomy, Geodesy, and Physics.

The following brief table of some of GAUSS' earlier mathematical discoveries is an indication of his striking achievements, which would continue unabated all through his life.

Age	Year	
18	1795	Series expansion for the arithmetic-geometric mean. The method of least squares. Conjecture: the prime number theorem. Non-euclidean geometry.
19	1796	Quadratic reciprocity law. Determination of regular polygons constructible with ruler and compass (includes the 17-gon).
20	1797	The fundamental theorem of algebra.
22	1799	Relation between the arithmetic-geometric mean and the length of the lemniscate.
23	1800	Doubly-periodic functions.
24	1801	Publication of *Disquisitiones Arithmeticae*

In *Disquisitiones Arithmeticae* (GAUSS (1801)), a landmark in number theory, widely translated and very influential, GAUSS presented in an organized way his discoveries of the preceding years, completing and clarifying the work of his predecessors FERMAT, EULER, LAGRANGE, and LEGENDRE.

This book contains: the theory of congruences (with the happy introduction of the notation $a \equiv b \pmod{n}$); indeterminate linear equations; binary quadratic forms; indeterminate quadratic equations; cyclotomy; and the construction of regular polygons with ruler and compass.

While still young, GAUSS' attention veered to other subjects in mathematics, astronomy, and physics. Since this chapter concerns binary quadratic forms, I abstain from discussing his contributions to other topics. For a recent and enlightening critical study of GAUSS, see KAUFMANN-BÜHLER (1981), where attention is given to other facets of GAUSS' work. It is also instructive to consult GAUSS' diary (see Gauss' Werke (1870)) and the commentary by GRAY (1984).

3   Brief historical background

One of the famous discoveries of FERMAT concerns the representation of prime numbers as sums of squares.

Let p be a prime number. Then $p = x^2 + y^2$ (with integers x, y) if and only if $p = 2$ or $p \equiv 1 \pmod 4$. In this event, the representation of p is unique (with $0 < x < y$ when $p \neq 2$). With the Legendre symbol, the condition is rephrased as follows: $p = 2$ or $(-1/p) = +1$.

Similarly, $p = x^2 + 2y^2$ (with integers x, y) if and only if $p = 2$ or $p \equiv 1$ or $3 \pmod 8$; equivalently, $p = 2$ or $(-2/p) = +1$. Again, the representation is unique (with $0 < x, y$).

In the same way, $p = x^2 + 3y^2$ (with integers x, y) if and only if $p = 3$ or $p \equiv 1 \pmod 3$, or equivalently, $p = 3$ or $(-3/p) = +1$. Again, the representation is unique (with $0 < x, y$).

Nevertheless, EULER knew that the condition $p = 5$ or $p \equiv 1, 3, 7$ or $9 \pmod{20}$, or, equivalently, $p = 5$ or $(-5/p) = +1$, expresses that the prime p is of the form $p = x^2 + 5y^2$ or $p = 2x^2 + 2xy + 3y^2$ (with integers x, y). Actually, EULER conjectured that $p = x^2 + 5y^2$ if and only if $p = 5$ or $p \equiv 1$ or $9 \pmod{20}$. However, the proof of this conjecture involves the theory of genera.

LAGRANGE and LEGENDRE studied the more general problem: given the integers a, b, c, represent the integer m in the form

$$m = ax^2 + bxy + cy^2,$$

where x, y are integers.

In *Disquisitiones Arithmeticae*, GAUSS presented systematically and in depth the results of EULER, LEGENDRE, and LAGRANGE, and developed the theory well beyond his predecessors.

The historical development of the fascinating theory of binary quadratic forms is thoroughly described in the excellent books of WEIL (1984) and EDWARDS (1977).

4   Binary quadratic forms

A *binary quadratic form* (or simply, a form) is a homogeneous polynomial of degree 2 in two indeterminates

$$Q = aX^2 + bXY + cY^2,$$

with coefficients a, b, $c \in \mathbb{Z}$.

A simple notation is $Q = \langle a, b, c \rangle$.

In his theory, GAUSS had good reasons to consider only the forms with b even; respecting the tradition still today, this restriction is observed in some presentations of the theory. The results for the more general theory, which was considered by EISENSTEIN, can be easily related to those obtained by GAUSS for the forms $\langle a, b, c \rangle$ with b even.

The form $Q = \langle a, b, c \rangle$ is said to be *primitive* if $\gcd(a, b, c) = 1$.

If $\langle a, b, c \rangle$ is any form and $d = \gcd(a, b, c)$, then the form $\langle \frac{a}{d}, \frac{b}{d}, \frac{c}{d} \rangle$ is primitive. This us allows to pass from arbitrary forms to the primitive ones.

The *discriminant* of $Q = \langle a, b, c \rangle$ is $D = D(Q) = b^2 - 4ac$. The discriminant is also denoted $\mathrm{Discr}(Q)$.

An integer D is the discriminant of some form if and only if $D \equiv 0$ or $1 \pmod 4$. Indeed, a discriminant satisfies one of the above congruences. Conversely,

$$\begin{cases} \text{if } D \equiv 0 \pmod 4, \text{ let } P = \langle 1, 0, \frac{-D}{4} \rangle, \\ \text{if } D \equiv 1 \pmod 4, \text{ let } P = \langle 1, 1, \frac{-D+1}{4} \rangle. \end{cases}$$

Then P has discriminant D, and it is called the *principal form* of discriminant D.

If $D = D(Q)$ is a square, then

$$aQ = \left[aX - \frac{-b + \sqrt{D}}{2} Y \right] \left[aX - \frac{-b - \sqrt{D}}{2} Y \right],$$

so it is the product of linear factors with integral coefficients. This case is degenerate, and therefore it will be always assumed that the discriminant is not a square. Thus, $ac \neq 0$.

An integer m is a *value* of Q, or is *represented* by Q, if there exist integers x, y such that $m = Q(x, y) = ax^2 + bxy + cy^2$; each such relation is said to be a *representation* of m by Q. If, moreover, $\gcd(x, y) = 1$, then one speaks of *primitive values* and *primitive representations*.

The set of values of Q is

$$\{\text{values of } Q\} = \{mt^2 \mid m \text{ is a primitive value of } Q \text{ and } t \in \mathbb{Z}\}.$$

The forms are classified as follows:

definite forms when $D < 0$;

indefinite forms when $D > 0$.

If the form Q is indefinite, then it clearly assumes positive as well as negative values.

It is easy to see that if $Q = \langle a, b, c \rangle$ and $D = b^2 - 4ac < 0$, then the following conditions are equivalent:

1. $a > 0$;
2. $c > 0$;
3. $Q(x, y) > 0$ for all non-zero integers x, y.

In this case, Q is said to be *positive definite* when $a > 0$, and *negative definite* when $a < 0$.

Since positive definite forms $\langle a, b, c \rangle$ correspond to negative definite forms $\langle -a, -b, -c \rangle$, it suffices to study positive definite forms. Thus, unless stated to the contrary, all forms with negative discriminant will be positive definite forms.

The following notions will be useful later.

It is convenient to say that the *conjugate* of the form $Q = \langle a, b, c \rangle$ is the form $\bar{Q} = \langle a, -b, c \rangle$. Clearly, $Q, \bar{Q}$ have the same discriminant and Q is positive definite if and only if $\bar{Q}$ is also.

The roots of the form $Q = \langle a, b, c \rangle$ are

$$\omega = \frac{-b + \sqrt{D}}{2a} \quad \text{(the } \textit{first root}\text{) and}$$

$$\eta = \frac{-b - \sqrt{D}}{2a} \quad \text{(the } \textit{second root}\text{).}$$

The following notation will be used.

If D is any discriminant, let $\mathcal{Q}_D$ be the set of all forms when $D > 0$ (respectively, all positive definite forms when $D < 0$); similarly, let $\mathrm{Prim}(\mathcal{Q}_D)$ be the subset of $\mathcal{Q}_D$ consisting of primitive forms. Let

$$\mathcal{Q} = \bigcup \{ \mathcal{Q}_D \mid D \equiv 0 \text{ or } 1 \pmod 4 \} \quad \text{and}$$
$$\mathrm{Prim}(\mathcal{Q}) = \bigcup \{ \mathrm{Prim}(\mathcal{Q}_D) \mid D \equiv 0 \text{ or } 1 \pmod 4 \}.$$

Let $D \equiv 0$ or $1 \pmod 4$. D is said to be a fundamental discriminant when every form with discriminant D is primitive.

It is easy to see that this happens if and only if

1. whenever $D \equiv 1 \pmod 4$, then D is square-free,
2. whenever $D \equiv 0 \pmod 4$, then $D = 4D'$ with D' square-free and $D' \equiv 2$ or $3 \pmod 4$.

Thus, the fundamental discriminants are of the form

$$D = \pm q_1 q_2 \cdots q_r, \quad \text{or} \quad D = \pm 4 q_2 \cdots q_r, \quad \text{or} \quad D = \pm 8 q_2 \cdots q_r,$$

where $q_1, q_2, \ldots, q_r$ are distinct odd primes.

On the other hand, not every discriminant D of one of the above forms is fundamental, e.g., $D = -4 \times 3$.

Every discriminant D is uniquely expressible in the form $D_0 f^2$ where D_0 is a fundamental discriminant and $f \geq 1$. D_0 is called the *fundamental discriminant associated to the discriminant D*.

The following bijection is easy to establish:

$$\mathcal{Q}_D \longrightarrow \bigcup_{e \mid f} \mathrm{Prim}(\mathcal{Q}_{D/e^2}).$$

In particular, if $D = D_0$ is a fundamental discriminant, then $\mathcal{Q}_D = \mathrm{Prim}(\mathcal{Q}_D)$.

5   The fundamental problems

The theory of binary quadratic forms deals with the following problems:

Problem 1. Given integers m and D, $D \equiv 0$ or $1 \pmod 4$, to find if there exists a primitive representation of m by some form of discriminant D.

In the affirmative, consider the next questions:

Problem 2. To enumerate all the forms $Q \in \mathcal{Q}_D$ such that m has a primitive representation by Q.

Problem 3. For each $Q \in \mathcal{Q}_D$ such that m has a primitive representation by Q, to determine all the representations of m by Q.

In order to solve the problems, it is necessary to study the equivalence of forms.

6   Equivalence of forms

Let $\mathrm{GL}_2(\mathbb{Z})$ denote the *linear group of rank 2 over* $\mathbb{Z}$; it consists of all matrices

$$A = \begin{pmatrix} \alpha & \beta \\ \gamma & \delta \end{pmatrix}$$

with α, β, γ, $\delta \in \mathbb{Z}$ and $\alpha\delta - \beta\gamma = \pm 1$.

Let $\mathrm{SL}_2(\mathbb{Z})$ consist of those matrices $A \in \mathrm{GL}_2(\mathbb{Z})$ such that $\alpha\delta - \beta\gamma = 1$. It is a normal subgroup of $\mathrm{GL}_2(\mathbb{Z})$ of index 2, called the *special linear group of rank* 2 *over* $\mathbb{Z}$.

The group $\mathrm{GL}_2(\mathbb{Z})$ acts on the set $\mathcal{Q}$ of forms in the following way. If $A = \begin{pmatrix} \alpha & \beta \\ \gamma & \delta \end{pmatrix} \in \mathrm{GL}_2(\mathbb{Z})$, let $T_A : \mathcal{Q} \to \mathcal{Q}$ be the map defined as follows.

If $Q = \langle a, b, c \rangle$, $Q' = \langle a', b', c' \rangle$, then

$$T_A(\langle a, b, c \rangle) = \langle a', b', c' \rangle$$

where

$$(*) \begin{cases} a' = a\alpha^2 + b\alpha\gamma + c\gamma^2 = Q(\alpha, \gamma), \\ b' = 2a\alpha\beta + b(\alpha\delta + \beta\gamma) + 2c\gamma\delta, \\ c' = a\beta^2 + b\beta\delta + c\delta^2 = Q(\beta, \delta). \end{cases}$$

Thus,

$$Q'(X, Y) = Q(\alpha X + \beta Y, \gamma X + \delta Y).$$

It is easy to see that if A, $A' \in \mathrm{GL}_2(\mathbb{Z})$, then $T_{AA'}(Q) = T'_A(T_A(Q))$. Q is a primitive form if and only if $T_A(Q)$ is a primitive form.

The mapping T_A is the identity map if and only if $A = \pm I$, where

$$I = \begin{pmatrix} 1 & 0 \\ 0 & 1 \end{pmatrix}$$

is the identity matrix.

The forms Q, Q' are said to be *equivalent* if there exists $A \in \mathrm{GL}_2(\mathbb{Z})$ such that $Q' = T_A(Q)$. This fact is denoted by $Q \sim Q'$, and it is easy to see that this relation is an equivalence relation.

The forms Q, Q are *properly equivalent* if there exists $A \in \mathrm{SL}_2(\mathbb{Z})$ such that $Q' = T_A(Q)$; the notation is $Q \approx Q'$, and, again, this is an equivalence relation. The proper equivalence class of $Q = \langle a, b, c \rangle$ shall be denoted $\mathbb{Q} = \langle a, b, c \rangle$.

Clearly, if $Q \approx Q'$, then also $Q \sim Q'$.

Each equivalence class is either the union of two proper equivalence classes or just a proper equivalence class. For example, for every integer $a \neq 0$, every form equivalent to $\langle a, 0, a \rangle$ is also properly equivalent to $\langle a, 0, a \rangle$.

It is easy to note that if $Q \sim Q'$, then Q, Q' have the same set of values; thus an integer is represented by Q if and only if it is represented by Q'. Similarly, Q and Q' have the same set of primitive values.

Furthermore, if $Q \sim Q'$ then $\mathrm{Discr}(Q) = \mathrm{Discr}(Q')$.

If $D \equiv 0, 1 \pmod 4$, let $\mathrm{Cl}_+(\mathcal{Q}_D)$ denote the set of proper equivalence classes of forms with discriminant D, and let $\mathrm{Cl}_+(\mathrm{Prim}\,\mathcal{Q}_D)$ be the subset of proper equivalence classes of primitive forms. The similar notations $\mathrm{Cl}(\mathcal{Q}_D)$, $\mathrm{Cl}(\mathrm{Prim}\,\mathcal{Q}_D)$ are used for sets of equivalence classes.

It is a fundamental fact that in general, for a given discriminant, there exist more than one equivalence class of primitive forms.

Thus for example, the forms $\langle 1, 0, 5\rangle$, $\langle 2, 2, 3\rangle$ are primitive with discriminant -20. They cannot be equivalent, since 5 is a value of the first, but not of the second form, as easily verified by observing that $2x^2 + 2xy + 3y^2 \not\equiv 1 \pmod 4$ for all integers x, y.

7 Conditional solution of the fundamental problems

Solution of Problem 1. Given m and D, $D \equiv 0$ or $1 \pmod 4$, there exists a primitive representation of m by some form of discriminant D if and only if there exists n such that $D \equiv n^2 \pmod{4m}$.

The proof offers no difficulty. First, observe the following:

If m has a primitive representation by $Q \in \mathcal{Q}_D$, then there exists $n \in \mathbb{Z}$ such that $D \equiv n^2 \pmod{4m}$ and $Q \sim \langle m, n, l\rangle$, with $l = \frac{n^2 - D}{4m}$.

Indeed, there exist integers α, γ such that $\gcd(\alpha, \gamma) = 1$ and $m = Q(\alpha, \gamma)$. Let β, δ be integers such that $\alpha\delta - \beta\gamma = 1$, and let $A = \begin{pmatrix} \alpha & \beta \\ \gamma & \delta \end{pmatrix} \in \mathrm{SL}_2(\mathbb{Z})$; let $Q' = T_A(Q)$, so $Q' = \langle m, n, l\rangle \in \mathcal{Q}_D$ (for some n, l); thus $D = n^2 - 4ml$, hence $D \equiv n^2 \pmod{4m}$ and $l = \frac{n^2 - D}{4m}$.

Conversely, if there exists n such that $D \equiv n^2 \pmod{4m}$, let $l = \frac{n^2 - D}{4m}$, so $\langle m, n, l\rangle \in \mathcal{Q}_D$ and m has a primitive representation by $\langle m, n, l\rangle$.

For example, if $m = 4$, $D = 17$, then 4 has a primitive representation by the form $\langle 4, 1, -1\rangle$, which has discriminant $D = 17$.

Solution of Problem 2. Suppose that $n_1, \ldots, n_k$ are the integers such that $1 \leq n_i \leq 2m$ and $D \equiv n_i^2 \pmod{4m}$. Then m has a primitive representation by the form $Q \in \mathcal{Q}_D$ if and only if $Q \approx \langle m, n, l \rangle$, where $n \equiv n_i \pmod{2m}$ for some i, and $l = \frac{n^2 - D}{4m}$.

One implication is clear. Conversely, as indicated in the solution of Problem 1, if m has a primitive representation by $Q \in \mathcal{Q}_D$ then there exists n such that $D \equiv n^2 \pmod{4m}$ and $Q \approx \langle m, n, l \rangle$ where $l = \frac{n^2 - D}{4m}$. If $n \equiv n' \pmod{2m}$, with $1 \leq n' \leq 2m$, then $D \equiv n^2 \equiv n'^2 \pmod{4m}$, so $n' = n_i$ (for some i).

This gives a procedure to find the forms Q which provide a primitive representation of m (once Problem 4 is solved).

For example, 4 has a primitive representation by a form Q of discriminant 17 if and only if Q is properly equivalent to one of the forms $\langle 4, 1, -1 \rangle$, $\langle 4, 7, 2 \rangle$.

Solution of Problem 3. Let $m = Q(\alpha, \gamma)$, with $\gcd(a, \gamma) = 1$, be a primitive representation of m by the form Q. Then there exist integers β, δ which are unique such that $A = \begin{pmatrix} \alpha & \beta \\ \gamma & \delta \end{pmatrix} \in \mathrm{SL}_2(\mathbb{Z})$ and $T_A(Q) = \langle m, n, l \rangle$ with $D \equiv n^2 \pmod{4m}$, $1 \leq n \leq 2m$, $l = \frac{n^2 - D}{4m}$. This is not hard to show by noting that if β_0, δ_0 are such that $\alpha \delta_0 - \gamma \beta_0 = 1$, then all the possible pairs of integers (β, δ) such that $\alpha \delta - \beta \gamma = 1$ are given by

$$\begin{cases} \beta = \beta_0 + k\alpha \\ \delta = \delta_0 + k\gamma \end{cases} \qquad \text{for any } k \in \mathbb{Z}.$$

It is possible to choose k in a unique way, such that

$$n = 2a\alpha\beta + b(\alpha\delta + \beta\gamma) + 2c\gamma\delta$$

where $1 \leq n < 2m$ and $D \equiv n^2 \pmod{4m}$.

The representation $m = Q(\alpha, \gamma)$ is said to belong to n, when n is determined as above. Conversely, to every n, such that $1 \leq n < 2m$, $D \equiv n^2 \pmod{4m}$, and $Q \approx \langle m, n, l \rangle$, it corresponds, for example, to the following representation: if $T_A(Q) = \langle m, n, l \rangle$, with $A = \begin{pmatrix} \alpha & \beta \\ \gamma & \delta \end{pmatrix} \in \mathrm{SL}_2(\mathbb{Z})$, then $m = T_A(Q)(1, 0) = Q(\alpha, \gamma)$. Clearly, this representation belongs to n.

It remains to describe all primitive representations belonging to the same value n. If $m = Q(\alpha, \gamma) = Q(\alpha', \gamma')$ are primitive representations, if (β, δ), (β', δ') are the unique pairs of integers such that $A = \begin{pmatrix} \alpha & \beta \\ \gamma & \delta \end{pmatrix}$, $A' = \begin{pmatrix} \alpha' & \beta' \\ \gamma' & \delta' \end{pmatrix} \in \mathrm{SL}_2(\mathbb{Z})$, with $T_A(Q) = \langle m, n, l \rangle$,

$T_{A'}(Q) = \langle m, n', l' \rangle$ and $1 \le n$, $n' < 2m$, $l = \frac{n^2 - D}{4m}$, $l' = \frac{n'^2 - D}{4m}$, then $n = n'$ if and only if there exists $B \in SL_2(\mathbb{Z})$ such that $A' = BA$ and $T_B(Q) = Q$.

So, the enumeration of all possible primitive representations of m by Q requires the solutions of the following problems:

Problem 4. If Q, Q' are forms of a given discriminant, to decide whether Q and Q' are equivalent (respectively properly equivalent) or not.

More precisely, it is required to find a finite algorithm to solve this problem.

Problem 5. For any form Q, to determine the set

$$\{B \in SL_2(\mathbb{Z}) \mid T_B(Q) = Q\}.$$

This set, which is clearly a subgroup of $SL_2(\mathbb{Z})$, is called the *automorph* of Q. It is, in fact, the stabilizer of Q by the action of $SL_2(\mathbb{Z})$ on the set of forms.

Note that if $Q \approx Q'$ and if $A_0 \in SL_2(\mathbb{Z})$ is such that $T_{A_0}(Q) = Q'$, then $\{A \in SL_2(\mathbb{Z}) \mid T_A(Q) = Q'\} = \{BA_0 \mid B \text{ in the automorph of } Q\}$.

Indeed, if $T_B(Q) = Q$, then $T_{BA_0}(Q) = Q'$. Conversely, if $T_A(Q) = Q'$, then $T_{AA_0^{-1}}(Q) = Q$, so $AA_0^{-1} = B$ is in the automorph of Q and $A = BA_0$.

8   Proper equivalence classes of definite forms

Let $D < 0$, $D \equiv 0$ or $1 \pmod 4$.

The main idea in this study of proper equivalence classes of positive definite forms, is to select, in an appropriate way, special reduced forms, as I shall indicate now.

Lemma 1. If $Q \in \mathcal{Q}_D$, then there exists $\langle a, b, c \rangle$ such that $Q \approx \langle a, b, c \rangle$ and $|b| \le a \le c$.

PROOF.   Let $Q = \langle m, n, l \rangle$, let $\epsilon = \pm 1$ be such that $\epsilon n = |n|$. If

$$A = \begin{pmatrix} 1 & -\epsilon \\ 0 & 1 \end{pmatrix} \quad \text{and} \quad b = \begin{pmatrix} 1 & 0 \\ -\epsilon & 1 \end{pmatrix},$$

then

$$T_A(Q) = \langle m, n - 2\epsilon m, m + l - |n| \rangle = \langle m', n', l' \rangle,$$
$$T_B(Q) = \langle m + l - |n|, n - 2\epsilon l, l \rangle = \langle m', n', l' \rangle.$$

If $|n| > m$, then in $T_A(Q)$, $m' + l' < m + l$.

If $|n| > l$, then in $T_B(Q)$, $m' + l' < m + l$.

By repeating this process, one reaches a form $\langle a, b, c \rangle$ in the same class, with $|b| \leq a$, $|b| \leq c$.

If $a \leq c$, stop. If $c < a$, let $C = \begin{pmatrix} 0 & 1 \\ -1 & 0 \end{pmatrix}$, then $T_C(\langle a, b, c \rangle) = \langle c, -b, a \rangle$. $\qquad \square$

It is important to observe that if $Q \approx \langle a, b, c \rangle$ with $|b| \leq a \leq c$, then

$$a = \inf\{Q(\alpha, \beta) \mid \alpha, \beta \text{ integers}, Q(\alpha, \beta) \neq 0\},$$
$$ac = \inf\{Q(\alpha, \beta)Q(\gamma, \delta) \mid \alpha, \beta, \gamma, \delta \text{ integers}, \alpha\delta - \beta\gamma \neq 0\}.$$

Thus a, c, and hence also $|b|$, are uniquely defined by Q. It follows that if $\langle a, b, c \rangle \approx \langle a', b', c' \rangle$ with $|b| \leq a \leq c$, and $|b'| \leq a' \leq c'$, then $a = a'$, $c = c'$, $b = \pm b'$. Moreover in this situation, $\langle a, b, c \rangle \approx \langle a, -b, c \rangle$ if and only if $a = |b|$, or $a = c$, or $b = 0$.

Combining these facts, the main result is the following:

If $Q \in \mathcal{Q}_D$, there exists a unique form $\langle a, b, c \rangle \in \mathcal{Q}_D$ such that $Q \approx \langle a, b, c \rangle$ and

$$\text{(Red)} \quad \begin{cases} |b| \leq a \leq c \\ \text{if } |b| = a \text{ or } c = a, \text{ then } b \geq 0. \end{cases}$$

The forms satisfying the condition (Red) above, are called the *reduced positive definite forms* with discriminant D.

It should also be noted that if $Q, Q' \in \mathcal{Q}_D$ then the following conditions are equivalent.

(1) $Q \approx Q'$ or $Q \approx \bar{Q}'$ ($\bar{Q}'$ denotes the conjugate of Q')
(2) $Q \sim Q'$
(3) {values of Q} = {values of Q'}.

The only non-trivial implication is (3) $\Rightarrow$ (1). Let $Q_0 = \langle a, b, c \rangle$, $Q'_0 = \langle a', b', c' \rangle$ be reduced forms such that $Q \approx Q_0$, $Q' \approx Q'_0$, so Q_0,

Q'_0 have the same sets of values. By the characterization of a, c, a', c' as infima of values, then it follows that $a = a'$, $c = c'$, and since these forms have the same discriminant, then $|b| = |b'|$. Therefore, $Q \approx Q'$ or $Q \approx \bar{Q}'$.

It should be noted that SCHINZEL (1980) showed that there exist forms Q, Q' with different discriminants but with the same sets of values; for example, $Q = \langle 1, 0, 3 \rangle$ and $Q' = \langle 1, 1, 1 \rangle$.

Numerical example of reduction of a definite form

Let $\langle 2, 5, 4 \rangle \in Q_{-7}$. Then

$$\langle 2, 5, 4 \rangle \approx \langle 2, 1, 1 \rangle \approx \langle 1, -1, 2 \rangle \approx \langle 1, 1, 2 \rangle.$$

It is easy to see that if $\langle a, b, c \rangle \in \mathcal{Q}_D$ is a reduced form, then

$$\begin{cases} 0 \leq |b| \leq \sqrt{\frac{|D|}{3}}, \\ b^2 \equiv D \pmod 4, \\ a \text{ divides } \frac{b^2 - D}{4}, \\ |b| \leq a \leq \frac{b^2 - D}{4a}. \end{cases}$$

Thus, the number of reduced forms is finite and equal to the number of proper equivalence classes; it follows also that the number of equivalence classes is finite.

It is useful to fix the following notations.
Let

$$\begin{aligned} \tilde{h}(D) &= \text{number of equivalence classes of } \mathcal{Q}_D, \\ h(D) &= \text{number of equivalence classes of } \mathrm{Prim}(\mathcal{Q}_D), \\ \tilde{h}_+(D) &= \text{number of proper equivalence classes of } \mathcal{Q}_D, \\ h_+(D) &= \text{number of proper equivalence classes of } \mathrm{Prim}(\mathcal{Q}_D). \end{aligned}$$

The following inequalities are trivial: $h(D) \leq \tilde{h}(D)$, $h_+(D) \leq \tilde{h}_+(D)$ and $h(D) \leq h_+(D) \leq 2h(D)$, $\tilde{h}(D) \leq \tilde{h}_+(D) \leq 2\tilde{h}(D)$. As examples will show, the above numbers may actually be distinct.

From the bijection given in §4, it is easy to show that if $D = D_0 f^2$, where D_0 is the fundamental discriminant associated to D, then

$$\tilde{h}(D) = \sum_{e|f} h\left(\frac{D}{e^2}\right),$$

$$\tilde{h}_+(D) = \sum_{e|f} h_+\left(\frac{D}{e^2}\right);$$

also, if D is a fundamental discriminant,

$$\tilde{h}(D) = h(D) \quad \text{and} \quad \tilde{h}_+(D) = h_+(D).$$

In view of the preceding considerations, it is easy to write a formula for the number $\tilde{h}_+(D)$.

Put

$$n(a, b) = \begin{cases} 1 & \text{when } b = 0, \\ 1 & \text{when } b = a, \\ 1 & \text{when } a = \sqrt{\frac{b^2 - D}{4}}, \\ 2 & \text{otherwise.} \end{cases}$$

Then $\tilde{h}_+(D) = \sum_{b \in \mathcal{B}} \sum_{a \in \mathcal{A}_b} n(a, b)$, where

$$\mathcal{B} = \{b \mid 0 \le b < \sqrt{\frac{|D|}{3}}, b \equiv D \pmod 2\}$$

$$\mathcal{A}_b = \{a \mid a \text{ divides } \frac{b^2 - D}{4} \text{ and } b \le a \le \sqrt{\frac{b^2 - D}{4}}\}$$

This formula is easy to use when $|D|$ is small.

Numerical example

To calculate $h_+(-303)$ and to determine all reduced forms with discriminant $D = -303$:

$b \in \mathcal{B}$ exactly when b is odd and $0 < b < 10$, thus $\mathcal{B} = \{1, 3, 5, 7, 9\}$.

It is easy to see that $\mathcal{A}_1 = \{1, 2, 4\}$, $\mathcal{A}_3 = \{3, 6\}$, $\mathcal{A}_5 = 0$, $\mathcal{A}_7 = \{8\}$, and $\mathcal{A}_9 = 0$. According to the definition of $n(a, b)$,

$$\tilde{h}_+(-303) = n(1, 1) + n(1, 2) + n(1, 4) + n(3, 3) + n(3, 6) + n(7, 8)$$
$$= 1 + 2 + 2 + 1 + 2 + 2 = 10.$$

Since -303 is a fundamental discriminant, then $h_+(-303) = \tilde{h}_+(-303) = 10$.

Here is the explicit determination of the reduced forms of discriminant -303:

b	$\frac{b^2-D}{4}$	a	b
± 1	76	1 2 4 1̸9̸ 3̸8̸ 7̸6̸	76 38 19 4̸ 2̸ 1̸
± 3	78	3 6 1̸3̸ 2̸6̸ 3̸9̸ 7̸8̸	26 13 6 3 2 1
± 5	82	4̸1̸ 8̸2̸	2̸ 1̸
± 7	88	8 1̸1̸ 2̸2̸ 4̸4̸ 8̸8̸	11 8̸ 4̸ 2̸ 1̸
± 9	96	1̸2̸ 1̸6̸ 3̸2̸ 4̸8̸ 9̸6̸	8̸ 6̸ 3̸ 2̸ 1̸

So, the reduced forms are $\langle 1, 1, 76 \rangle$, $\langle 2, 1, 38 \rangle$, $\langle 4, 1, 19 \rangle$, $\langle 1, -1, 76 \rangle$, $\langle 2, -1, 38 \rangle$, $\langle 4, -1, 19 \rangle$, $\langle 3, 3, 26 \rangle$, $\langle 6, 3, 13 \rangle$, $\langle 6, -3, 13 \rangle$, $\langle 8, 7, 11 \rangle$, and $\langle 8, -7, 11 \rangle$.

From this, it follows that $\tilde{h}(-303) = h(-303) = 6$.

A   Another numerical example

If $D = -72 = 9 \times (-8)$, the associated fundamental discriminant is $D_0 = -8$.

The same method yields

$$h(-72) = h_+(-72) = 2,$$
$$h(-8) = h_+(-8) = 1.$$

Thus $\tilde{h}(-72) = \tilde{h}_+(-72) = 3$, with the reduced forms $\langle 1, 0, 18 \rangle$, $\langle 2, 0, 9 \rangle$, and $\langle 3, 0, 6 \rangle$ (this last one not primitive).

9   Proper equivalence classes of indefinite forms

Let $D > 0$ (not a square).

The form $Q = \langle a, b, c \rangle \in \mathcal{Q}_D$ is said to be a *reduced indefinite form* with discriminant D if the following conditions are satisfied:

$$(\text{Red}) \quad \begin{cases} 0 < b < \sqrt{D}, \\ \sqrt{D} - b < 2|a| < \sqrt{D} + b. \end{cases}$$

It follows (as easily seen) that

$$ac < 0, \; |a| < \sqrt{D}, \; |c| < \sqrt{D} \text{ and also } \sqrt{D} - b < 2|c| < \sqrt{D} + b.$$

Also, if $|a| \leq |c|$ and $\sqrt{D} - 2|a| < b < \sqrt{D}$, then $\langle a, b, c \rangle$ is a reduced form.

Here is the easy algorithm to enumerate all the reduced forms in $\mathcal{Q}_D$.

For each integer b, $0 < b < \sqrt{D}$, such that 4 divides $D - b^2$, factor the integer $(D - b^2)/4$ in all possible ways.

For each factorization $(D - b^2)/4 = ac$, with a, $c > 0$, check if the conditions

$$(*) \begin{cases} \sqrt{D} - b < 2a < \sqrt{D} + b \\ \sqrt{D} - b < 2c < \sqrt{D} + b \end{cases}$$

are satisfied. If not, reject the factorization. If $(*)$ is satisfied, then the forms $\langle a, b, -c \rangle$, $\langle -a, b, c \rangle$, $\langle c, b, -a \rangle$, $\langle -c, b, a \rangle$ are reduced with discriminant D; note that there are only two distinct forms when $a = c$.

Numerical example

The reduced forms with discriminant $D = 52$ must have b even, $0 < b < 52$, 4 dividing $52 - b^2$; thus $b = 2$, 4, or 6. If $b = 2$, then $-ac = \frac{52-4}{4} = 12$; moreover, $\sqrt{52} - 2 < 2|a|$, $2|c| < \sqrt{52} + 2$, so $(a, c) = (\pm 3, \mp 4)$, $(\pm 4, \mp 3)$. If $b = 4$, then $-ac = \frac{52-16}{4} = 9$; moreover, $\sqrt{82} - 4 < 2|a|$, $2|c| < \sqrt{52} + 4$, so $(a, c) = (\pm 3, \mp 3)$. If $b = 6$, then $ac < 0$, $ac = \frac{52-36}{4} = 4$; moreover $\sqrt{52} - 6 < 2|a|$, $2|c| < \sqrt{52} + 6$, so $(a, c) = (\pm 1, \mp 4)$, $(\pm 2, \mp 2)$, $(\pm 4, \mp 1)$.

Thus the reduced forms are $\langle \pm 3, 2, \mp 4 \rangle$, $\langle \pm 4, 2, \mp 3 \rangle$, $\langle \pm 1, 6, \mp 4 \rangle$, $\langle \pm 2, 6, \mp 2 \rangle$, and $\langle \pm 4, 6, \mp 1 \rangle$.

Lemma 2. *Every $Q \in \mathcal{Q}_D$ is equivalent to a reduced form.*

PROOF. Let $Q = \langle m, n, l \rangle$. It will be shown that there exists $\langle a, b, c \rangle \approx Q$, such that $|a| \leq |c|$, $\sqrt{D} - 2|a| < b < \sqrt{D}$; by a previous remark, $\langle a, b, c \rangle$ is a reduced form.

Let $\lambda = [\sqrt{D}]$ and consider the set of $2|l|$ integers $\{\lambda + 1 - 2|l|, \lambda + 2 - 2|l|, \ldots, \lambda\}$. Then there exists a unique n', $\lambda + 1 - 2|l| \leq n' \leq \lambda$ such that $n \equiv -n' \pmod{2|l|}$. Let

$$\delta = -\frac{n + n'}{2|l|} \quad \text{and} \quad A = \begin{pmatrix} 0 & 1 \\ -1 & \delta \end{pmatrix}.$$

Then $T_A \langle m, n, l \rangle = \langle l, n', l' \rangle$, with $l' = m + n\delta + l\delta^2$ and $\sqrt{D} - 2|l| < n' < \sqrt{D}$.

If $|l| > |l'|$, repeat the argument. Since it cannot always be $|l| > |l'| > |l''| > \cdots$, one reaches a form $\langle a, b, c \rangle \approx Q$, with $|c| \leq |a|$ and $\sqrt{D} - 2|a| < b < \sqrt{D}$, which is reduced.   □

Numerical example of reduction of an indefinite form

Let $\langle 76, 58, 11 \rangle \in \mathcal{Q}_{20}$. Then

$$\langle 76, 58, 11 \rangle \approx \langle 11, -14, 4 \rangle \approx \langle 4, -2, -1 \rangle \approx \langle -1, 4, 5 \rangle.$$

It may be shown that there exist only finitely many forms $\langle a, b, c \rangle \in \mathcal{Q}_D$ with $ac < 0$. Hence, there are only finitely many reduced forms, and it may be concluded:

The number of proper equivalence classes of indefinite forms with discriminant D is finite. Therefore, the number of proper equivalence classes of $\mathrm{Prim}(\mathcal{Q}_D)$ is finite, and so are the numbers of equivalence classes of $\mathcal{Q}_D$ and of $\mathrm{Prim}(\mathcal{Q}_D)$.

As in the case of positive definite forms, the following notation will be used:

$\tilde{h}_+(D)$, $h_+(D)$ for the numbers of proper equivalence classes of $\mathcal{Q}_D$, $\mathrm{Prim}(\mathcal{Q}_D)$ respectively;

$\tilde{h}(D)$, $h(D)$ for the numbers of equivalence classes of $\mathcal{Q}_D$, $\mathrm{Prim}(\mathcal{Q}_D)$.

And again, the following inequalities are trivial:

$$h_+(D) \le \tilde{h}_+(D), \quad h(D) \le \tilde{h}(D),$$

$$h(D) \le h_+(D) \le 2h(D), \quad \tilde{h}(D) \le \tilde{h}_+(D) \le 2\tilde{h}(D).$$

If $D = D_0 f^2$, where D_0 is a fundamental discriminant, then

$$\tilde{h}(D) = \sum_{e \mid f} h\left(\frac{D}{e^2}\right) \quad \text{and} \quad \tilde{h}_+(D) = \sum_{e \mid f} h_+\left(\frac{D}{e^2}\right).$$

Hence, if D is a fundamental discriminant, then $\tilde{h}_+(D) = h_+(D)$ and $\tilde{h}(D) = h(D)$. As examples show, $h(D)$, $h_+(D)$ (respectively, $\tilde{h}(D)$, $\tilde{h}_+(D)$) may actually be distinct.

It is essential to study the case when two reduced forms are properly equivalent.

Let $\mathcal{R}_D$ denote the set of reduced forms in $\mathcal{Q}_D$.

If $Q = \langle a, b, c \rangle \in \mathcal{R}_D$, there exists a unique properly equivalent form $Q' = \langle c, b', c' \rangle \in \mathcal{R}_D$ such that $2c \mid b + b'$, and there exists a unique properly equivalent form $Q'' = \langle a'', b'', a \rangle \in \mathcal{R}_D$ such that $2a \mid b + b''$. The form Q' is said to be *right-adjacent*, and Q'' is said to be *left-adjacent* to Q. Necessarily $2|c|$ divides $b + b'$ and $2|a|$ divides

$b+b''$. Let $Q' = p(Q)$, $Q'' = \lambda(Q)$; then $Q = \lambda(Q')$, $Q = \rho(Q'')$. Note that $\rho(Q) = T_A(Q)$ where

$$A = \begin{pmatrix} 0 & 1 \\ -1 & \delta \end{pmatrix},$$

and δ is unique. The actual determination of δ will be indicated later, but it should already be noted that $a\delta > 0$.

Since $\mathcal{R}_D$ is finite, for every $Q \in \mathcal{R}_D$ there exist indices $i < i+k$, such that if $Q_1 = \rho(Q)$, $Q_2 = \rho(Q_1)$, $\ldots$, then $Q_i = Q_{i+k}$; but $Q_{i-1} = \lambda(Q_i) = \lambda(Q_{i+k}) = Q_{i+k-1}$, $\ldots$, and this gives $Q = Q_k$. Let $k \geq 1$ be the minimal possible. Then the set $\{Q = Q_0, Q_1, \ldots, Q_k\}$ is called the *period* of Q. Clearly, this set is also the period of each $Q_i (1 \leq i \leq k)$. Thus, $\mathcal{R}_D$ is partitioned into periods; forms in the same period are clearly properly equivalent. The converse is one of the main results in the theory:

If $Q, Q' \in \mathcal{R}_D$ and $Q \approx Q'$, then Q, Q' are in the same period. Thus $h_+(D)$ is equal to the number of periods and, similarly, $h_+(D)$ is the number of periods of primitive reduced forms.

It is not difficult to see that the number of forms in each period is even.

There is a very interesting relation between periods and continued fraction expansions.

The fact that a form $\langle a, b, c \rangle$ is reduced may be expressed in terms of the roots ω, η of the form. Namely, $\langle a, b, c \rangle$ is reduced if and only if $|w| > 1$, $|\eta| < 1$, and $\omega\eta < 0$.

If $\{Q = Q_0, Q_1, \ldots, Q_{2r-1}\}$ is the period of the reduced form Q, let $\rho(Q_i) = T_{A_i}(Q_i) = Q_{i+1}$, where $A_i = \begin{pmatrix} 0 & 1 \\ -1 & \delta_i \end{pmatrix}$. Let ω_i, η_i be the roots of the form Q_i. Then

$$|\omega_i| = \frac{1}{|\delta_i| + |\omega_{i+1}|} \qquad \text{for } i = 0, 1, \ldots, 2r - 1 \text{ and } |\omega_r| = |\omega_0|.$$

So,

$$|\omega| = [|\overline{\delta_0}|, |\delta_1|, \ldots, |\delta_{r-1}|],$$

that is, $|\delta_0|$, $|\delta_1|$, $\ldots$, $|\delta_{r-1}|$ are the partial quotients in the regular continued fraction expansion of the quadratic irrational number ω; moreover, this expansion is purely periodic.

It is also true that if $Q = \langle a, b, c \rangle$ is a form with discriminant $D > 0$, and if $\left| \frac{-b+\sqrt{D}}{2a} \right|$ has regular continued fraction expansion which is purely periodic with period having length r, and partial quotients $|\delta_0|, |\delta_1|, \ldots, |\delta_{r-1}|$, then Q is a reduced form, its period has $2r$ elements when r is odd, r elements when r is even and it is equal to $\{Q = Q_0, Q_1, \ldots, Q_{2r-1}\}$, with

$$Q_{i+1} = T_{A_i}(Q_i), \quad A_i = \begin{pmatrix} 0 & 1 \\ -1 & \delta_i \end{pmatrix}, \quad A_{i+r} = \begin{pmatrix} 0 & 1 \\ -1 & -\delta_i \end{pmatrix},$$

for $0 \le i \le r-1$; if $Q_i = \langle a_i, b_i, c_i \rangle$, then $a_i \delta_i > 0$ (for $i = 0, 1, \ldots, r-1$).

Numerical example

It is easy to calculate $\tilde{h}_+(D)$, $h_+(D)$ for small values of $D > 0$. For example, if $D = 68$ (not a fundamental discriminant), then $\sqrt{D} = 2\sqrt{17} = 8.24\ldots$.

Determination of the reduced forms with discriminant 68: Since 4 divides $D - b^2$ and $0 < b < \sqrt{D}$, it follows that $b = 2, 4, 6, 8$. Also, $\sqrt{D} - b < 2|a| < \sqrt{D} + b$. This gives the possibilities:

b	$\sqrt{D}-b$	$\sqrt{D}+b$	$|ac|$	$|a|$	$|c|$
2	6.24	10.24	16	4	4
4	4.24	12.24	13	—	—
6	2.24	14.24	8	$\begin{cases} 2 \\ 4 \end{cases}$	$\begin{matrix} 4 \\ 2 \end{matrix}$
8	0.24	16.24	1	1	1

Thus $\mathcal{R}_{68}$ consists of the 8 forms

$$\langle \pm 4, 2, \mp 4 \rangle, \quad \langle \pm 2, 6, \mp 4 \rangle, \quad \langle \pm 4, 6, \mp 2 \rangle, \quad \langle \pm 1, 8, \mp 1 \rangle.$$

Calculation of the period of $\langle 4, 2, -4 \rangle$: its first root is $\omega = \frac{\sqrt{17}-1}{4}$, which has the following continued fraction expansion: $\omega = [\overline{1, 3, 1}]$. For $0 \le i \le 5$, let $A_i = \begin{pmatrix} 0 & 1 \\ -1 & -\delta_i \end{pmatrix}$ with $\delta_0 = 1$, $\delta_1 = -3$, $\delta_2 = 1$, $\delta_3 = -1$, $\delta_4 = 3$, $\delta_5 = -1$. Let $Q_{i+1} = T_{A_i}(Q_i)$ (for $i = 0, 1, \ldots, 5$); then the period of $\langle 4, 2, -4 \rangle$ consists of this form and the forms $\langle -4, 6, 2 \rangle$, $\langle 2, 6, -4 \rangle$, $\langle -4, 2, 4 \rangle$, $\langle 4, 6, -2 \rangle$, and $\langle -2, 6, 4 \rangle$.

The other period is $\{\langle 1, 8, -1 \rangle, \langle -1, 8, 1 \rangle\}$. Thus $\tilde{h}_+(68) = 2$, while $h_+(68) = 1$. Also, $\tilde{h}(68) = 2$, $h(68) = 1$.

A   Another numerical example

Let $D = 76 = 4 \times 19$ (it is a fundamental discriminant). The reduced forms are $\langle\pm3, 4, \mp5\rangle$, $\langle\pm5, 4, \mp3\rangle$, $\langle\pm2, 6, \mp5\rangle$, $\langle\pm5, 6, \mp2\rangle$, $\langle\pm1, 8, \mp3\rangle$, and $\langle\pm3, 8, \mp1\rangle$.

To calculate the period of $\langle3, 4, -5\rangle$, its first root is $\omega = \frac{-2+\sqrt{19}}{3}$. The regular continued fraction expansion of ω is $\omega = [\overline{1, 3, 1, 2, 8, 2}]$ and the period of $\langle3, 4, -5\rangle$ is $\{\langle3, 4, -5\rangle, \langle-5, 6, 2\rangle, \langle2, 6, -5\rangle, \langle-5, 4, 3\rangle, \langle3, 8, -1\rangle, \langle-1, 8, 3\rangle\}$.

There is another period consisting of the other six reduced forms. Thus $h_+(76) = 2$, while $h(76) = 1$.

10   The automorph of a primitive form

Recall that the automorph of a form $Q = \langle a, b, c\rangle$ consists of all the $A \in \mathrm{SL}_2(\mathbb{Z})$ such that $T_A(Q) = Q$.

The description of the automorph requires a preliminary study of the behavior of the roots of Q, under the action of any $A \in \mathrm{GL}_2(\mathbb{Z})$.

Thus, let $Q = \langle a, b, c\rangle$, with roots ω, η; let $A \in \mathrm{GL}_2(\mathbb{Z})$ and let $Q' = T_A(Q)$ have roots ω', η'.

If $\zeta \in \{\omega', \eta'\}$, then

$$(\gamma\zeta' + \delta)^2 Q\left(\frac{\alpha\zeta' + \beta}{\gamma\zeta' + \delta}, 1\right) = Q(\alpha\zeta' + \beta, \gamma\zeta' + \delta) = Q'(\zeta', 1) = 0;$$

since $\gamma\zeta' + \delta \neq 0$, then $\frac{\alpha\zeta' + \beta}{\gamma\zeta' + \delta}$ is a root of Q.

A simple calculation shows that if $A \in \mathrm{SL}_2(\mathbb{Z})$, then

$$\frac{\alpha\omega' + \beta}{\gamma\omega' + \beta} = \omega \quad \text{and} \quad \frac{\alpha\eta' + \beta}{\gamma\eta' + \delta} = \eta,$$

while if $A \notin \mathrm{SL}_2(\mathbb{Z})$, then ω', η' correspond respectively to η, ω.

If $A \in \mathrm{SL}_2(\mathbb{Z})$ is in the automorph of the primitive form $Q = \langle a, b, c\rangle$, then from $T_A(Q) = Q$ it follows that $\omega = \frac{\alpha\omega+\beta}{\gamma\omega+\delta}$, hence

$$\gamma\omega^2 + (\delta - \alpha)\omega - \beta = 0.$$

But also

$$a\omega^2 + b\omega + c = 0,$$

and since $\gcd(a, b, c) = 1$, there exists an integer $u \neq 0$ such that $\gamma = au$, $\delta - \alpha = bu$, $-\beta = cu$. Let $t = \delta + \alpha$; then $t \equiv bu \pmod{2}$,

$\alpha = \frac{t-bu}{2}$, $\delta = \frac{t+bu}{2}$. From $\alpha\delta - \beta\gamma = 1$, it follows that $\frac{t^2-b^2u^2}{4} + acu^2 = 1$, and finally $t^2 - Du^2 = 4$.

This calculation points out to a description of the automorph of $Q = \langle a, b, c \rangle$:

A is in the automorph of Q if and only if there exist integers t, u such that

$$\begin{cases} t^2 - Du^2 = 4 \\ \alpha = \frac{t-bu}{2} \\ \beta = -cu \\ \gamma = au \\ \delta = \frac{t+bu}{2}. \end{cases}$$

Thus, the automorph of Q is in one-to-one correspondence with the solutions in integers of the equation

$$T^2 - DU^2 = 4.$$

If $D < 0$, this equation has only the solutions

$$(t, u) = \begin{cases} (\pm 2, 0) & \text{if } D \neq -4, -3, \\ (\pm 2, 0), (\pm 1, \pm 1) & \text{if } D = -3, \\ (\pm 2, 0), (0, \pm 1) & \text{if } D = -4. \end{cases}$$

Thus, if $D < 0$, then the number of elements in the automorph of Q is

$$w = \begin{cases} 2 & \text{if } D \neq -3, -4, \\ 6 & \text{if } D = -3, \\ 4 & \text{if } D = -4. \end{cases}$$

If $D > 0$, LAGRANGE proved what FERMAT already knew: there are infinitely many pairs (t, u) of integers such that $t^2 - Du^2 = 4$. If (t_1, u_1) is such that $t_1, u_1 > 0$ and $t_1 + u_1\sqrt{D}$ is minimal, then all the solutions (t, u) are given by the relations

$$\frac{t + u\sqrt{D}}{2} = \left(\frac{t_1 + u_1\sqrt{D}}{2} \right)^n \qquad \text{for } n = 0, \pm 1, \pm 2, \ldots.$$

Thus, if $D > 0$, then the automorph of Q is infinite.

LAGRANGE has also indicated an algorithm, involving continued fractions, to determine t_1, u_1; this is, of course, well-known.

Numerical example

To determine all the primitive representations of $m = 17$ by the form $Q = \langle 2, 6, 5 \rangle$ with discriminant -4.

First, note that -4 is a square modulo 68 because -1 is a square modulo 17. The integers $n = 4, 13$ are the only solutions of $-1 \equiv n^2 \pmod{17}$, $1 \leq n < 17$, so 8, 26 are the only integers n such that $-4 \equiv n^2 \pmod{68}$, $1 \leq n < 34$.

To know whether there is a primitive representation of 17 by Q, belonging to 8, it must be verified that $Q \approx \langle 17, 8, 1 \rangle$. But $h(-4) = 1$, so necessarily Q, $\langle 17, 8, 1 \rangle$ are equivalent to the only reduced form $\langle 1, 0, 1 \rangle$ of discriminant -4. The reduction is performed as follows:

$$\langle 2, 6, 5 \rangle \xrightarrow[\left(\begin{smallmatrix} 1 & -1 \\ 0 & 1 \end{smallmatrix}\right)]{\approx} \langle 2, 2, 1 \rangle \xrightarrow[\left(\begin{smallmatrix} 1 & 0 \\ -1 & 1 \end{smallmatrix}\right)]{\approx} \langle 1, 0, 1 \rangle,$$

$$\langle 17, 8, 1 \rangle \xrightarrow[\left(\begin{smallmatrix} 1 & 0 \\ -1 & 1 \end{smallmatrix}\right)]{\approx} \langle 10, 6, 1 \rangle \xrightarrow[\left(\begin{smallmatrix} 1 & 0 \\ -1 & 1 \end{smallmatrix}\right)]{\approx} \langle 5, 4, 1 \rangle$$

$$\langle 5, 4, 1 \rangle \xrightarrow[\left(\begin{smallmatrix} 1 & 0 \\ -1 & 1 \end{smallmatrix}\right)]{\approx} \langle 2, 2, 1 \rangle \xrightarrow[\left(\begin{smallmatrix} 1 & 0 \\ -1 & 1 \end{smallmatrix}\right)]{\approx} \langle 1, 0, 1 \rangle$$

Let

$$A = \begin{pmatrix} 1 & -1 \\ 0 & 1 \end{pmatrix} \begin{pmatrix} 1 & 0 \\ -1 & 1 \end{pmatrix} \begin{pmatrix} 1 & 0 \\ 1 & 1 \end{pmatrix} \begin{pmatrix} 1 & 0 \\ 1 & 1 \end{pmatrix} \begin{pmatrix} 1 & 0 \\ 1 & 1 \end{pmatrix} \begin{pmatrix} 1 & 0 \\ 1 & 1 \end{pmatrix}$$
$$= \begin{pmatrix} -2 & -1 \\ 3 & 1 \end{pmatrix}$$

Then,
$$T_A(\langle 2, 6, 5 \rangle) = \langle 17, 8, 1 \rangle.$$

The automorph of $Q = \langle 2, 6, 5 \rangle$ consists of the matrices

$$\begin{pmatrix} \pm 1 & 0 \\ 0 & \pm 1 \end{pmatrix} \quad \text{and} \quad \begin{pmatrix} \mp 3 & \mp 5 \\ \pm 2 & \pm 3 \end{pmatrix}.$$

All the primitive representations are obtained from the matrices BA, where B is in the automorph of Q:

$$\begin{pmatrix} \mp 2 & \mp 1 \\ \pm 3 & \pm 1 \end{pmatrix} \quad \text{and} \quad \begin{pmatrix} \mp 9 & \mp 2 \\ \pm 5 & \pm 1 \end{pmatrix}.$$

This gives the four representations:

$$17 = Q(\mp 2, \pm 3) = Q(\mp 9, \pm 5).$$

A similar calculation gives the primitive representations of 17 by Q, which belong to 26:

$$T_{A'}(\langle 2, 6, 5 \rangle) = \langle 17, 26, 10 \rangle, \quad \text{where } A' = \begin{pmatrix} 7 & 5 \\ -3 & -2 \end{pmatrix}.$$

The matrices BA', with B in the automorph of Q, are

$$\begin{pmatrix} \pm 7 & \pm 5 \\ \mp 3 & \mp 2 \end{pmatrix} \quad \text{and} \quad \begin{pmatrix} \mp 6 & \mp 5 \\ \pm 5 & \pm 4 \end{pmatrix};$$

this gives the four representations:

$$17 = Q(\pm 7, \mp 3) = Q(\mp 6, \pm 5).$$

The classical case, which concerns the form $Q = \langle 1, 0, 1 \rangle$, was studied by FERMAT and is treated in elementary books with a direct approach. But the results of FERMAT may also be obtained as a special case of GAUSS' theory:

An integer $m \geq 1$ has a primitive representation as the sum of two squares if and only if $m = m_1 m_2^2$, where $\gcd(m_1, m_2) = 1$ and the primes dividing m_1 are either 2 or primes $p \equiv 1 \pmod 4$. In this situation, the number of primitive representations of m as a sum of two squares is equal to

$$\rho(m) = 4(d_1(m) - d_3(m)),$$

where

$$d_1(m) = \#\{d > 0 \mid d \equiv 1 \pmod 4, d \mid m\},$$
$$d_3(m) = \#\{d > 0 \mid d \equiv 3 \pmod 4, d \mid m\}.$$

To arrive at this conclusion, the main points to observe are the following:

m is a sum of two squares if and only if m is representable by a form of discriminant -4, because $h_+(-4) = 1$. This happens if and only if $-4 \equiv n^2 \pmod{4m}$, for some n; equivalently, -1 is a square modulo m. An easy computation with the Jacobi symbol leads to the condition stated for m.

Each n, $1 \leq n < 2m$, such that $-4 \equiv n^2 \pmod{4m}$ corresponds to $\omega = 4$ primitive representations of m as the sum of two squares. From the theory of congruences, the number of solutions n as above is $\sum_{k|m} \left(\frac{-1}{k}\right) = d_1(m) - d_3(m)$. Hence, the number of primitive representations is indeed $4(d_1(m) - d_3(m))$.

11   Composition of proper equivalence classes of primitive forms

One of the very important and deepest contributions of GAUSS in the study of binary quadratic forms is the theory of composition. The idea had been already sketched by LEGENDRE, in a particular case.

Let D be any discriminant. GAUSS defined a binary operation in the set $\mathrm{Cl}_+(\mathrm{Prim}(\mathcal{Q}_D))$ of proper equivalence classes of primitive forms with discriminant D. As it turns out, this operation, called *composition*, satisfies nice properties.

The theory is presented here in a simplified form due to DIRICHLET.

Let $Q = \langle a, b, c \rangle$, $Q' = \langle a', b', c' \rangle$ be primitive forms with discriminant D. The new form $Q'' = \langle a'', b'', c'' \rangle$ shall be defined as follows.

Let $\delta = \gcd\left(a, a', \frac{b+b'}{2}\right)$ and let u, v, w be any integers such that

$$au + a'v + \frac{b+b'}{2}w = \delta.$$

Note that there are infinitely many possible choices for u, v, w. Define

$$\begin{cases} a'' = \frac{aa'}{\delta^2}, \\ b'' = \frac{1}{\delta}\left[aub' + a'vb + \frac{bb'+D}{2}w\right], \\ c'' = \frac{(b'')^2 - D}{4a''}. \end{cases}$$

Then, $Q'' = \langle a'', b'', c'' \rangle$ is also a primitive form with discriminant D, which depends on the choice of u, v, w—this is stressed by denoting $Q'' = Q''_{(u,v,w)}$. It may be shown that if u_1, v_1, w_1 are also integers satisfying $au_1 + av_1 + \frac{b+b'}{2}w_1 = \delta$, then the form $Q''_{(u_1,v_1,w_1)}$, although different from $Q''_{(u,v,w)}$, is nevertheless properly equivalent to $Q''_{(u,v,w)}$.

Thus, to the pair of primitive forms Q, Q', it may be assigned the proper equivalence class $\mathbf{Q}''_{(u,v,w)}$ of $Q''_{u,v,w}$. It is also true that if $Q \approx Q_1$ and $Q' \approx Q'_1$, if Q'', Q''_1 are defined as indicated above from Q, Q', respectively, Q_1, Q'_1, then $Q'' \approx Q''_1$. This allows us to define an operation of *composition* of proper equivalence classes of primitive forms of discriminant D:

$$\mathbf{Q} * \mathbf{Q}' = \mathbf{Q}'',$$

where Q'' was defined (with any u, v, w) as indicated above.

GAUSS showed that if Q, Q', $Q'' \in \mathrm{Prim}(\mathcal{Q}_D)$ and if $\mathbf{Q} * \mathbf{Q}' = \mathbf{Q}''$, then for any integers x, y, x', y', there exist integers x'', y'' which are linear combinations with coefficients in $\mathbb{Z}$ of xx', xy', yx', yy', such that $Q''(x'', y'') = Q(x,y)Q'(x', y')$.

The proof of this important property of the composition is rather sophisticated, and I shall sketch it now.

First, it may be shown that given any two classes $\mathbf{Q}$, $\mathbf{Q}'$ of primitive forms, it is possible to choose primitive forms $Q_1 \approx Q$, $Q'_1 \approx Q'$ such that

$$Q_1 = \langle a, b, a'c \rangle$$

$$Q'_1 = \langle a', b, ac \rangle$$

with a, $a' \geq 1$, $\gcd(a, a') = 1$.

Let u, v be such that $au + a'v = 1$, let $w = 0$ and consider the primitive form Q''_1, obtained from Q_1, Q'_1 and $(u, v, 0)$ as already indicated in the definition of composition. A simple calculation gives $Q''_1 = \langle aa', b, c \rangle$, and $\mathbf{Q}''_1 = \mathbf{Q}_1 * \mathbf{Q}_1 = \mathbf{Q} * \mathbf{Q}'$.

Then, it may be shown that $Q_1(x, y) \cdot Q'_1(x', y') = Q''_1(x'', y'')$, where

$$\begin{cases} x'' = xx' - cyy', \\ y'' = axy' + a'yx' + byy'. \end{cases}$$

This suffices to prove the statement.

The composition of proper equivalence classes of primitive forms satisfies the associative, and commutative laws. The proper equivalence class $\mathbf{P}$ of the principal form P is such that $\mathbf{Q} * \mathbf{P} = \mathbf{Q}$ for every class $\mathbf{Q}$. The class $\mathbf{P}$ is called the *principal class*. Finally, for every class $\mathbf{Q}$, there exists a class $\mathbf{Q}'$, necessarily unique, such that $\mathbf{Q} * \mathbf{Q}' = \mathbf{P}$; the class $\mathbf{Q}'$ is the *inverse of* $\mathbf{Q}$ under composition.

Thus, $\mathrm{Cl}_+(\mathrm{Prim}(\mathcal{Q}_D))$ is a finite abelian group under the operation of composition—a fact which was discovered by GAUSS, although obviously not phrased in this language.

The inverse of the class $\langle a, b, c \rangle$ is $\langle a, -b, c \rangle$; it is the class of the associated forms.

A class equal to its inverse is called an *ambiguous class*. These are exactly the elements of order 1 or 2 in the group of classes of primitive forms, with operation of composition. It is easy to see that if $\langle a, b, c \rangle$ is such that a divides b, then $\langle a, b, c \rangle \approx \langle a, -b, c \rangle$, hence the class $\langle a, b, c \rangle$ is ambiguous.

Numerical example

If $D = -20$ and $P = \langle 1, 0, 5 \rangle$, $Q = \langle 2, 2, 3 \rangle$, then $\mathbf{Q} * \mathbf{Q} = \mathbf{P}$, as it is easy to verify.

The structure theorem for finite abelian groups tells us that the group under composition $\mathrm{Cl}_+(\mathrm{Prim}(\mathcal{Q}_D))$ is the direct product, in a unique way (up to isomorphism), of primary cyclic groups.

The structure of $\mathrm{Cl}_+(\mathrm{Prim}(\mathcal{Q}_D))$ will be discussed in §13.

12   The theory of genera

The theory of genera was created to try to characterize in a simple manner those primes that may be represented by a given form with a fundamental discriminant.

Explicitly, let D be a discriminant, let p be a prime not dividing $2D$, and assume that p may be represented by some primitive form of discriminant D; in other words, D is a square modulo $4p$, so $\left(\frac{D}{p}\right) = +1$. The problem is to decide which primitive forms of discriminant D represent p, just by computing the values at p of certain quadratic characters associated to D.

As it will be seen, the theory of genera does not quite succeed to attain its aim.

As was said in §3, if $D = -4$ and $Q = X^2 + Y^2$, the prime p is a sum of two squares if and only if $p = 2$ or $p \equiv 1 \pmod 4$. Similarly, if $D = -12$, $Q = X^2 + 3Y^2$, then p is represented by Q if and only if $p = 3$ or $p \equiv 1 \pmod 3$.

However, as already indicated, if $D = -20$, the class number $h_+(-20) = 2$ and there are two reduced forms of discriminant -20

which are not properly equivalent, namely

$$Q_1 = X^2 + 5Y^2 \quad \text{and} \quad Q_2 = 2X^2 + 2XY + 3Y^2.$$

Now:

$p = 5$ is represented by Q_1;

if $p \equiv 11, 13, 17, 19 \pmod{20}$, that is $\left(\frac{-5}{p}\right) = -1$, then p is neither represented by Q_1, nor by Q_2;

if $p \equiv 1, 3, 7, 9 \pmod{20}$, that is $\left(\frac{-5}{p}\right) = +1$, then p is represented by Q_1 or by Q_2.

The question that remains is to decide, for a given prime p of the last type, which one of Q_1 or Q_2 represents it.

The same problem arises also for any discriminant D such that $h_+(D) > 1$, and the theory of genera is a serious attempt to solve it.

Let D be any discriminant (not necessarily fundamental). Let q_1, $\ldots$, q_r, $(r \geq 0)$, be the distinct odd primes dividing the square-free kernel of D. The numbering is such that $q_1 \equiv \cdots \equiv q_s \equiv 1 \pmod{4}$ and $q_{s+1} \equiv \cdots \equiv q_r \equiv -1 \pmod{4}$, with $0 \leq s \leq r$.

For every m such that q_i does not divide m, let $\chi_i(m) = \left(\frac{m}{q_i}\right)$ $(i = 1, \ldots, r)$, so χ_i is a character modulo q_i.

For every odd integer m, let

$$\delta(m) = (-1)^{(m-1)/2} \quad \text{and} \quad \eta(m) = (-1)^{(m^2-1)/8}.$$

Again, δ is a character modulo 4 and η is a character modulo 8.

To each type of discriminant, there will be assigned a set of characters, according to the following rule:

Discriminant	Assigned characters	$(\#, \#')$
$D \equiv 1 \pmod 4$	$\chi_1, \ldots, \chi_r$	(r, r)
$D = 4D', D' \equiv 1 \pmod 4$	$\chi_1, \ldots, \chi_r$	$(r, r+1)$
$D = 4D', D' \equiv 3 \pmod 4$	$\chi_1, \ldots, \chi_r, \delta$	$(r+1, r+1)$
$D = 4D', D' \equiv 2 \pmod 8$	$\chi_1, \ldots, \chi_r, \eta$	$(r+1, r+1)$
$D = 4D', D' \equiv 6 \pmod 8$	$\chi_1, \ldots, \chi_r, \delta\eta$	$(r+1, r+1)$
$D = 4D', D' = 4E^2 q_1 \cdots q_r$	$\chi_1, \ldots, \chi_s, \chi_{s+1}\delta, \ldots, \chi_r\delta$	$(r, r+1)$
$D = 4D', D' = 8E^2 q_1 \cdots q_r$	$\chi_1, \ldots, \chi_s, \chi_{s+1}\delta, \ldots, \chi_r\delta, \eta$	$(r+1, r+1)$

$(\#, \#') = $ (number of assigned characters, number of primes dividing D)

It is convenient to denote in each case by Θ the set of all assigned characters, and by t their number.

By the quadratic reciprocity law, if $\gcd(m, 2D) = 1$, then

$$\prod_{\theta \in \Theta} \theta(m) = \left(\frac{D}{m}\right).$$

The following facts may be proved. If $\gcd(m, 2D) = 1$ and m is any value of $Q \in \mathrm{Prim}(\mathcal{Q}_D)$, then $\left(\frac{D}{m}\right) = 1$. If $\gcd(m', 2D) = 1$ and m' is any other value of Q, then $\theta(m) = \theta(m')$ for every assigned character θ.

This allows us to define $\theta(Q) = \theta(m)$ for every integer m represented by Q and such that $\gcd(m, 2D) = 1$, and for every assigned character θ.

It follows at once that if $Q \approx Q'$, then $\theta(Q) = \theta(Q')$ for every assigned character θ, and so it is possible to define for every proper equivalence class: $\theta(\mathbf{Q}) = \theta(Q)$, for every assigned character θ.

Let $\{+1, -1\}^t$ be the multiplicative group of t-tuples of integers $+1$ or -1, and define the map

$$\Xi : \mathrm{Cl}_+(\mathrm{Prim}(\mathcal{Q}_D)) \longrightarrow \{+1, -1\}^t$$

by

$$\Xi(\mathbf{Q}) = (\theta(\mathbf{Q}))_{\theta \in \Theta}.$$

It is easy to show that the map Ξ is a homomorphism of the group $\mathrm{Cl}_+(\mathrm{Prim}(\mathcal{Q}_D))$, under composition, to $\{+1, -1\}^t$.

Note that $\prod_{\theta \in \Theta} \theta(\mathbf{Q}) = 1$.

It may be shown that the image of Ξ is the set of all $\sigma = (\sigma_1, \ldots, \sigma_t)$ with each $\sigma_i \in \{+1, -1\}$, such that $\prod_{i=1}^{t} \sigma_i = 1$. Hence, the image has 2^{t-1} elements.

For every σ in the image of Ξ, the inverse image of σ,

$$\Xi^{-1}(\sigma) = \{\mathbf{Q} \mid \Xi(\mathbf{Q}) = \sigma\},$$

is called the *genus* of $\mathrm{Cl}_+(\mathrm{Prim}(\mathcal{Q}_D))$ with generic character σ.

The genus with generic character $\sigma_1 = \{+1, \ldots, +1\}$ is called the principal genus and it is a subgroup of $\mathrm{Cl}_+(\mathrm{Prim}(\mathcal{Q}_D))$. Each genus is a coset of the principal genus.

The following notation will sometimes be adopted: $[\mathbf{Q}]$ is the genus of the class $\mathbf{Q}$.

The number of genera is $g(D) = 2^{t-1}$ and the number of classes in each genus is $f(D) = \frac{h_+(D)}{2^{t-1}}$; in particular, 2^{t-1} divides $h_+(D)$.

Note that if D is a fundamental discriminant, then t is equal to the number of distinct prime divisors of D (including 2, if D is even).

GAUSS proved the following important theorem, called the squaring or duplication theorem:

The principal genus consists of the squares (under composition) of the proper equivalence classes of primitive forms.

From the above considerations, one obtains the following criterion. Let p be a prime, not dividing $2D$, and let $\mathcal{G}$ be a genus of proper equivalence classes of primitive forms of discriminant D, say

$$\mathcal{G} = \{\mathbf{Q}_1, \ldots, \mathbf{Q}_k\}.$$

Then p is represented by some form Q belonging to one of the classes $Q_i \in \mathcal{G}(1 \leq i \leq k)$ if and only if $(\theta(p))_{\theta \in \Theta} = \Xi(\mathbf{Q}_i)$.

Note that if the genus has more than one proper equivalence class, the above criterion does not tell which form represents p, among those whose proper equivalence class is in the genus.

Returning to the previous example of discriminant $D = -20$, with the two classes $\mathbf{Q}_1 = \langle 1, 0, 5 \rangle$ (the principal class) and $\mathbf{Q}_2 = \langle 2, 2, 3 \rangle$; note that it has $g(-20) = 2$ genera, so $\mathbf{Q}_1, \mathbf{Q}_2$ are in different genera. If p is an odd prime $p \neq 5$, then p is represented by Q_1 if and only if $\left(\frac{p}{5}\right) = 1$ and $(-1)^{(p-1)/2} = 1$, so $p \equiv 1 \pmod 4$ and $p \equiv \pm 1 \pmod 5$, or equivalently, $p \equiv 1$ or $9 \pmod{20}$.

Similarly, p is represented by Q_2 if and only if

$$(-1)^{(p-1)/2} = \chi_1(Q_2) = \chi_1(3) = (-1)^{(3-1)/2} = -1,$$

and

$$\left(\frac{p}{5}\right) = \chi_2(p) = \chi_2(Q_2) = \chi_2(2) = \left(\frac{2}{5}\right) = -1.$$

Thus $p \equiv 3 \pmod 4$ and $p \equiv \pm 2 \pmod 5$, or equivalently, $p \equiv 3, 7 \pmod{20}$.

Note that the theory of genera establishes the conjecture of EULER concerning the form $X^2 + 5Y^2$.

The treatment with the theory of genera is not so conclusive when there is more than one form in each genus.

Numerical example

Let $D = -56 = -8 \times 7$.

The number of genera is 2.

The reduced forms are $P = \langle 1, 0, 14 \rangle$, $Q_1 = \langle 3, 2, 5 \rangle$, $Q_2 = \langle 2, 0, 7 \rangle$, $Q_3 = \langle 3, -2, 5 \rangle$.

A simple calculation shows that $\mathrm{Cl}_+(\mathrm{Prim}(\mathcal{Q}_{-56}))$ is a cyclic group, with $\mathbf{Q}_1^2 = \mathbf{Q}_2$, $\mathbf{Q}_1^3 = \mathbf{Q}_3$, $\mathbf{Q}_1^4 = \mathbf{P}$.

The principal genus is $\{\mathbf{P}, \mathbf{Q}_2\}$ and the non-principal genus is $\{\mathbf{Q}_1, \mathbf{Q}_3\}$; it has generic character $\{-1, -1\}$. From this, it follows that a prime $p \neq 2, 7$ is represented by P or Q_2 if and only if $\chi_1(p) = (-1)^{(P^2-1)/2} = 1$, that is $\left(\frac{2}{p}\right) = 1$ and $\left(\frac{p}{7}\right) = 1$; a simple calculation gives $p \equiv 1, 9, 15, 23, 25,$ or $39 \pmod{56}$. But, it cannot be obtained a condition stating that p is represented by P (respectively, by Q_2).

In the same way, if $p \neq 2, 7$, then p is represented by Q_1 or by Q_3 if and only if $p \equiv 3, 5, 13, 19, 27, 45 \pmod{56}$.

As will be explained below, the work of EULER on the "numeri idonei" (also called "convenient numbers") and the work of GAUSS generated interest in fundamental discriminants $D < 0$ whose principal genus consists only of the principal class.

Here is a list of the 65 known fundamental discriminants $D < 0$, such that the principal genus consists only of the principal class:

$h_+(D)$	$-D$
1	3 4 7 8 11 19 43 67 163
2	15 20 24 35 40 51 52 88 91 115 123 148 187 232
	235 267 403 427
4	84 120 132 168 195 228 280 312 340 372 408 435
	483 520 532 555 595 627 708 715 760 795 1012 1435
8	420 660 840 1092 1155 1320 1380 1428 1540 1848
	1995 3003 3315
16	5460

Up to now, no other such fundamental discriminant is known! (See §18 for a further discussion on this point.)

There are also the following 36 known non-fundamental discriminants with principal genus consisting only of the principal class:

$$-D = 3 \times 2^2, 3 \times 3^2, 3 \times 4^2, 3 \times 5^2, 3 \times 7^2, 3 \times 8^2, 4 \times 2^2, 4 \times 3^2,$$
$$4 \times 4^2, 4 \times 5^2, 7 \times 2^2, 7 \times 4^2, 7 \times 8^2, 8 \times 2^2, 8 \times 3^2, 8 \times 6^2,$$
$$11 \times 3^2, 15 \times 2^2, 15 \times 4^2, 15 \times 8^2, 20 \times 3^2, 24 \times 2^2, 35 \times 3^2,$$
$$40 \times 2^2, 88 \times 2^2, 120 \times 2^2, 168 \times 2^2, 232 \times 2^2,$$
$$280 \times 2^2, 312 \times 2^2, 408 \times 2^2, 520 \times 2^2, 760 \times 2^2,$$

$$840 \times 2^2, 1320 \times 2^2, 1848 \times 2^2.$$

From the theory of genera, for each of the above discriminants, it is possible to decide with the method indicated if an odd prime may or may not be represented by any one of the primitive forms of this discriminant.

There is an interesting connection, discovered by GAUSS, between negative discriminants with one class in each genus, and EULER's convenient numbers, which were defined in order to find large primes.

The definition of convenient numbers involves odd integers $m \geq 1$ satisfying the following properties:

(i) if x, y, x', y' are non-negative integers such that $x^2 + ny^2 = x'^2 + n(y')^2$, then $(x', y') = (x, y)$ or (y, x);

(ii) if x, y are non-negative integers such that $m = x^2 + ny^2$, then $\gcd(x, y) = 1$.

The integer $n \geq 1$ is a *convenient number* when it satisfies the following property: every odd integer $m \geq 1$ relatively prime to n which satisfies the above conditions (i) and (ii), is a prime.

GAUSS showed:

Let $n \geq 1$ be an integer. Then the principal genus of the fundamental discriminant $D = -4n$ consists of only one class if and only if n is a convenient number.

Thus, the 65 known convenient numbers are: 1, 2, 3, 4, 5, 6, 7, 8, 9, 10, 12, 13, 15, 16, 18, 21, 22, 24, 25, 28, 30, 33, 37, 40, 42, 45, 48, 57, 58, 60, 70, 72, 78, 85, 88, 93, 102, 105, 112, 120, 130, 133, 165, 168, 177, 190, 210, 232, 240, 253, 273, 280, 312, 330, 345, 357, 385, 408, 462, 520, 760, 840, 1320, 1365, 1848.

More information about convenient numbers may be found in FREI (1985, 1984); STEINIG (1966).

There are other ways to indicate that two forms are in the same genus. To be able to express these conditions, the notion of equivalence is extended as follows.

Let R be any of the following rings:

(1) $R = \mathbb{Z}_{(n)}$ (the ring of rational numbers, with denominators prime to n);

(2) $R = \mathbb{Z}_p$ (the ring of p-adic integers, with p prime);

(3) $R = \mathbb{Z}/m\mathbb{Z}$ (the ring of residue classes modulo $m \geq 2$).

In each case, let $\lambda : \mathbb{Z} \to R$ be the natural ring homomorphism; so in cases (1), (2), λ is the embedding, and in case (3), λ is the residue map. If $Q = \langle a, b, c \rangle$ let $\lambda Q = \langle \lambda(a), \lambda(b), \lambda(c) \rangle = \lambda(a)X^2 +$

$\lambda(b)XY + \lambda(c)Y^2$ be the associated binary quadratic form over the ring R. Thus, in cases (1), (2), Q and λQ may be identified.

The forms $Q = \langle a, b, c \rangle$, $Q' = \langle a', b', c' \rangle$ are said to be *R-equivalent* if there exits $A = \begin{pmatrix} \alpha & \beta \\ \gamma & \delta \end{pmatrix} \in \mathrm{GL}_2(R)$ such that $\lambda Q' = T_A(\lambda Q)$. In cases (1), (2), this means (after the canonical embedding) that the conditions (*) of §6 are satisfied. In case (3), those equalities become congruences modulo m. The notation is $Q \sim Q'$ (over R).

Similarly, $\rho \in R$ is said to be a *value* of Q if there exist α, $\gamma \in R$ such that $(\lambda Q)(\alpha, \gamma) = \rho$. For example, $r \bmod m \in \mathbb{Z}/m\mathbb{Z}$ is a value of Q if there exist integers x, y such that $Q(x, y) \equiv r \pmod m$.

Each of the following equivalent conditions characterizes when two forms Q, $Q' \in \mathcal{Q}_D$ are in the same genus:

(i) Q, Q' have the same set of values in each ring $\mathbb{Z}/m\mathbb{Z}$, for all $m \geq 2$;

(ii) $Q \sim Q'$ (over $\mathbb{Z}/m\mathbb{Z}$) for all $m \geq 2$;

(iii) $Q \sim Q'$ (over $\mathbb{Z}_p$) for all primes p;

(iv) $Q \sim Q'$ (over $\mathbb{Z}_{(n)}$) for every $n \geq 2$.

These results are the cornerstone of a local-global theory of quadratic forms which shall not be developed here.

13 The structure of the group of proper equivalence classes of primitive forms

Let D be any discriminant.

Recall that $\mathrm{Cl}_+(\mathrm{Prim}(\mathcal{Q}_D))$ is a finite abelian group. As such, it is the direct product of its p-Sylow subgroups S_p, for every prime p dividing $h_+(D)$. In turn, each non-trivial p-Sylow subgroup is the direct product of $k(p) \geq 1$ cyclic p-groups. The integer $k(p)$, which is uniquely defined, is called the *p-rank* of the group $\mathrm{Cl}_+(\mathrm{Prim}(\mathcal{Q}_D))$.

First, consider the prime $p = 2$. The 2-Sylow subgroup S_2 contains the subgroup $\mathcal{A}$ of ambiguous classes (the classes of order dividing 2).

GAUSS showed that the order of $\mathcal{A}$ is equal to the number $g(D) = 2^{t-1}$ of genera (t denotes the number of assigned characters of D, which is the same as the number of primes dividing D, when D is a fundamental discriminant). A detailed and clear presentation of this proof can be found, for example, in the book of FLATH (1989).

The only ambiguous class is the principal class if and only if $D = p$ or $4p$, where p is an odd prime, $p \equiv 1 \pmod 4$.

It is easy to see that if a non-principal genus contains an ambiguous class $\mathbf{Q}$, there is a bijection between the set $\mathcal{A} \cap [\mathbf{P}]$ of ambiguous classes in the principal genus and the set $\mathcal{A} \cap [\mathbf{Q}]$.

Concerning the principal genus $[\mathbf{P}]$, it may or may not be cyclic. If it is cyclic of even order, then $\mathcal{A} \cap [\mathbf{P}]$ has only two classes, and so every genus either has no ambiguous class, or exactly two ambiguous classes—this happens for exactly one-half of the genera. If, however, $\mathbf{P}$ is the only ambiguous class in $\mathcal{A} \cap [\mathbf{P}]$, then each genus has exactly one ambiguous class.

If the principal genus is not cyclic, let $e(D)$ be the maximum of the orders of its classes; so $e(D) < f(D)$ (the order of the principal genus) and, in fact, $e(D)$ divides $f(D)$.

GAUSS called D a *regular* discriminant when the principal genus is cyclic; otherwise, D is called *irregular* and $f(D)/e(D)$ is its *irregularity index*.

For example, if the principal genus contains 3 or more ambiguous classes, then D is irregular and the irregularity index is even. If the number of ambiguous classes in the principal genus is 1 or 2, then $f(D)/e(D)$ is odd (but not necessarily equal to 1)

In article 306 of *Disquisitiones Arithmeticae*, GAUSS indicated infinitely many negative discriminants with irregularity index multiple of 3, namely,

$$D = -(216k + 27), \quad \text{with } k \geq 1,$$
$$D = -(1000k + 75), \quad \text{with } k \geq 1,$$
$$\text{etc.} \ldots$$

He also gave the following examples:

$-D = 576, 580, 820, 884, 900,$ with irregularity index 2,
$-D = 243, 307, 339, 459, 675, 755, 891, 974,$ with irregularity index 3.

GAUSS gave just one example of an irregular positive discriminant: $D = 3026$; it has irregularity index 2.

Now, let p be an odd prime. Nothing was said by GAUSS concerning the p-rank of the group of classes. I shall return to this matter in §20.

14   Calculations and conjectures

GAUSS made many calculations concerning the forms $\langle a, 2b, c \rangle$, to which he had restricted his attention. But his results may be easily reinterpreted for arbitrary forms.

For $-3000 < D < 0$, and D a fundamental discriminant, he obtained:

$h_+(D) = 1$ if and only if $-D = 3, 4, 7, 8, 11, 19, 43, 67, 163$ (these are 9 values);

$h_+(D) = 2$ if and only if $-D$ assumes 18 values, of which the largest is 427;

$h_+(D) = 3$ if and only if $-D$ assumes 16 values, of which the largest is 907; etc. . . .

Based on these calculations, GAUSS conjectured (see *Disquisitiones Arithmeticae*, article 303):

Conjecture 1. There exist only 9 fundamental discriminants $D < 0$, such that $h_+(D) = 1$.

Conjecture 2. For every $n \geq 2$, there exist only finitely many fundamental discriminants $D < 0$, such that $h_+(D) = n$. In particular, for $n = 2$ only 18 values, for $n = 3$ only 16 values, etc. . . .

More specifically, the conjecture will be established if an algorithm is devised to find all discriminants $D < 0$ such that $h_+(D) = n$.

Concerning proper equivalence classes of indefinite forms with fundamental discriminant $D > 0$, the following is the ongoing belief (see also Gauss, loc. cit., article 304):

Conjecture 3. There exist infinitely many fundamental discriminants $D > 0$ such that $h_+(D) = 1$.

With respect to the number of classes in the principal genus, GAUSS has conjectured (article 303):

Conjecture 4. For each integer $m \geq 1$ there exist only finitely many discriminants $D < 0$ such that the number of classes in the principal genus of D is equal to m.

This may also be expressed as follows:

$$\lim_{|D| \to \infty} f(D) = \infty$$

($f(D)$ is the number of classes in the principal genus of D).

GAUSS noted by numerical computation that the number of positive discriminants D for which $f(D) = 1$ becomes increasingly rare as D increases. He ventures that there are infinitely many such discriminants, and poses the problem to study the behavior of

$$\frac{\{\#\{D \mid 1 \le D \le N, f(D) = 1\}}{N}$$

as $N \to \infty$.

It will be seen that Conjectures 1, 2, and 4 have now been established, and only Conjecture 3 remains open.

15   The aftermath of Gauss (or the "math" after Gauss)

The rich theory developed by GAUSS, and published in *Disquisitiones Arithmeticae* when he was only 24 years old, has had a lasting impact. Its presentation required the whole section V of the book, over 250 pages long. GAUSS' text is full of numerical examples and algorithms, clarifying and completing the results previously obtained by FERMAT, EULER, LEGENDRE, and especially by LAGRANGE. Here I have touched on only a few aspects of his study.

The period following the publication of GAUSS' theory saw the analytical work of DIRICHLET on the computation of the number of proper equivalence classes of primitive forms of a given discriminant, and later, the geometric theory of forms as developed by KLEIN.

It also saw the far-reaching interpretation of DEDEKIND. In order to provide a transparent explanation for the composition of proper equivalence classes of primitive forms. DEDEKIND established a connection between forms and ideals in quadratic number fields; see DEDEKIND's supplements to DIRICHLET's *Vorlesungen über Zahlentheorie*.

16   Forms versus ideals in quadratic fields

To explain the correspondence between forms and ideals in quadratic number fields, I begin by briefly recalling some facts.

Let $d \neq 0, 1$ be a square-free integer and let $K = \mathbf{Q}(\sqrt{d})$ be the associated quadratic field consisting of the elements $\alpha = x + y\sqrt{d}$, where $x, y \in \mathbf{Q}$.

The *discriminant* of K is defined to be

$$D_K = \begin{cases} d & \text{if } d \equiv 1 \pmod 4, \\ 4d & \text{if } d \equiv 2 \text{ or } 3 \pmod 4. \end{cases}$$

So, D_K is a fundamental discriminant.

This defines bijections between the set of square-free integers, $d \neq 0$, 1, the set of quadratic fields, and the set of fundamental discriminants.

Let

$$\omega = \begin{cases} \frac{1+\sqrt{d}}{2} & \text{if } d \equiv 1 \pmod 4, \\ \sqrt{d} & \text{if } d \equiv 2 \text{ or } 3 \pmod 4. \end{cases}$$

Then $\{1, \omega\}$ is also a basis of the $\mathbf{Q}$-vector space K, so every element α of K may be written in unique way as $\alpha = x + y\omega$, with $x, y \in \mathbf{Q}$.

The *conjugate* of $\alpha = x + y\sqrt{d}$ $(x, y \in \mathbf{Q})$ is $\bar{\alpha} = x - y\sqrt{d}$, and the norm of α is $N(\alpha) = \alpha\bar{\alpha} = x^2 - y^2 d \in \mathbf{Q}$. In particular,

$$\bar{\omega} = \begin{cases} \frac{1-\sqrt{d}}{2} & \text{if } d \equiv 1 \pmod 4, \\ -\sqrt{d} & \text{if } \equiv 2 \text{ or } 3 \pmod 4, \end{cases}$$

and

$$N(\omega) = \begin{cases} \frac{1-d}{4} & \text{if } d \equiv 1 \pmod 4, \\ -d & \text{if } d \equiv 2 \text{ or } 3 \pmod 4. \end{cases}$$

If $\alpha \in K$ is written in terms of the basis $\{1, \omega\}$, then

$$N(x + y\omega) = \begin{cases} x^2 + xy + \frac{1-d}{4}y^2 & \text{if } d \equiv 1 \pmod 4, \\ x^2 - y^2 d & \text{if } d \equiv 2 \text{ or } 3 \pmod 4. \end{cases}$$

The element $\alpha \in K$ is said to be an *algebraic integer* if it is the root of a quadratic monic polynomial $X^2 - aX + b \in \mathbf{Z}[X]$. In this situation $a = \alpha + \bar{\alpha}$, and $N(\alpha) = \alpha\bar{\alpha} = b \in \mathbf{Z}$.

The set of algebraic integers of K is a subring of K which will be denoted by $\mathcal{O}_K$. Clearly, $\mathbf{Z} \subset \mathcal{O}_K$, K is the field of quotients of $\mathcal{O}_K$, and $\mathcal{O}_K = \mathbf{Z} \oplus \mathbf{Z}\omega$, so $\mathcal{O}_K$ is a free $\mathbf{Z}$-module of rank 2.

The correspondence indicated above for fundamental discriminants may be extended to all possible discriminants $D \equiv 0$ or $1 \pmod 4$; they will correspond bijectively to orders in quadratic fields, which I shall now introduce.

An *order* of K is a subring $\mathcal{O}$ of K which is a free $\mathbb{Z}$-module of rank 2. Thus $\mathbb{Z} \subset \mathcal{O}$, K is the field of quotients of $\mathcal{O}$, and there exist two elements α, $\beta \in \mathcal{O}$ such that every $\gamma \in \mathcal{O}$ may be written uniquely in the form $\gamma = x\alpha + y\beta$, with x, $y \in \mathbb{Z}$. This is written as $\mathcal{O} = \mathbb{Z}\alpha \oplus \mathbb{Z}\beta$.

In particular, the ring of algebraic integers $\mathcal{O}_K$ is an order of K.

The *discriminant* of any free $\mathbb{Z}$-module $\mathbb{Z}\alpha \oplus \mathbb{Z}\beta$ is, by definition, equal to

$$\det \begin{pmatrix} \alpha & \beta \\ \bar{\alpha} & \bar{\beta} \end{pmatrix}^2 .$$

It is independent of the choice of the basis.

The discriminant of an order $\mathcal{O}$ is denoted by $\mathrm{Discr}(\mathcal{O})$ and is an integer congruent to 0 or 1 modulo 4. In particular, the discriminant of the order $\mathcal{O}_K$ of all algebraic integers of K is $\mathrm{Discr}(\mathcal{O}_K) = D_K$.

The discriminant establishes a map from the set of orders of quadratic fields to the set of integers congruent to 0 or 1 modulo 4 (which are not squares). Conversely, if $D \equiv 0$ or $1 \pmod 4$ (D not a square), let $D = f^2 D_0$, where $f \geq 1$ and D_0 is a fundamental discriminant. Let K be the quadratic field with discriminant $D_K = D_0$. Then $\mathcal{O}(D) = \mathbb{Z} \oplus \mathbb{Z}\frac{D+\sqrt{D}}{2}$ is an order of K with $\mathrm{Discr}(\mathcal{O}(D)) = D$; f is called the *conductor* of the order $\mathcal{O}(D)$. Thus, $\mathcal{O}_K = \mathcal{O}(D_K)$.

Note that

$$\mathcal{O}(D) = \left\{ \frac{x + y\sqrt{D}}{2} \mid x, y \in \mathbb{Z}, \quad x \equiv yD \pmod 2 \right\}.$$

This establishes a bijection between the set of orders of quadratic fields and the set of integers (not a square) congruent to 0 or 1 modulo 4.

Moreover, if $D = f^2 D_0$, $D' = e^2 D_0$ and e divides f with $e < f$, then $\mathcal{O}(D) \subset \mathcal{O}(D')$. In particular, $\mathcal{O}_K$ is the only maximal order in the field K with discriminant $D_K = D_0$.

For every $D = f^2 D_K$, the additive quotient group $\mathcal{O}_K/\mathcal{O}(D)$ is finite, having f elements.

A *fractional ideal* I of the order $\mathcal{O} = \mathcal{O}(D)$ is an additive subgroup of K such that

(1) $\alpha I \subseteq I$ for every $\alpha \in \mathcal{O}$,

(2) there exists a non-zero element $\delta \in \mathcal{O}$ such that $\delta I \subseteq \mathcal{O}$.

Each non-zero fractional ideal I of $\mathcal{O}$ admits a basis consisting of two numbers α, $\beta \in I$, that is, $I = \mathbb{Z}\alpha \oplus \mathbb{Z}\beta$.

For every $\alpha \in K$, the set $\mathcal{O}\alpha = \{\beta\alpha \mid \beta \in \mathcal{O}\}$ is a fractional ideal of the order $\mathcal{O}$, called the *principal ideal* defined by α. In particular, $\mathcal{O} = \mathcal{O}1$ is the *unit ideal*, $0 = \mathcal{O}0$ is the *zero ideal*.

If I is a fractional ideal of $\mathcal{O}$, then so is its *conjugate* $\bar{I} = \{\bar{\alpha} \mid \alpha \in I\}$. If I, J are fractional ideals of $\mathcal{O}$ let $I \cdot J = \{\sum_{i=1}^{n} \alpha_i\beta_i \mid \alpha_i \in I, \beta_i \in I, n \geq 1\}$. Then $I \cdot J$ is also a fractional ideal of $\mathcal{O}$. The multiplication of fractional ideals is an associative and commutative operation, withthe unit ideal as the unit element. Also $\mathcal{O}\alpha \cdot \mathcal{O}\beta = \mathcal{O}\alpha\beta$ for any $\alpha, \beta \in K$.

The product $I \cdot \bar{I}$ is a principal fractional ideal of $\mathbb{Z}$, generated by a unique positive rational number $l > 0$: $I \cdot \bar{I} = \mathcal{O}l$; by definition, the *norm* of I is $N(I) = l$. It is clear that $N(I \cdot J) = N(I)N(J)$ and $N(\mathcal{O}\alpha) = |N(\alpha)|$.

If $\{\alpha, \beta\}$ is any basis of I, then $(\alpha\bar{\beta} - \bar{\alpha}\beta)^2 = N(I)^2 D$.

A fractional ideal I of $\mathcal{O} = \mathcal{O}(D)$ is said to be *invertible* if there exists a fractional ideal J such that $I \cdot J = \mathcal{O}$.

The following conditions on a non-zero fractional ideal I of $\mathcal{O}$ are equivalent:

(1) I is invertible.

(2) $\mathcal{O} = \{\alpha \in K \mid \alpha I \subseteq I\}$.

If $\gcd(N(I), f) = 1$, then I is invertible.

In particular, if $f = 1$, then all the non-zero fractional ideals of $\mathcal{O}(D_K)$ are invertible.

If I, J are invertible fractional ideals of $\mathcal{O}$ and $I \subseteq J$, then there exists a fractional ideal $J' \subseteq \mathcal{O}$ such that $I = JJ'$ and hence $N(J)$ divides $N(I)$. In particular, for every $\alpha \in K$, $\alpha \neq 0$, if $\alpha \in J$, then $N(J)$ divides $N(\alpha)$.

Let $\mathcal{I} = \mathcal{I}(\mathcal{O}(D))$ denote the set of all invertible fractional ideals of $\mathcal{O}(D)$; thus $\mathcal{I}$ is a multiplicative group which contains the subgroups

$$\mathcal{P} = \mathcal{P}(\mathcal{O}(D)) = \{\mathcal{O}\alpha \mid \alpha \in K, \alpha \neq 0\}$$

and

$$\mathcal{P}_+ = \mathcal{P}_+(\mathcal{O}(D)) = \{\mathcal{O}\alpha \mid \alpha \in K, \alpha \neq 0, N(\alpha) > 0\}.$$

The *equivalence* of invertible ideals I, $J \in \mathcal{I}$ is defined as follows: $I \sim J$ if there exists $\alpha \in K$, $\alpha \neq 0$, such that $I = J \cdot \mathcal{O}\alpha$. The set of equivalence classes of invertible ideals of $\mathcal{O}$ is denoted by $\mathrm{Cl}(\mathcal{O}(D))$, and the equivalence class of I is denoted by $\mathrm{cl}(I)$. If $I \sim I'$ and $J \sim J'$, then $I \cdot J \sim I' \cdot J'$. This allows us to define the operation $\mathrm{cl}(I) \cdot \mathrm{cl}(J) = \mathrm{cl}(I \cdot J)$.

Endowed with this operation, $\mathrm{Cl}(\mathcal{O}(D))$ is an abelian group isomorphic to the quotient group $\mathcal{I}/\mathcal{P}$.

The *strict equivalence* invertible ideals I, $J \in \mathcal{I}$ is defined as follows: $I \approx J$ if there exists $\alpha \in K$, with $N(\alpha) > 0$, such that $I = J \cdot \mathcal{O}\alpha$. The set of strict equivalence classes of invertible ideals of $\mathcal{O}$ is denoted by $\mathrm{Cl}_+(\mathcal{O}(D))$ and the strict equivalence class of I is denoted by $cl_+(I)$. Again, if $I \approx I'$, $J \approx J'$, then $I \cdot J \approx I' \cdot J'$, which allows us to define the operation $\mathrm{cl}_+(I) \cdot \mathrm{cl}_+(J) = \mathrm{cl}_+(I \cdot J)$. With this operation, $\mathrm{Cl}_+(\mathcal{O}(D))$ is an abelian group, isomorphic to $\mathcal{T}/\mathcal{P}_+$.

The mapping $\mathrm{cl}_+(I) \mapsto \mathrm{cl}(I)$ from $\mathrm{Cl}_+(\mathcal{O})$ to $\mathrm{Cl}(\mathcal{O})$ is a surjective homomorphism, with a kernel consisting of one or two elements.

Whereas $\mathcal{T}(\mathcal{O}(D))$ is an infinite group, the group $\mathrm{Cl}_+(\mathcal{O}(D))$ is finite, hence also $\mathrm{Cl}(\mathcal{O}(D))$ is finite. This important result is the counterpart, in DEDEKIND's interpretation, of the finiteness of the group $\mathrm{Cl}_+(\mathrm{Prim}(D))$, as I shall soon explain.

The number of elements in $\mathrm{Cl}(\mathcal{O}(D))$ is denoted $h(\mathcal{O}(D))$ and called the *class number* of the order $\mathcal{O}(D)$. The number of elements of $\mathrm{Cl}_+(\mathcal{O}(D))$ is called the *strict class number* of $\mathcal{O}(D)$ and denoted by $h_+(\mathcal{O}(D))$.

From the above homomorphism, $h((\mathcal{O}(D)) \leq h_+((\mathcal{O}(D)) \leq 2h((\mathcal{O}(D))$. The exact relation between the class number and strict class number of an order will be made more precise.

The following facts about orders will also be needed later.

An element $\alpha \in \mathcal{O} = \mathcal{O}(D)$ such that $\alpha^{-1} \in \mathcal{O}$ is called a *unit* of $\mathcal{O}$. If $\alpha \in \mathcal{O}$ and there exists $k \geq 1$ such that $\alpha^k = 1$, then α is a root of unity and also a unit. The set $U = U(\mathcal{O}(D))$ of units of $\mathcal{O}$ forms a multiplicative group. If α is a unit, then so is $\bar{\alpha}$ and $N(\alpha\bar{\alpha}) = \pm 1$.

Consider the situation in the case of the maximal order $\mathcal{O}(D_K)$.

If $d < 0$, the group of units of $\mathcal{O}(D_K)$ is finite, so every unit is a root of unity. Let w denote the number of units. Then

$$
w = \begin{cases}
4 & \text{if } d = -1; & \text{the units are } \pm 1, \pm\sqrt{-1} \\
6 & \text{if } d = -3; & \text{the units are } \pm 1, (\pm 1 \pm \sqrt{-3})/2 \\
2 & \text{if } d \neq -1, -3; & \text{the units are } \pm 1.
\end{cases}
$$

If $D_K > 0$, there exists a unit $\epsilon > 1$, unique such that

$$
U = \{\pm \epsilon^k \mid k \in \mathbb{Z}\}.
$$

The unit ϵ is called the *fundamental unit* of $\mathcal{O}(D_K)$. The only roots of unity are ± 1.

Since $U(\mathcal{O}(D)) = U(\mathcal{O}(D_K)) \cap \mathcal{O}(D)$, if $D_K < 0$, then $U(\mathcal{O}(D))$ is finite, consisting only of roots of units. If $D_K > 0$, then there exists a smallest $t \geq 1$ such that $\epsilon^t \in \mathcal{O}(D)$ and $U(\mathcal{O}(D)) = \{\pm \epsilon^{tk} \mid k \geq 1\}$; ϵ^t is the fundamental unit of $\mathcal{O}(D)$.

The fundamental unit may have norm equal to 1 or to -1, both cases being possible.

If $D_K \equiv 1 \pmod 4$, then $D_K = d$ is square-free; the fundamental unit $\epsilon = \frac{x_1 + y_1 \sqrt{d}}{2}$ (with $x_1, y_1 \geq 1$, $x_1 \equiv y_1 \pmod 2$) is such that $x_1^2 - y_1^2 d = \pm 4$; moreover, for every pair (x, y), $x, y \geq 1$, such that $x^2 - y^2 d = \pm 4$, necessarily $x_1 + x + y_1 \sqrt{d} < x + y\sqrt{d}$.

If $D_K \equiv 0 \pmod 4$, then $D_K = 4d$, with $d \equiv 2, 3 \pmod 4$; the fundamental unit $\epsilon = x_1 + y_1 \sqrt{d}$ (with $x_1, y_1 \geq 1$) is such that $x_1^2 - y_1^2 d = \pm 1$ and $x_1 + y_1 \sqrt{d} < x + y\sqrt{d}$ whenever $x, y \geq 1$, $x^2 - y^2 d = \pm 1$.

This theory was developed by LAGRANGE.

The relationship between the class number and the strict class number of the order $\mathcal{O}(D)$ is the following:

(1) If $D < 0$ or $D > 0$ and the fundamental unit of $\mathcal{O}(D)$ has norm -1, then $h_+(\mathcal{O}(D)) = h(\mathcal{O}(D))$.

(2) If $D > 0$ and the fundamental unit has norm 1, then $h_+(\mathcal{O}(D)) = 2h(\mathcal{O}(D))$.

Now I shall indicate the important correspondence between proper equivalence classes of primitive forms and strict equivalence classes of invertible fractional ideals of orders.

Let $D \equiv 0, 1 \pmod 4$ (D not a square), so $D = f^2 D_0$, where D_0 is a fundamental discriminant. Let $K = \mathbb{Q}(\sqrt{D_0})$, so its discriminant is $D_K = D_0$. Let $\mathcal{O} = \mathcal{O}(D)$ and let $I \in \mathcal{I}(\mathcal{O})$, so I is a non-zero invertible fractional ideal of the order $\mathcal{O}$. Thus I has a basis $\{\alpha, \beta\}$, and therefore

$$\det \begin{pmatrix} \alpha & \beta \\ \bar{\alpha} & \bar{\beta} \end{pmatrix} = \alpha\bar{\beta} - \bar{\alpha}\beta \neq 0.$$

Since $\frac{\alpha\bar{\beta} - \bar{\alpha}\beta}{\sqrt{d}}$ is equal to its conjugate, then it is a rational number. Therefore, either $\alpha\bar{\beta} - \bar{\alpha}\beta > 0$ or $\beta\bar{\alpha} - \bar{\beta}\alpha > 0$. Thus, it is possible to choose a pair (α, β) such that $I = \mathbb{Z}\alpha \oplus \mathbb{Z}\beta$ and $\alpha\bar{\beta} - \bar{\alpha}\beta > 0$; (α, β) is called a *positively oriented* basis for I.

Since $\mathcal{O}\alpha \subseteq I$ and $\mathcal{O}\beta \subseteq I$, it follows that $N(I)$ divides $N(\alpha)$ and $N(\beta)$. But $N(I)^2$ divides

$$(\alpha\bar{\beta} - \bar{\alpha}\beta)^2 = (\alpha\bar{\beta} + \bar{\alpha}\beta)^2 - 4N(\alpha)N(\beta),$$

hence $N(I)$ divides $\alpha\bar{\beta} + \bar{\alpha}\beta$.

Let

$$Q = \left\langle \frac{N(\alpha)}{N(I)}, \frac{\alpha\bar{\beta} + \bar{\alpha}\beta}{N(I)}, \frac{N(\beta)}{N(I)} \right\rangle,$$

so Q has discriminant equal to D.

Note that Q depends on the choice of the positively oriented basis $\{\alpha, \beta\}$; this is denoted by writing $Q = Q_{(\alpha,\beta)}$. If $\{\alpha', \beta'\}$ is another positively oriented basis of I, it may be shown that $Q_{(\alpha,\beta)}$ and $Q_{(\alpha',\beta')}$ are properly equivalent. Similarly, if the ideals $I, I' \in \mathcal{I}$ are strictly equivalent, then the associated forms Q, Q' (using any positively oriented bases of I, I') are properly equivalent. This defines the map $\mathrm{cl}_+(I) \mapsto \mathbf{Q}$ from $\mathrm{Cl}_+(\mathcal{O}(D))$ to $\mathrm{Cl}_+(\mathrm{Prim}(\mathcal{Q}_D))$.

Conversely, let $Q = \langle a, b, c \rangle$ be a primitive form with discriminant $D = f^2 D_0$, where D_0 is a fundamental discriminant. Let $K = \mathcal{Q}(\sqrt{D_0})$, so $D_K = D_0$.

If $a > 0$, let $I = \mathbb{Z}a \oplus \mathbb{Z}\left(\frac{b - \sqrt{D}}{2}\right)$.

If $a < 0$ (hence $D > 0$), let $I = \mathbb{Z}a\sqrt{D} \oplus \mathbb{Z}\left(\frac{b - \sqrt{D}}{2}\right)\sqrt{D}$.

It is easy to see that, in both cases, I is an invertible fractional ideal of the order $\mathcal{O}(D)$.

Once again, if $Q \approx Q'$, then $I \approx I'$. Note also that if $a > 0$, then the basis $\left(a, \frac{b - \sqrt{D}}{2}\right)$, (respectively, if $a < 0$, then the basis $\left(a\sqrt{D}, \frac{b - \sqrt{D}}{2}\sqrt{D}\right)$) of I is positively oriented.

This defines a mapping $\mathbf{Q} \mapsto \mathrm{cl}_+(I)$ from $\mathrm{Cl}_+(\mathcal{Q}_D)$ to $\mathrm{Cl}_+(\mathcal{O}(D))$. The two mappings are inverse to each other, as may be verified.

Moreover, if $\mathbf{Q} * \mathbf{Q}' = \mathbf{Q}''$ (composition of proper equivalence classes), then the corresponding strict classes of ideals satisfy $\mathrm{cl}_+(I) \cdot \mathrm{cl}_+(I') = \mathrm{cl}_+(I'')$.

In other words, the group (under composition) $\mathrm{Cl}_+(\mathrm{Prim}(\mathcal{Q}_D))$ and the group $\mathrm{Cl}_+(\mathcal{O}(D))$ are isomorphic.

In particular, it follows also that $h_+(D) = h_+(\mathcal{O}(D))$.

Here it should be observed that there is no isomorphism in general between $\mathrm{Cl}(\mathrm{Prim}(\mathcal{Q}_D))$ and $\mathrm{Cl}(\mathcal{O}(D))$. For example, it was shown that $h(-303) = 6$, $h_+(-303) = 10$, so $h(\mathcal{O}(-303)) = h_+(\mathcal{O}(-303)) = 10$.

17 Dirichlet's class number formula

Using analytical methods, in 1839 DIRICHLET gave a formula for the number of proper equivalence classes of primitive forms of a given discriminant D.

First, I recall the definition and main properties of the Kronecker symbol, which will be used in the sequel.

Let $D \equiv 0$ or $1 \pmod 4$, D not a square.

The Kronecker symbol is defined as follows:

$$(1) \ \left(\frac{D}{2}\right) = \begin{cases} 0 & \text{if } D \equiv 0 \pmod 4, \\ 1 & \text{if } D \equiv 1 \pmod 8, \\ -1 & \text{if } D \equiv 5 \pmod 8; \end{cases}$$

(2) if p is an odd prime, then $\left(\frac{D}{p}\right)$ is the Legendre symbol; in particular, $\left(\frac{D}{p}\right) = 0$ when $p \mid D$;

(3) if $n = \prod_{i=1}^r p_i^{e_i}$ (with p_i prime, $e_i \geq 1$), then $\left(\frac{D}{n}\right) = \prod_{i=1}^r \left(\frac{D}{p_i}\right)^{e_i}$; in particular $\left(\frac{D}{1}\right) = 1$.

The computation of the Kronecker symbol is reduced to that of Legendre symbols, and this may be speedily done using Gauss' reciprocity law.

It is also necessary to use the well-known fact that, given m and D (as above), the number of integers n, such that $1 \leq n < 2m$ and $D \equiv n^2 \pmod{4m}$, is equal to

$$\sum_{k \mid m, \, 1 \leq k} \left(\frac{D}{k}\right).$$

Let $D > 0$ and denote by ϵ the fundamental unit of the real quadratic field $\mathbf{Q}(\sqrt{D})$ associated to D. Let $Q = \langle a, b, c \rangle \in \mathrm{Prim}(\mathcal{Q}_D)$. The primitive representation $m = Q(\alpha, \beta)$ is said to be a *primary representation* if

$$2a\alpha + (b - \sqrt{D})\beta > 0$$

and

$$1 \le \left| \frac{2a\alpha + (b + \sqrt{D})\beta}{2a\alpha + (b - \sqrt{D})\beta} \right| \le (\epsilon')^2$$

where

$$\epsilon' = \begin{cases} \epsilon & \text{if } N(\epsilon) = +1, \\ \epsilon^2 & \text{if } N(\epsilon) = -1. \end{cases}$$

If $D > 0$, define $w = 1$.

If $D < 0$, it is convenient to say that every primitive representation is primary, and for $D < 0$, w has already been defined, as being the number of roots of unity of the quadratic field $\mathbf{Q}(\sqrt{D})$.

Then, for any $Q \in \mathrm{Prim}(\mathcal{Q}_D)$, the number of primitive primary representations of $m \ge 1$ by Q and belonging to n (where $1 \le n < 2m$, $D \equiv n^2 \pmod{4m}$) is equal to 0 or to w.

Let $Q \in \mathrm{Prim}(\mathcal{Q}_D)$, $m \ge 1$, and denote by $\psi(m, Q)$ the number of primitive primary representations of m by Q.

Let $\{Q_1, \ldots, Q_{h_+(D)}\}$ be a set of $h_+(D)$ pairwise non-properly equivalent primitive forms with discriminant D.

Let $\psi(m) = \sum_{i=1}^{h_+(D)} \psi(m, Q_i)$.

Then $\psi(m) = w \sum_{k|m} \left(\frac{D}{k} \right)$; this equality reflects the fact that every primitive representation belongs to some n, $1 \le n < 2m$, and $D \equiv n^2 \pmod{4m}$.

For each $Q \in \mathrm{Prim}(\mathcal{Q}_D)$ and real number $t > 1$, let

$$\Psi(t, Q) = \sum_{\substack{1 \le m \le t \\ \gcd(m, D) = 1}} \psi(m, Q).$$

The limiting average of $\Psi(t, Q)$ exists and may be computed:

$$\lim_{t \to \infty} \frac{1}{t} \Psi(t, Q) = \begin{cases} \dfrac{2\pi}{\sqrt{|D|}} \cdot \dfrac{\phi(|D|)}{|D|} & \text{if } D < 0, \\[2ex] \dfrac{\log \epsilon'}{\sqrt{D}} \cdot \dfrac{\phi(D)}{D} & \text{if } D > 0. \end{cases}$$

This average is independent of the choice of Q.

The class number $h_+(D)$ appears as follows:

$$\sum_{i=1}^{h_+(D)} \Psi(t, Q_i) = \sum_{\substack{1 \le m \le t \\ \gcd(m,D)=1}} \sum_{i=1}^{h_+(D)} \psi(m, Q_i)$$

$$= \sum_{\substack{1 \le m \le t \\ \gcd(m,D)=1}} \psi(m)$$

$$= w \sum_{\substack{1 \le m \le t \\ \gcd(m,D)=1}} \sum_{k|m} \left(\frac{D}{k}\right).$$

Dividing by t and considering the limit as t tends to infinity, the left-hand side yields

$$h_+(D) \cdot C_D \cdot \frac{\phi(|D|)}{|D|},$$

where

$$C_D = \begin{cases} \dfrac{2\pi}{\sqrt{|D|}} & \text{if } D < 0, \\[2mm] \dfrac{\log \epsilon'}{\sqrt{D}} & \text{if } D > 0. \end{cases}$$

The calculation of the right-hand side is less obvious. For details, the excellent books of HUA (1982) or BOREVICH and SHAFAREVICH (1966) should be consulted. At any rate,

$$\lim_{t \to \infty} \frac{1}{t} \left[w \sum_{\substack{1 \le m \le t \\ \gcd(m,D)=1}} \sum_{k|m} \left(\frac{D}{k}\right) \right] = w \frac{\psi(|D|)}{|D|} L(D)$$

where

$$L(D) = \sum_{k=1}^{\infty} \frac{1}{k} \left(\frac{D}{k}\right).$$

Note that if the mapping $n \mapsto \chi(n) = (D/n)$ is a modular character, then $L(D)$ is nothing more than $L(1|\chi)$, the value at $s = 1$ of the L-series of χ:

$$L(s|\chi) = \sum_{n=1}^{\infty} \frac{\chi(n)}{n^s} \quad \text{(convergent for } \operatorname{Re}(s) > 1\text{)}.$$

This series $L(D)$ converges, and it follows that

$$h_+(D) = \frac{w}{C_D} L(D) = \begin{cases} \frac{w\sqrt{|D|}}{2\pi} L(D) & \text{if } D < 0, \\ \frac{\sqrt{D}}{\log \epsilon'} L(D) & \text{if } D > 0. \end{cases}$$

The computation of $L(D)$ is delicate. For fundamental discriminants D, it yields (see HUA (1982)):

$$L(D) = \begin{cases} -\frac{\pi}{|D|^{3/2}} \sum_{k=1}^{|D|-1} \left(\frac{D}{k}\right) k & \text{if } D < 0, \\ -\frac{1}{\sqrt{D}} \sum_{k=1}^{D-1} \left(\frac{D}{k}\right) \log \sin \frac{k\pi}{D} & \text{if } D > 0, \end{cases}$$

and finally, Dirichlet's formula for the strict class number (for fundamental discriminants) is:

$$h_+(D) = \begin{cases} -\frac{w}{2|D|} \sum_{k=1}^{|D|-1} \left(\frac{D}{k}\right) k & \text{if } D < 0, \\ -\frac{1}{\log \epsilon'} \sum_{k=1}^{D-1} \left(\frac{D}{k}\right) \log \sin \frac{k\pi}{D} & \text{if } D > 0. \end{cases}$$

For the class number $h(D)$, noting the relation between the fundamental unit and ϵ', as well as between $h_+(D)$ and $h(D)$, the formula may be rewritten as follows:

$$h(D) = \begin{cases} -\frac{w}{2|D|} \sum_{k=1}^{|D|-1} \left(\frac{D}{k}\right) k & \text{if } D < 0, \\ -\frac{1}{2\log \epsilon} \sum_{k=1}^{D-1} \left(\frac{D}{k}\right) \log \sin \frac{k\pi}{D} & \text{if } D > 0. \end{cases}$$

More generally, if $D = f^2 D_0$, where D_0 is a fundamental discriminant, then

$$L(D) = \prod_{p | f} \frac{1 - \left(\frac{D_0}{p}\right)}{p} L(D_0),$$

and this value leads at once to the formulas for $h_+(D)$ and $h(D)$, for arbitrary discriminants.

Another expression for $h(D)$, when D is a fundamental discriminant, $D < -4$, is the following:

$$h(D) = \frac{1}{2 - \left(\frac{D}{2}\right)} \sum_{\substack{1 \leq k < |D|/2 \\ \gcd(k,D)=1}} \left(\frac{D}{k}\right).$$

From this, it follows that if p is a prime number, $p \equiv 3 \pmod 4$ and $p \neq 3$, then

$$
h(-p) = \begin{cases} R - N & \text{if } p \equiv 7 \pmod 8, \\ \frac{1}{3}(R - N) & \text{if } p \equiv 3 \pmod 8, \end{cases}
$$

where

$$
R = \#\{k \mid 1 \leq k < \frac{P}{2}, \left(\frac{k}{p}\right) = -1\},
$$

$$
N = \#\{k \mid 1 \leq k < \frac{P}{2}, \left(\frac{k}{p}\right) = -1\}.
$$

For example,

$$
h(-43) = \frac{1}{3}(R - N) = \frac{1}{3}(12 - 9) = 1.
$$

It is interesting to note that in article 303 of *Disquisitiones Arithmeticae*, GAUSS stated that for $D > 0$, the product $h(D) \log \epsilon$ has a role similar to $h_+(D)$ when $D < 0$. This fact recurs later in the work of SIEGEL (1936) and reflects the crucial importance of the L-series of the character χ.

18 Solution of the class number problem for definite forms

Even though the class number formula indeed allows the computation of $h(D)$ (see BUELL's tables of $h(D)$ (BUELL (1976) and BUELL (1987)) for $|D| < 25 \times 10^6$), no inference can be made concerning the growth of $h(D)$. Thus, the formula does not suffice to decide whether the conjectures of GAUSS (see §14) are true. These questions are much more difficult.

The two excellent articles of GOLDFELD (1985) and OESTERLÉ (1988) should be read, as they contain a lucid account of the work which culminated in the solution of the problem for proper equivalence classes of positive definite forms with fundamental discriminant, or equivalently, for imaginary quadratic fields.

I shall borrow unashamedly from these authoritative accounts—after my sincere kudos, what else can the authors do but forgive me?

Before considering the question in its full generality, it is worth pointing out that a special case had been settled by LANDAU, already in 1903:

If D is any fundamental discriminant such that 4 divides D, then $h(-D) = 1$ if and only if $-D = 4$ or 8.

It is not difficult to give upper estimates for $L(D) = L(1|\chi)$. If $D \leq -5$, then $L(D) \leq \log|D|$, hence

$$h(D) < \frac{\sqrt{D}\log|D|}{\pi}.$$

For the class number problem, what is important is the determination of lower bounds for $h(D)$.

Consider the zeta function of the field $K = \mathbf{Q}(\sqrt{D})$:

$$\zeta_K(s) = \sum \frac{1}{N(I)^s} \text{(summation over all non-zero ideals } I \text{ of } \mathcal{O}),$$

where $N(I)$ indicates the norm of the ideal I. This series converges absolutely for $\mathrm{Re}(s) > 1$.

If $K = \mathbf{Q}$, then $\zeta_{\mathbf{Q}}(s)$ is the Riemann zeta function

$$\zeta(s) = \sum_{n=1}^{\infty} \frac{1}{n^s} \quad \text{(for } \mathrm{Re}(s) > 1\text{)}.$$

Classical calculations give

$$\zeta_K(s) = \zeta(s)L(s|\chi) \quad \text{(for } \mathrm{Re}(s) > 1\text{)},$$

where $\chi(n) = \left(\frac{D}{n}\right)$ for every $n \geq 1$.

Riemann's hypothesis states that all non-real zeroes $\sigma + it$ of $\zeta(s)$ are such that $\sigma = 1/2$.

In the present context, the generalized Riemann's hypothesis is the analogous statement for the L-series $L(s|\chi)$.

Everyone knows that both the Riemann and the generalized Riemann hypothesis, however plausible they may be, have yet to be proved. It is a common practice in analytic number theory to deduce consequences from these hypotheses—as in the last century was the case for non-euclidean geometry.

HECKE (see LANDAU (1913)) proved that under a hypothesis similar to, but weaker than, the generalized Riemann hypothesis for the

L-series of the character χ, it follows that there is a constant $c > 0$ such that

$$h(D) \geq \frac{1}{c} \cdot \frac{\sqrt{|D|}}{\log |D|}.$$

This implies that $\lim_{D \to -\infty} h(D) = \infty$, and also if $h(D) = 1$, then

$$c \geq \frac{\sqrt{|D|}}{\log |D|},$$

and, therefore,

$$|D| \leq (c \log |D|)^2.$$

What has been proved, without assuming the generalized Riemann's hypothesis?

DEURING (1933) showed:

Assuming that the classical Riemann hypothesis is false, there exist only finitely many discriminants $D < 0$ such that $h(D) = 1$.

Soon after, MORDELL showed, assuming the classical Riemann hypothesis to be false, that $\lim_{D \to -\infty} h(D) = \infty$.

In the same year of 1934, assuming that the generalized Riemann hypothesis is false, HEILBRONN (1934a) concluded also that $\lim_{D \to -\infty} h(D) = \infty$.

As GOLDFELD says: "Here was the first known instance of a proof which first assumed that the generalized Riemann hypothesis was true and then that it was false, giving the right answer in both cases!"

SIEGEL (1936) showed, in a different way, that

$$\log h(D) \sim \log \sqrt{|D|} \quad (\text{asymptotically, as } D \to -\infty);$$

in particular, $\lim_{D \to -\infty} h(D) = \infty$.

The above proofs did not provide any effective bound for the discriminants $D < 0$ such that $h(D)$ is less than any given value.

A refined proof by HEILBRONN and LINFOOT (1934b), led to the conclusion that, apart from at most one extra tenth discriminant, all values of $D < 0$ such that $h(D) = 1$ are those already mentioned: $|D| = 3, 4, 7, 8, 11, 19, 43, 67, 163$.

It took a rather long time to rule out this extra discriminant; the story is quite interesting.

HEEGNER (1952)—who incidentally, as a high school teacher, was an outsider—published a paper showing that the extra tenth discriminant does not in fact exist. HEEGNER's proof, using the theory of modular forms, was discounted as being incorrect; in fact, there were errors in it, as well as obscure passages.

BAKER (1966) put to good use his effective minoration of linear forms of three logarithms, and showed that the extra discriminant does not exist.

STARK (1967) gave another proof, similar to HEEGNER's. Still another proof was due to SIEGEL (1968).

A reexamination of HEEGNER's proof by DEURING (1968) sufficed to put it back on solid ground. And STARK (1969) compounded the embarrassment by showing how the theorem could have been proved by the effective minoration of a linear form in two logarithms—and this was already fully possible using the transcendence results of GEL'FOND and LINNIK, known in 1949.

The road was paved to deal with the imaginary quadratic fields of class number 2. Using effective minorations of logarithms, BAKER (1971) and STARK (1971), independently, showed that $h(D) = 2$ exactly when

$$|D| = 5, 6, 10, 13, 15, 22, 35, 37, 51, 58, 91, 115, 123, 187, 235, 267,$$
$$403, 427.$$

Still a long way was ahead before GAUSS' conjectures for the class number of imaginary quadratic fields could be settled. It was essential to obtain effective minorations for the class number. The way to this achievement was convoluted, and involved the theory of modular forms and elliptic functions.

The culmination of the work of GOLDFELD (1977) and GROSS and ZAGIER (1986) gave (in 1983) the following effective minoration for $h(D)$:

For every $\delta > 0$, there exists an effectively computable number $C = C(\epsilon) > 0$ such that

$$h(D) > C(\log |D|)^{1-\epsilon}.$$

This is sufficient to imply that for every given number n there is an effective bound $B(n)$, depending on n, such that if $h(D) = n$, then $|D| \leq B(n)$. So, GAUSS' class number conjectures for definite forms is true.

Explicit computations by OESTERLÉ (1983) led to the minoration:

$$h(D) > \frac{1}{55}(\log |D|) \prod_{p|D, p \neq 2} \left(1 - \frac{[2\sqrt{p}]}{p+1}\right).$$

This minoration holds for discriminants prime to 5077.

This, and similar estimates, have allowed the determination of all fields with class number 3 (there are 16 such fields, and for these, $D \leq 907$), with class number 4 (there are 54 such fields, and for these, $D < 1555$), and more is still to come along these lines.

The same method of HEILBRONN allowed CHOWLA to give a (non-effective) lower bound for the number of classes in the principal genus of any discriminant $D < 0$. Recall that this number is $h(D)/g(D)$, where $g(D)$ is the number of genera.

CHOWLA (1934) showed that

$$\lim_{|D| \to \infty} \frac{h(D)}{g(D)} = \infty.$$

In particular, for every $n \geq 1$ there exist only finitely many discriminants $D < 0$ such that the number of classes in the principal genus is n. This gives a solution, albeit non-effective, of the fourth conjecture of GAUSS (see §14).

Further work by CHOWLA and BRIGGS (1954) and WEINBERGER (1973a) led to the following interesting conclusion:

Apart from the known discriminants with only one class in the principal genus, listed in §12, there exists at most one other D, and $|D| > 10^{60}$. Whether or not such a discriminant actually exists is still unknown. However, the existence is denied, as soon as an appropriate weak hypothesis is made about the zeroes of the associated L-series; see CHOWLA and BRIGGS (1954) and GROSSWALD (1963).

19   The class number problem for indefinite forms

Recall that GAUSS had conjectured, on the basis of numerical calculations, that there should exist infinitely many fundamental discriminants $D > 0$ such that $h(D) = 1$, or equivalently, there exist infinitely many real quadratic fields with class number one.

The problem is very much tied to the size of the fundamental unit ϵ_D. Indeed, SIEGEL showed in 1936:

$$\log(h(D) \log \epsilon_D) \sim \log \sqrt{D} \quad \text{(asymptotically, as } D \to \infty\text{)}.$$

Extensive computations of the class number by WADA (1981), MOLLIN and WILLIAMS (1992), and, more recently, by JACOBSON

(1998) give weight to this conjecture. Yet, its proof is elusive and of great difficulty.

I wish to indicate some recent approaches to the problem and related studies.

First, there are very interesting heuristic considerations by COHEN and LENSTRA (1984) (or COHEN (1993)) that involve the automorphism group of the class group and lead to the conclusion that the proportion of real quadratic fields with class number one ought to be 75.466%. This is indeed very close to the proportion observed in the tables. I shall return in the next section to these conjectures, which have a much wider scope.

An interesting notion, studied by LACHAUD (1986, 1987), is the *caliber* of a fundamental discriminant D, or of the corresponding field $K = \mathbf{Q}(\sqrt{D})$. By definition, the caliber $c(D)$ is the number of reduced primitive forms, under proper equivalence. Note that if $D < 0$, then $c(D) = h(D)$, but if $D > 0$, then in general $c(D) > h(D)$, and how much greater depends on the periods of the roots associated to the reduced forms.

For each class $\mathbf{Q}$ of reduced forms, let $m(\mathbf{Q})$ denote the number of forms in its period. Then

$$m(\mathbf{Q}) \log \alpha \leq \log \epsilon'_D < m(\mathbf{Q}) \log \sqrt{D}$$

where

$$\epsilon'_D = \begin{cases} \epsilon_D & \text{if } N(\epsilon_D) = +1, \\ \epsilon_D^2 & \text{if } N(\epsilon_D) = -1, \end{cases}$$

ϵ_D is the fundamental unit, and $\alpha = (1+\sqrt{5})/2$ is the golden number. Then

$$c(D) \log \alpha \leq h_+(D) \log \epsilon'_D < c(D) \log \sqrt{D},$$

and this may be rewritten as

$$c(D) \log \alpha \leq h(D) \log \epsilon_D < c(D) \log \sqrt{D}.$$

From SIEGEL's result,

$$\log(h(D) \log \epsilon_D) \sim \log \sqrt{D},$$

and it follows that

$$\log c(D) \sim \log \sqrt{D}.$$

Hence, for every $n \geq 1$, there exist only finitely many real quadratic fields $K = \mathbf{Q}(\sqrt{D})$ such that $c(D) \leq n$.

This is the analog to the result indicated for the class number of imaginary quadratic fields, but just the contrary of what is expected for the class number of real quadratic fields.

As a consequence, for every $n \geq 1$ and $m \geq 1$, the set $\{D > 0 \mid D$ is a fundamental discriminant, $h(D) \leq n$, and the maximum $m(D)$ of the lengths of the periods of reduced forms of discrinimant D is at most $m\}$ is finite—because for each such D, $c(D) \leq mn$. See SASAKI (1986).

In particular, for every $m \geq 1$, there exist only finitely many fundamental discriminants $D > 0$ such that $h(D) = 1$, and the periods of the reduced forms are of length at most equal to m. However, the preceding assertions are not effective results.

As usual, with a weaker form of the generalized Riemann hypothesis about the L-series of the character χ of D, it is possible to obtain an effective result, namely,

$$h(D) \log \epsilon_D < 4.23 c(D).$$

Then, with the same assumption, LACHAUD showed that the only real quadratic fields $\mathbf{Q}(\sqrt{D})$ with caliber one are the seven fields $\mathbf{Q}(\sqrt{D})$ with $D = 2, 5, 13, 29, 53, 173, 293$. Moreover, SASAKI showed that if also $m(D) = 1$, then $D = 2$.

Other types of results have the following flavor: if $D > 0$ is a fundamental discriminant of a given "shape", there are only finitely many real quadratic fields $\mathbf{Q}(\sqrt{D})$ with class number one.

Thus, CHOWLA and FRIEDLANDER (1976) had conjectured that if p is a prime, $p = m^2 + 1$, and $\mathbf{Q}(\sqrt{p})$ has class number one, then $p = 2, 5, 17, 37, 101, 197, 677$. Analogously, if p is a prime, $p = m^2 + 4$ and $\mathbf{Q}(\sqrt{p})$ has class number one, then $p = 5, 13, 29, 173, 293$.

This was proved, under the generalized Riemann hypothesis, independently by LACHAUD (1987) and by MOLLIN and WILLIAMS (1988).

I want also to highlight another theorem of the same family, proved by MOLLIN and WILLIAMS (1989):

A square-free positive integer $d = n^2 + r$, where r divides $4n$, is said to be of *extended Richaud-Degert type*.

There are 43 (and possibly 44) integers d of extended Richaud-Degert type whose corresponding field $\mathbf{Q}(\sqrt{d})$ has class number one;

a complete list is:

$$d = 2, 3, 5, 6, 7, 11, 13, 14, 17, 21, 23, 29, 33, 37, 38, 47, 53, 62, 69, 77,$$
$$83, 93, 101, 141, 167, 173, 197, 213, 227, 237, 293, 398, 413, 437,$$
$$453, 573, 677, 717, 1077, 1133, 1253, 1293,$$
$$1757 \text{ and possibly another value.}$$

Now, let $d > 0$, $d \neq 1$ be a square-free integer and, as previously, let

$$\omega = \begin{cases} \sqrt{d} & \text{if } d \equiv 2 \text{ or } 3 \pmod 4, \\ \frac{1+\sqrt{d}}{2} & \text{if } d \equiv 1 \pmod 4. \end{cases}$$

Denote by k the length of the period of the continued fraction expansion of ω.

MOLLIN and WILLIAMS (1989) have also determined explicitly (with possibly one exception) all the finitely many real quadratic fields $K = \mathbf{Q}(\sqrt{d})$ having class number one or two and such that the length of the period of ω is $k \leq 24$. This time not all integers d are of extended Richaud-Degert type.

For a unified presentation of the results thus far obtained by MOLLIN and WILLIAMS, the reader may wish to consult their paper MOLLIN and WILLIAMS (1990) and the book *Quadratics* by MOLLIN (1996).

There are, there were, and there will be, more partial results about this problem before a real insight will allow us to find the right way to approach it.

20   More questions and conjectures

The study of the conjectures of §14 has led to more embracing and deeper problems, all interrelated and, most likely, very difficult. Even though, at the present, and to my knowledge, there is no method to attack these questions with any significant success, I think it is nevertheless worthwhile to explicitly state the problems.

Problem 1. Is every natural number equal to the class number of some quadratic field with negative discriminant, respectively positive discriminant, D?

The following question is intimately related and even more difficult:

Problem 2. Is every finite abelian group G isomorphic to the class group of a quadratic field with discriminant $D < 0$, respectively $D > 0$? If so, are there infinitely many number fields $\mathbf{Q}(\sqrt{D})$, with $D > 0$, such that the class group of $\mathbf{Q}(\sqrt{D})$ is isomorphic to G?

BOYD and KISILEVSKY (1972) showed that there are only finitely many imaginary quadratic fields with class group isomorphic to a product of cyclic groups of order 3; they showed, under the generalized Riemann hypothesis, the corresponding result for class groups products of cyclic groups of order $n > 3$.

The next problem concerns the p-rank $r_p(D)$ (for p prime) of the class group of the quadratic field with discriminant D.

Problem 3. Let $p \geq 3$. Is every natural number equal to the p-rank $r_p(D)$ for some negative discriminant, respectively positive discriminant, D? In the affirmative, are there infinitely many discriminants D such that $r_3(D)$ is greater or equal to a given natural number n?

CRAIG (1977) showed that there exist infinitely many negative discriminants D, such that $r_3(D) \geq 4$.

Can one at least decide:

Problem 4. Is $\sup\{r_p(D) : |D| \geq 1\} = \infty$? (for $D < 0$, respectively $D > 0$).

In this respect, it is important to learn how to determine discriminants for which the p-rank is likely to be large.

It would also be very relevant to obtain estimates for the p-rank. Perhaps this could be feasible for negative discriminants.

Concerning this question, I wish to report that now there are known discriminants $D < 0$ for which the 3-rank is n, for every $n \leq 6$, and also discriminants $D < 0$ for which the 5-rank is n, for every $n \leq 4$.

QUER (1987) showed

$$r_3(-408368221541174183) = 6$$

and SCHOOF (1983) computed

$$r_5(-258559351511807) = 4;$$

see also LLORENTE and QUER (1988).

It is not necessary to discuss the 2-rank because it follows from the theory of genera that if D is a fundamental discriminant with r distinct prime factors, then the 2-rank is $r_2(D) = r - 1$ because the number of ambiguous classes is 2^{r-1}.

This is the place to mention some results of divisibility of the class number, which are, however, not strong enough to settle any of the above problems.

NAGELL (1929) showed that for every $n > 1$ there exist infinitely many imaginary quadratic fields $\mathbf{Q}(\sqrt{D})$, $D < 0$, with class number divisible by n.

This is a theorem which was rediscovered by HUMBERT (1939) and by ANKENY and CHOWLA (1955). In 1986, MOLLIN extended the result with a simpler proof (see MOLLIN (1986)).

Similarly, for real quadratic fields, first HONDA (1968) showed that there exist infinitely many real quadratic fields with class number divisible by 3. In 1970, it was shown by YAMAMOTO (1970) and also by WEINBERGER (1973b): for every $n > 1$ there exist infinitely many real quadratic fields with class number divisible by n.

Now I return to the heuristic arguments of COHEN and LENSTRA (1984); see also the book of COHEN (1993). From inspection of the tables of class groups (see BUELL (1976, 1987), SAITO and WADA (1988a,b)) for negative discriminants, it is apparent that the 3-Sylow subgroup is eight times more often isomorphic to C_9 than to $C_3 \times C_3$ (here C_n denotes the multiplicative cyclic group of order n). This is exactly the ratio

$$\frac{\#\operatorname{Aut}(C_9)}{\#\operatorname{Aut}(C_3 \times C_3)}.$$

This, and similar facts, suggest that probabilities of occurrence of a type of p-Sylow subgroup should be computed by weighing the groups G with weights $1/\#\operatorname{Aut}(G)$. With this simple idea, COHEN and LENSTRA arrived at probabilities which are amazingly close to observed values.

First, let $D < 0$.

The probability that the odd part of the class group is a cyclic group is equal to

$$\frac{\zeta(2)\zeta(3)}{\zeta(6)C_\infty \prod_{i=1}^\infty (1 - \frac{1}{2^i})} = 97.757\%$$

where

$$C_\infty = \prod_{n=2}^{\infty} \zeta(n) = 2.2948\ldots.$$

If p is an odd prime, the probability that the class number is divisible by p is equal to

$$l(p) = 1 - \prod_{i=1}^{\infty} \left(1 - \frac{1}{p_i}\right) = \frac{1}{p} + \frac{1}{p^2} - \frac{1}{p^5} - \frac{1}{p^7} + \frac{1}{p^{12}} + \frac{1}{p^{15}} + \cdots.$$

Explicitly,

$$l(3) \simeq 44\%,$$
$$l(5) \simeq 24\%,$$
$$l(7) \simeq 16\%, \text{etc}\ldots.$$

If p is an odd prime, the probability that the p-rank of the class group is equal to $n \geq 1$ is equal to

$$t_p(n) = \frac{\prod_{i=1}^{\infty} \left(1 - \frac{1}{p^i}\right)}{p^{n^2} \prod_{j=1}^{n} \left(1 - \frac{1}{p^j}\right)^2}.$$

The probability that the p-Sylow subgroup ($p > 2$) of the class group be equal to a given group is:

$S_3 = C_9$	: 9.33%
$S_3 = C_3 \times C_3$	: 1.17%
$S_3 = C_3 \times C_3 \times C_3$	: 0.005%
$S_3 = C_3 \times C_3 \times C_3 \times C_3$	: $2.3 \times 10^{-8}\%$
$S_5 = C_{25}$	: 3.80%
$S_5 = C_5 \times C_5$	: 0.16%, etc....

Now, let $D > 0$.

The probability that the order of the odd part of the class group be equal to n is $u(n)$, where

$$u(1) = 75.5\%$$
$$u(3) = 12.6\%$$
$$u(5) = 3.8\%$$
$$u(7) = 1.8\%$$
$$u(9) = 1.6\%, \text{ etc}\ldots.$$

$u(n)$ is also the probability that the class number of $\mathbf{Q}(\sqrt{p})$ (with p prime) is equal to n.

The probability that the odd prime p divides the class number is

$$1 - \prod_{k=2}^{\infty} \left(1 - \frac{1}{p^k}\right).$$

The probability that the p-rank of the class group be equal to $n \geq 1$ is

$$t_p'(n) = \frac{\prod_{i=1}^{\infty}\left(1 - \frac{1}{p^i}\right)}{p^{n(n+1)} \prod_{j=1}^{n}\left(1 - \frac{1}{p^j}\right) \prod_{j=1}^{n+1}\left(1 - \frac{1}{p^j}\right)},$$

etc. . . .

The above heuristic results suggest, of course, what *should* be the answers to the problems stated in the beginning of this section.

The reader may wish to consult the recent paper of JACOBSON (1998). This paper contains tables requiring extensive calculations for $D < 10^9$. The numerical results confirm the amazing conjectures of COHEN and LENSTRA and provide lists of discriminants for which the class group contains non-cyclic p-Sylow subgroups (for all $p \leq 23$), etc.

21   Many topics have not been discussed

This extended version of my lecture is already much longer than intended. Yet, many topics of no lesser importance could not and will not be discussed. Among these topics, the geometric theory of quadratic forms, as developed by KLEIN, leading to an intimate connection with modular forms; see the expository paper of SERRE (1985).

The problem of representation of integers by quadratic forms, which cannot be completely solved by the methods presented here when there is more than one class in the principal genus, can however be dealt with using class field theory, more specifically with the Hilbert symbol. This development is very well presented in the book of COX (1989).

SHANKS made good use of the class group and even of its infrastructure, to invent clever algorithms for factorization and primality; see SHANKS (1969, 1976, 1989).

References

1801　C. F. Gauss. *Disquisitiones Arithmeticae.* G. Fleischer, Leipzig. Translated by A. A. Clarke, Yale Univ. Press, New Haven, 1966.

1870　C. F. Gauss. *Werke.* Königl. Ges. d. Wiss., Göttingen.

1892/94　P. Bachmann. *Zahlentheorie, Vol. I and II.* B. G. Teubner, Leipzig.

1907　J. Sommer. *Vorlesungen über Zahlentheorie.* B. G. Teubner, Leipzig.

1913　E. Landau. Über die Klassenzahl imaginär-quadratischer Zahlkörper. *Göttinger Nachr.,* 285–295.

1929　T. Nagell. Über die Klassenzahl imaginär-quadratischer Zahlkörper. *Abh. Math. Sem. Univ. Hamburg,* 1:140–150.

1933　M. Deuring. Imaginäre quadratische Zahlkörper mit der Klassenzahl 1. *Math. Z.,* 37:405–415.

1934　S. Chowla. An extension of Heilbronn's class-number theorem. *Quart. J. Math. Oxford,* 5:304–307.

1934a　H. Heilbronn. On the class number of imaginary quadratic fields. *Quart. J. Math. Oxford,* 5(2):150–160.

1934b　H. Heilbronn and E. H. Linfoot. On the imaginary quadratic corpora of class number one. *Quart. J. Math. Oxford,* 5 (2):293–301.

1936　C. L. Siegel. Über die Classenzahl quadratischer Zahlkörper. *Acta Arith.,* 1:83–86. Reprinted in *Gesammelte Abhandlungen, Vol. I,* 406–409. Springer-Verlag, Berlin, 1966.

1939　P. Humbert. Sur les nombres de classes de certains corps quadratiques. *Comm. Math. Helvetici,* 12:233–245 and 13:67 (1940).

1952　K. Heegner. Diophantische Analysis und Modulfunktionen. *Math. Z.,* 56:227–253.

1954　S. Chowla and W. E. Briggs. On discriminants of binary quadratic forms with a single class in each genus. *Can. J. Math.,* 6:463–470.

1955　N. C. Ankeny and S. Chowla. On the divisibility of the class number of quadratic fields. *Pacific J. Math.,* 5:321–324.

1961　G. B. Mathews. *Theory of Numbers.* Reprinted by Chelsea Publ. Co., Bronx, NY.

1962　H. Cohn. *Advanced Number Theory.* Dover, New York.

1963 E. Grosswald. Negative discriminants of binary quadratic forms with one class in each genus. *Acta Arith.*, 8:295–306.

1966 A. Baker. Linear forms in the logarithms of algebraic numbers. *Mathematika*, 13:204–216.

1966 Z. I. Borevich and I. R. Shafarevich. *Number Theory.* Academic Press, New York.

1966 J. Steinig. On Euler's idoneal numbers. *Elem. of Math.*, 21:73–96.

1967 H. M. Stark. A complete determination of the complex quadratic fields of class number one. *Michigan Math. J.*, 14:1–27.

1968 M. Deuring. Imaginäre-quadratische Zahlkörper mit der Klassenzahl Eins. *Invent. Math.*, 5:169–179.

1968 T. Honda. On real quadratic fields whose class numbers are multiples of 3. *J. reine u. angew. Math.*, 233:101–102.

1968 P. G. Lejeune-Dirichlet. *Vorlesungen über Zahlentheorie (mit Zusätzen versehen von R. Dedekind).* Chelsea Publ. Co., New York. Reprint. First edition in 1863.

1968 C. L. Siegel. Zum Beweise des Starkschen Satz. *Invent. Math.*, 5:180–191.

1969 D. Shanks. Class number, a theory of factorization, and genera. In *1969 Number Theory Institute (Proc. Sympos. Pure Math., Vol. XX, State Univ. New York, Stony Brook, N.Y., 1969)*, 415–440, Providence, R.I. Amer. Math. Soc.

1969 H. M. Stark. On the "gap" in a theorem of Heegner. *J. Nb. Th.*, 1:16–27.

1970 B. A. Venkov. *Elementary Number Theory.* Wolters-Noordhoff Publishing, Gröningen. Translated from the Russian and edited by H. Alderson.

1970 Y. Yamamoto. On unramified Galois extensions of quadratic number fields. *Osaka J. Math.*, 7:57–76.

1971 A. Baker. Imaginary quadratic fields with class number 2. *Ann. of Math. (2)*, 94:139–152.

1971 H. M. Stark. A transcendence theorem for class number problems. *Ann. Math. (2)*, 94:153–173.

1972 D. W. Boyd and H. Kisilevsky. On the exponent of the ideal class groups of complex quadratic fields. *Proc. Amer. Math. Soc.*, 31:433–436.

1973a P. J. Weinberger. Exponents of the class groups of complex quadratic fields. *Acta Arith.*, 22:117–124.

1973b P. J. Weinberger. Real quadratic fields with class numbers divisible by *n*. *J. Nb. Th.*, 5:237–241.

1975 A. Baker. *Transcendental Number Theory.* Cambridge Univ. Press, Cambridge.

1976 D. A. Buell. Class groups of quadratic fields. *Math. of Comp.*, 30:610–623.

1976 S. Chowla and J. B. Friedlander. Some remarks on *L*-functions and class numbers. *Acta Arith.*, 28:414–417.

1976 D. Shanks. A survey of quadratic, cubic and quartic algebraic number fields (from a computational point of view). In *Proceedings of the Seventh Southeastern Conference on Combinatorics, Graph Theory, and Computing (Louisiana State Univ., Baton Rouge, LA)*, 15–40. Utilitas Math., Winnipeg, Manitoba.

1977 M. Craig. A construction for irregular discriminants. *Osaka J. Math.*, 14:365–402.

1977 H. M. Edwards. *Fermat's Last Theorem: A Genetic Introduction to Algebraic Number Theory.* Springer-Verlag, New York.

1977 D. M. Goldfeld. *The conjectures of Birch and Swinnerton-Dyer and the class numbers of quadratic fields.* Astérisque 41–42, 219–227.

1980 H. Davenport. *Multiplicative Number Theory.* Springer-Verlag, New York, 2nd edition.

1980 A. Schinzel. On the relation between two conjectures on polynomials. *Acta Arith.*, 38:285–322.

1981 W. Kaufmann-Bühler. *Gauss: A Biographical Study.* Springer-Verlag, Berlin-Heidelberg-New York.

1981 H. Wada. A table of ideal class numbers of real quadratic fields. *Sophia Kokyoroku in Mathematics.* Number 10.

1981 D. B. Zagier. *Zetafunktionen und quadratische Körper.* Springer-Verlag, Berlin.

1982 L. K. Hua. *Introduction to Number Theory.* Springer-Verlag, Berlin.

1983 B. Gross and D. B. Zagier. Points de Heegner et dérivées de fonctions *L. C. R. Acad. Sci. Paris*, 297:85–87.

1983 J. Oesterlé. Nombres de classes des corps quadratiques imaginaires. *Séminaire Bourbaki, exp. 631.*

1983 R. J. Schoof. Class groups of complex quadratic fields. *Math. of Comp.*, 41:295–302.

1984 H. Cohen and H. W. Lenstra, Jr. Heuristics on class groups of number fields. In *Number Theory, Noordwijkerhout 1983, Lect. Notes in Math., 1068*, 33–62. Springer-Verlag, Berlin.

1984 G. Frei. Les nombres convenables de Leonhard Euler. *Sém. Th. des Nombres), Besançon, (1983–84).* 58 pages.

1984 J. J. Gray. A commentary on Gauss's mathematical diary, 1796–1814, with an English translation. *Expo. Math.*, 2: 97–130.

1984 A. Weil. *Number theory, an Approach through History, from Hammurapi to Legendre.* Birkhäuser, Boston.

1984 D. B. Zagier. *L*-series of elliptic curves, the Birch-Swinnerton-Dyer conjecture, and the class number problem of Gauss. *Notices Amer. Math. Soc.*, 31(7):739–743.

1985 G. Frei. Leonhard Euler's convenient numbers. *Math. Intelligencer*, 7(3):55–58, 64.

1985 D. M. Goldfeld. Gauss's class number problem for imaginary quadratic fields. *Bull. Amer. Math. Soc.*, 13: 23–37.

1985 J. P. Serre. $\Delta = b^2 - 4ac$. *Mathematical Medley*, 13(1): 1–10. See also the Appendix in Flath (1989).

1986 B. Gross and D. B. Zagier. Heegner points and derivatives of *L*-series. *Invent. Math.*, 84:225–320.

1986 G. Lachaud. Sur les corps quadratiques réels principaux. In *Séminaire de Théorie des Nombres, Paris 1984–85. Progress in Math. #63*, 165–175. Birkhäuser Boston, Boston, MA.

1986 R. A. Mollin. On class numbers of quadratic extensions of algebraic number fields. *Proc. Japan Acad., Ser. A*, 62: 33–36.

1986 R. Sasaki. A characterization of certain real quadratic fields. *Proc. Japan Acad. Ser. A Math. Sci.*, 62:97–100.

1987 J. M. Borwein and P. B. Borwein. *Pi and the AGM.* John Wiley & Sons, New York.

1987 D. A. Buell. Class groups of quadratic fields, II. *Math. of Comp.*, 48:85–93.

1987 G. Lachaud. On real quadratic fields. *Bull. Amer. Math. Soc.*, 17:307–311.

1987 J. Quer. Corps quadratiques de 3-rang 6 et courbes elliptiques de rang 12. *C. R. Acad. Sci. Paris*, 305(6): 215–218.

1988 P. Llorente and J. Quer. On the 3-Sylow subgroup of the class group of quadratic fields. *Math. of Comp.*, 50:321–333.

1988 R. A. Mollin and H. C. Williams. A conjecture of S. Chowla via the generalized Riemann hypothesis. *Proc. Amer. Math. Soc.*, 102:794–796.

1988 J. Oesterlé. Le problème de Gauss sur le nombres de classes. *L'Enseign. Math.*, 2e série, 34:43–67.

1988a M. Saito and H. Wada. A table of ideal class groups of imaginary quadratic fields. *Sophia Kokyoroku in Mathematics*. Number 28.

1988b M. Saito and H. Wada. Tables of ideal class groups of real quadratic fields. *Proc. Japan Acad., Ser. A*, 64:347–349.

1989 D. A. Buell. *Binary Quadratic Forms*. Springer-Verlag, New York.

1989 D. A. Cox. *Primes of the Form x^2+ny^2*. Wiley-Interscience, New York.

1989 D. E. Flath. *Introduction to Number Theory*. Wiley, New York.

1989 R. A. Mollin and H. C. Williams. Real quadratic fields of class number one and continued fraction period less than six. *C. R. Math. Reports Acad. Sci. Canada*, 11:51–56.

1989 D. Shanks. On Gauss and composition, I and II. In *Proc. Conf. Canadian Nb. Th. Assoc., Banff*, edited by R. A. Mollin, 163–204. Kluwer Acad. Publ., Dordrecht.

1990 R. A. Mollin and H. C. Williams. Class number problems for real quadratic fields. In *Number Theory and Cryptography (Sydney, 1989)*, 177–195. Cambridge Univ. Press, Cambridge. In London Math. Soc. Lecture Notes Ser., 154.

1992 R. A. Mollin and H. C. Williams. Computation of the class number of a real quadratic field. *Utilitas Math.*, 41:259–308.

1993 H. Cohen. *A Course in Computational Algebraic Number Theory*. Springer-Verlag, Berlin.

1996 R. A. Mollin. *Quadratics*. CRC Press, Boca Raton, FL.

1998 M. J. Jacobson, Jr. Experimental results on class groups of real quadratic fields (extended abstract). In *Algorithmic Number Theory (Portland, OR, 1998)*, 463–474. Springer-Verlag, Berlin.

7

Consecutive Powers

1 Introduction

(a) If we write the sequence of squares and cubes of integers in increasing order

$$4\ 8\ 9\ 16\ 25\ 27\ 36\ 49\ 64\ 81\ 100\ldots,$$

say,

$$z_1 < z_2 < z_3 < z_4 < \cdots < z_n < z_{n+1} < \cdots$$

we may ask many questions. For example:

(I) Are there consecutive integers in this sequence? Of course, yes: 8 and 9. Are there others? How many? Only finitely many?

If we examine a list of squares and cubes up to 1 000 000, we find no other example. Is this always true? Or will there be, perhaps by accident, other consecutive squares and cubes?

If a search with a computer is pushed further, we may observe that the differences appear to become larger (but not monotonically), that is, squares and cubes appear more sparsely. Yet we should not from this experimental observation conclude that no consecutive cube and square, other than 8 and 9, exist.

Consider the following situation where numbers more and more sparsely distributed are still sufficient for a certain representation. Namely, among the numbers up to 10 000 there are only 100 squares, so $\frac{1}{100}$; up to 1 000 000 there are only 1 000 squares, so $\frac{1}{1,000}$; up to 100 000 000 there are only 10 000 squares, so $\frac{1}{10,000}$; etc. So, the squares are less and less common. Yet, LAGRANGE proved that every natural number is the sum of (at most) *four* squares.

Thus, even though the squares are less and less present they occupy "strategic positions," so four squares are always enough to reproduce by addition any natural number.

This was mentioned just to prevent anyone to jump to false conclusions.

A second question is the following:

(II) Given k (now $k \geq 2$), for how many indices n is it true that $z_{n+1} - z_n \leq k$? Only finitely many?

We may also consider other sequences involving powers:

(b) The sequence $z_1 < z_2 < z_3 < \cdots$ of all proper powers of integers: squares, cubes, 5th powers, 7th powers, etc.

(c) If $a, b \geq 2$, $a \neq b$, we may consider the sequence $z_1 < z_2 < z_3 < \cdots$ of all powers of a or b.

For example, if $a = 2$, $b = 3$:

$$4 \ 5 \ 9 \ 16 \ 27 \ 32 \ 64 \ 81 \ 128 \ 243 \ 256 \ldots.$$

(d) If $E = \{p_1, \ldots, p_r\}$ where $r \geq 2$ and each p_i is a prime number, let S be the set of all natural numbers all of whose prime factors are in E:

$$S: z_1 < z_2 < z_3 < \cdots.$$

The sequence (c) is, of course, a subsequence of one of type (d).

For each of the sequences (b), (c), (d) we may ask the same questions (I) and (II). Also, for the sequence (b) of all powers, we may ask the question (which is of no interest for the sequences (a), (c), (d)):

(III) Are there three or more consecutive powers? How many? None? Finitely many?

Before we proceed, let us discuss whether these questions are just a curiosity. We may paraphrase GAUSS' point of view: "Any fool can ask questions about numbers, which even a thousand wise men cannot solve."

Are these questions of this kind? No! As they involve powers (therefore, multiplication in a rather special way) and differences, they combine the additive and multiplicative structure of the integers. Somewhat like a famous unsolved problem of FERMAT: is the sum of two nth powers again a nth power, when $n > 2$?

As it turns out, the study of these questions contributes substantially to the knowledge of the integers. And this amply justifies these investigations.

2 History

Our treatment of the problem will follow somewhat the historical development. So, we shall be very brief here, underlining only a few points.

(1) We may read in DICKSON's useful *History of the Theory of Numbers, Volume II*, that the first mention of this problem is in a question asked by PHILIPPE DE VITRY: Can $3^m \pm 1$ be a power of 2? This was solved by LEVI BEN GERSON (alias LEO HEBRACUS), who lived in Spain from 1288 to 1344. He showed that if $3^m \pm 1 = 2^n$, then $m = 2$, $n = 3$, so these numbers are 9 and 8.

(2) In 1657, in his "Deuxieme Deli aux Mathematiciens" (letter to FRÉNICLE DE BESSY), FERMAT proposed to show: if p is an odd prime and $n \geq 2$, then $p^n + 1$ is not a square; similarly, if $n \geq 4$, then $2^n + 1$ is not a square.

A proof, published by FRÉNICLE, was discovered in 1943 by HOFMANN.

(3) Using the method of infinite descent, which had been invented by FERMAT, EULER showed in 1738 that if the difference between a square and a cube is ± 1, then these numbers are 9 and 8.

(4) In 1844, in a letter to CRELLE (appearing in Volume I of CRELLE's journal), CATALAN asked for a proof that the only consecutive powers are 8 and 9. This assertion is now called "Catalan's conjecture". In other words, he proposed to prove that the equation $X^U - Y^V = 1$ in four unknown quantities, two of which are in the exponent, has only the solution $x = 3$, $u = 2$, $y = 2$, $v = 3$ in natural numbers bigger than 1. The only results of CATALAN on this equation are simple observations, which are in his *Mélanges Math-*

ematiques, XV, published much later in 1885. Among the various statements, CATALAN asserted, without proof, that if $x^y - y^x = 1$, then $x = 2$, $y = 3$—but this is rather a simple exercise to prove. For a biography of CATALAN, see JONGMANS (1996).

(5) In the next phase, various special cases, with powers having small exponents, were considered by LEBESGUE (1850) and, in this century, by NAGELL, OBLÁTH, S. SELBERG, CHAO KO, et al.

(6) Then came a series of results imposing divisibility constraints on any natural numbers x, y such that $x^m - y^n = 1$. The most important results of this kind refer to exponents m, n which are odd primes. They are due to CASSELS, INKERI, and HYYRÖ.

(7) Finally, there were the results concerning estimates on the number and size of possible consecutive powers. Here the most important contributions are first due to HYYRÖ and above all to TIJDEMAN who used high-powered methods from the theories of diophantine approximation and BAKER's estimates on linear forms of logarithms.

We shall discuss all these points in more detail.

3   Special cases

Unless stated to the contrary, the numbers appearing in the equations are natural numbers.

As is fitting, we begin with LEVI BEN GERSON'S result; the proof given here was provided by M. LANGEVIN, while another proof was published by FRANKLIN (1923).

(3.1) If $m, n \geq 2$ and $3^m - 2^n = \pm 1$, then $m = 2$, $n = 3$. Thus, in the sequence of powers of 2 or 3, the only consecutive integers are 8 and 9.

PROOF.   If $2^n - 3^m = 1$ then $2^n \equiv 1 \pmod{3}$, so n is even, $n = 2n'$. Then $3^m = 2^{2n'} - 1 = (2^{n'} - 1)(2^{n'} + 1)$, hence $2^{n'} - 1 = 3^{m'}$, $2^{n'} + 1 = 3^{m-m'}$ with $0 \leq m' < m - m'$. Subtracting, $2 = 3^{m'}(3^{m-2m'} - 1)$, hence $m' = 0$, $n' = 1$, $n = 2$, $m = 1$, against the hypothesis.

If $3^m - 2^n = 1$, if $n = 2$ this is impossible, so $n \geq 3$, hence $3^m \equiv 1 \pmod{8}$. Therefore, m is even, $m = 2m'$. Then $2^n = 3^{2m'} - 1 = (3^{m'} - 1)(3^{m'} + 1)$, hence $3^{m'} - 1 = 2^{n'}$, $3^{m'} + 1 = 2^{n-n'}$ with $0 \leq n' < n - n'$. Subtracting, $2 = 2^{n'}(2^{n-2n'} - 1)$, hence $n' = 1$, $n = 2n' + 1 = 3$, and $m = 2$.   □

The following observation is quite obvious: if $m, n \geq 2$, if there exist solutions in natural numbers of the equation $X^m - Y^n = 1$, and if p, q are primes such that $p \mid m$, $q \mid n$, then there exist solutions in natural numbers of the equation $X^p - Y^q = 1$. So, we are led to study the equation $X^p - Y^q = 1$ with p, q distinct primes.

EULER proved in 1738 the following basic lemma:

Lemma 1. *Let p, q be primes, and let $x, y \geq 2$ be such that $x^p - y^q = 1$. If p is odd, then*

$$x - 1 = a^q \qquad with \; y = aa', \; p \nmid aa',$$
$$\frac{x^p - 1}{x - 1} = (a')^q \qquad \gcd(a, a') = 1,$$

or

$$x - 1 = p^{q-1} a^q \qquad with \; y = paa', \; p \nmid a',$$
$$\frac{x^p - 1}{x - 1} = p(a')^q \qquad \gcd(a, a') = 1.$$

Similarly, if q is odd then

$$y + 1 = b^p \qquad with \; x = bb', \; q \nmid bb',$$
$$\frac{y^q + 1}{y + 1} = (b')^p \qquad \gcd(b, b') = 1,$$

or

$$y + 1 = q^{p-1} b^p \qquad with \; x = qbb', \; q \nmid b',$$
$$\frac{y^q + 1}{y + 1} = q(b')^p \qquad \gcd(b, b') = 1.$$

PROOF.   This proof is quite simple and so we shall indicate it. We have

$$y^q = x^p - 1 = (x - 1)\frac{x^p - 1}{x - 1}.$$

But $\gcd\left(x - 1, \frac{x^p - 1}{x - 1}\right) = 1$ or p, because

$$\frac{x^p - 1}{x - 1} = \frac{[(x - 1) + 1]^p - 1}{x - 1}$$

$$= (x - 1)^{p-1} + \binom{p}{1}(x - 1)^{p-2} + \ldots + \binom{p}{p - 2}(x - 1) + p.$$

Moreover, the greatest common divisor in question is equal to p exactly when $p \mid y$.

Next, we note that $p^2 \nmid \frac{x^p-1}{x-1}$. Indeed, if $p \mid \frac{x^p-1}{x-1}$, then $p \mid x-1$; but p^2 divides each summand $(x-1)^{p-1}, \ldots, \binom{p}{p-2}(x-1)$, so p^2 cannot divide $\frac{x^p-1}{x-1}$. Therefore we have shown the first assertion, concerning $x-1$ and $\frac{x^p-1}{x-1}$. The proof of the second assertion is similar.   □

With the same method, EULER proved:

Lemma 2. *If q is an odd prime, $x, y \geq 2$, and $x^2 - y^q = 1$, then*

$$x - 1 = 2a^q$$
$$x + 1 = 2^{q-1}(a')^q$$

or

$$\begin{cases} x + 1 = 2a^q \\ x - 1 = 2^{q-1}(a')^q \end{cases}$$

where $a, a' \geq 1$, a is odd, $\gcd(a, a') = 1$.

Using the method of infinite descent, EULER proved:

(3.2) If $x, y \geq 1$ and $x^2 - y^3 = \pm 1$, then $x = 3$, $y = 2$. So, in the sequence of squares and cubes, 8 and 9 are the only consecutive integers.

EULER had actually shown that the only solutions in positive rational numbers of $X^2 - Y^3 = \pm 1$ are $x = 3$, $y = 2$; his proof is rather tricky.

We also note here that EULER used the method of infinite descent to show that Fermat's equation $X^3 + Y^3 = Z^3$ has only trivial solutions in integers.

In 1921, NAGELL proposed another proof, reducing it to an earlier result of LEGENDRE (1830, Volume II, page 9): the equation $X^3 + Y^3 = 2Z^3$ has only the solutions $x = y = z$ or $x = -y$, $z = 0$ in integers. LEGENDRE's proof was also by infinite descent. BACHMANN gave in 1919 an incorrect proof of LEGENDRE's result without using the method of descent.

Another way of proving EULER's result, without the method of descent, uses the numbers in the cubic field $K = \mathbb{Q}(\sqrt[3]{2})$. From

$$\mp 1 = u^3 - 2v^3 = (u - \sqrt[3]{2}v)(u^2 + \sqrt[3]{2}uv + \sqrt[3]{4}v^2)$$

it follows that $u - \sqrt[3]{2}v$ is a unit of the field K. As is known (see LEVEQUE's book, Volume II, pages 108–109), $u - \sqrt[3]{2}v$ is, up to sign, a power of the fundamental unit $-1 + \sqrt[3]{2}$:

$$u - \sqrt[3]{2}v = \pm(-1 + \sqrt[3]{2})^n.$$

Then, it is shown that n cannot be negative, and $n \neq 2$. Finally it is shown that n cannot be greater than 2, by comparing coefficients in the two sides, and considering congruences modulo 3.

So, $u - \sqrt[3]{2}v = \pm(-1 + \sqrt[3]{2})$, leading to $x = 3$, $y = 2$.

After $X^2 - Y^3 = \pm 1$ was treated by EULER, next came the equations $X^2 - Y^m = \pm 1$ (with $n \geq 5$).

As it happens, and it is certainly surprising, one of these equations was rather easy to treat, while the other required 120 years to be solved!

Which is the easy one? Here is the answer. LEBESGUE used Gaussian integers to show, in 1850:

(3.3) The equation $X^m - Y^2 = 1$ has only trivial solutions in natural numbers.

PROOF.   Once more, we give a sketch of the proof, leaving the details to the reader. If $x, y \geq 2$ and $x^m = y^2 + 1 = (y+i)(y-i)$, then x is odd, y must be even, and there exist integers u, v such that

$$y + i = (u + iv)^m i^s \quad (\text{with } 0 \leq s \leq 3);$$

hence

$$y - i = (u - iv)^m (-i)^s.$$

So, $x = u^2 + v^2$ and since x is odd, then u or v is even. By subtracting, we have $2i = [(u + iv)^m - (u - iv)^m(-1)^s]i^s$ and this leads to

$$1 - \binom{m}{2}w^2 + \binom{m}{4}w^4 - \ldots \pm mw^{m-1} = \pm 1, \quad \text{where } w = u,$$

$v = \pm 1$ (when s is even), or $w = v$, $u = \pm 1$ (when s is odd); so w is even. The sign $-$ would imply that w^2 divides 2, which is impossible. The sign $+$ is also impossible, and this is seen by considering the 2-adic values of the summands in the above relation.   □

We shall postpone the study of the more difficult equation $X^2 - Y^m = 1$.

So, the next equations in line are $X^3 - Y^m = \pm 1$, which were studied by NAGELL in 1921. First, he showed:

(3.4) (a) If $m \geq 2$ is not a power of 3, the only non-zero solutions of $X^2 + X + 1 = Y^m$ are $(-1, 1)$ when m is odd, and $(-1, \pm 1)$ when m is even.

(b) If $m > 2$, the only non-zero solutions of $X^2 + X + 1 = 3Y^m$ are $x = 1$ and $x = -2$.

Moreover, if $m = 2$, there are also the solutions

$$\pm \frac{\sqrt{3}}{4} [(2 + \sqrt{3})^{2n+1} - (2 - \sqrt{3})^{2n+1}] - \frac{1}{2}$$

for $n = 0, 1, \ldots$.

The proof is much longer, so we just say that for (a) NAGELL worked in $\mathbb{Q}(\omega) = \mathbb{Q}(\sqrt{-3})$, where $\omega = \frac{-1 + \sqrt{-3}}{2}$ is a cube root of 1. He was led to the equations $X \pm \omega = (Z - \omega)^q$ where q is a prime, $q > 3$, the only solutions being $x = \pm 1, 0$. For (b), if $m = 2$, NAGELL worked in the field $\mathbb{Q}(\sqrt{3})$ which has the fundamental unit $2 + \sqrt{3}$. If $4 \mid m$, he was led to the equation $U^4 + V^4 = W^2$, which as FERMAT showed, has only trivial solutions. If m is a power of 3, NAGELL was led to the equation $X^3 + Y^3 = Z^3$. Finally, for all other values of m, he worked in the field $\mathbb{Q}(\omega)$.

Now, it was easy for NAGELL to show:

(3.5) The equations $X^3 \pm 1 = Y^m$ (with m not a power of 2) have only trivial solutions in integers.

PROOF. We may assume that $m = q$ is a prime, $q > 3$. If x, y are such that $y^q = x^3 \pm 1 = (x^2 \mp x + 1)(x \pm 1)$, then $x^2 \mp x + 1 = a^q$ or $3a^q$, where a is an integer. Replacing x by $-x$ (in the case of the minus sign), we have $x^2 + x + 1 = a^q$ or $3a^q$, and this leads to the result. $\square$

LJUNGGREN (1942, 1943) studied the equation $\frac{x^n - 1}{x - 1} = y^m$ and completed NAGELL's result **(3.4)**(a) above, showing that it holds also when m is a power of 3.

We return to the equation $X^2 - Y^n = 1$ (with $n > 3$), which resisted many attempts until it was finally completely solved. As we shall see, the solution was elementary, but certainly not straightforward. We shall present here several of the partial results. Even though they are now completely superseded, it is illuminating to see the ways mathematicians have tried to solve the equation, and the connection with other interesting problems.

As already indicated, the first mention of this equation was in FER-MAT's "second définux mathématiciens" of 1657. We give FRÉNICLE's result:

(3.6) If p is an odd prime, $n \geq 2$, then $p^n + 1$ is not a square. If $n \geq 4$, then $2^n + 1$ is not a square.

PROOF. If $p^n = x^2 - 1 = (x+1)(x-1)$, with $p \neq 2$, then $\gcd(x+1, x-1) = 1$, so

$$x + 1 = p^a$$
$$x - 1 = p^b$$

Thus $x - 1 = 1$, $x + 1 = p^n$, hence $p^n = 3$, $n \geq 2$, which is impossible. The proof of the other assertion is similar.   □

The weaker statement that for every integer y, the number $y^n + 1$ is not a 4th power was proved by S. SELBERG in 1932. His proof appealed to an older result of STØRMER (1899), which is linked with the speedy calculation of the decimal development of π.

Let us explain this unexpected connection, beginning with a quick history of the calculation of π.

Using the method of inscribed and circumscribed polygons, ARCHIMEDES gave, circa 250 B.C., the estimate

$$3.1408 = \frac{223}{71} < \pi < \frac{22}{7} = 3.1428.$$

In the XIIth century, FIBONACCI (LEONARDO DI PISA) calculated π to be about $\frac{864}{275} = 3.1418$. ADRIANUS METIUS (1571–1635) estimated

$$3.14150 = \frac{333}{106} < \pi < \frac{355}{113} = 3.14159.$$

In the XVIth century, F. VIÈTE (1540–1603) expressed π with an infinite product:

$$\pi = \cfrac{2}{\sqrt{\tfrac{1}{2}}\sqrt{\tfrac{1}{2} + \tfrac{1}{2}\sqrt{\tfrac{1}{2}}}\sqrt{\tfrac{1}{2} + \tfrac{1}{2}\sqrt{\tfrac{1}{2} + \tfrac{1}{2}\sqrt{\tfrac{1}{2}}} \cdots}$$

and this gave

$$3.1415926535 < \pi < 3.1415926537.$$

A. VAN ROOMAN calculated π to 15 decimal places in 1593. LU-DOLPH VAN CEULEN devoted his life to the calculation of π; first, in 1596 he computed 20 decimals, using a polygon with 60×2^{29} sides! And, after his death, his wife published in 1615 LUDOLPH's calculated value of π which had 32 correct decimals. His contemporaries were so impressed with this painstaking work that until recently German authors referred to π as the Ludolph number.

In the XVIIth century, there appeared infinite series, infinite products, and continued fractions: the series for the arcsine given by I. NEWTON (1642–1727) in 1676; the continued fraction expression for $\pi/4$ by LORD W. BROUNCKER (1620–1684), found in 1658:

$$\frac{\pi}{4} = \frac{1}{1+} \frac{1^2}{2+} \frac{3^2}{2+} \frac{5^2}{2+} \cdots ;$$

the infinite product for $\pi/2$ by J. WALLIS (1616–1703), in 1655:

$$\frac{\pi}{2} = \frac{2}{1} \cdot \frac{2}{3} \cdot \frac{4}{3} \cdot \frac{4}{5} \cdot \frac{6}{5} \cdot \frac{6}{7} \cdot \frac{8}{7} \cdot \frac{8}{9} \cdots ;$$

the power series for the arctangent function given by J. GREGORY (1638–1675) in 1671 and, independently, the alternating series for $\pi/4$ by G. W. LEIBNIZ (1646–1716) in 1674:

$$\frac{\pi}{4} = \arctan 1 = 1 - \frac{1}{3} + \frac{1}{5} - \frac{1}{7} + \cdots .$$

In 1699, A. SHARP (1651–1742) used the series for the arctangent, with $\lambda = \frac{1}{\sqrt{3}}$, and computed π to 72 decimals—a far cry from the lifelong efforts required with the polygon method.

In 1737 (published in 1744), EULER calculated π to 127 decimals. But, more interesting are the expressions which he discovered in 1755, for even powers of π in terms of the sums of series of inverses of even powers of natural numbers (values of the ζ-function):

$$\frac{\pi^2}{6} = \zeta(2) = \sum_{n=1}^{\infty} \frac{1}{n^2}$$

$$\frac{\pi^4}{90} = \zeta(4) = \sum_{n=1}^{\infty} \frac{1}{n^4}$$

$$\frac{\pi^6}{945} = \zeta(6) = \sum_{n=1}^{\infty} \frac{1}{n^6} .$$

The interest of the above formulas is more theoretical than computational, but we shall not enter into this matter here. EULER (published in 1783) obtained also the following generalization of VIÈTE's infinite product for π, namely, if $0 < \theta < 180°$, then

$$\theta = \frac{\sin \theta}{\cos \left(\frac{\theta}{2}\right) \cos \left(\frac{\theta}{2^2}\right) \cos \left(\frac{\theta}{2^3}\right) \cdots};$$

this gives Viète's formula when $\theta = 90°$.

The series for $\pi/4$ converges too slowly. However, it is possible to speed up the convergence by using the addition formula. Namely, from $\tan(x + y) = \frac{\tan x + \tan y}{1 - \tan x \tan y}$, taking $x = \arctan u$, $y = \arctan v$, then $\arctan u + \arctan v = \arctan \frac{u+v}{1-uv}$.

Choosing u, v such that $\frac{u+v}{1-uv} = 1$, then $\pi/4 = \arctan u + \arctan v$.

For example, let $u = \frac{1}{2}$, so $v = \frac{1}{3}$, hence (∗) $\pi/4 = \arctan \frac{1}{2} + \arctan \frac{1}{3}$ (formula by HUTTON, 1776); that is,

$$\frac{\pi}{4} = \left[\frac{1}{2} - \frac{1}{3}\left(\frac{1}{2}\right)^3 + \frac{1}{5}\left(\frac{1}{2}\right)^5 - \cdots\right] + \left[\frac{1}{3} - \frac{1}{3}\left(\frac{1}{3}\right)^3 + \frac{1}{5}\left(\frac{1}{3}\right)^5 - \cdots\right].$$

This series still converges too slowly. To improve the convergence, we determine v such that

$$\frac{1}{2} = \frac{\frac{1}{3} + v}{1 - \frac{v}{3}},$$

so $v = \frac{1}{7}$, and similarly v such that

$$\frac{1}{3} = \frac{\frac{1}{5} + v}{1 - \frac{v}{5}},$$

so $v = \frac{1}{8}$. Hence,

$$\arctan \frac{1}{2} = \arctan \frac{1}{3} + \arctan \frac{1}{7},$$

$$\arctan \frac{1}{3} = \arctan \frac{1}{5} + \arctan \frac{1}{8}.$$

It follows that

$$\frac{\pi}{4} = 2 \arctan \frac{1}{2} - \arctan \frac{1}{7} \qquad (∗)$$

and

$$\frac{\pi}{4} = 2\arctan\frac{1}{3} + \arctan\frac{1}{7} \qquad (*)$$
$$= 2\arctan\frac{1}{5} + \arctan\frac{1}{7} + 2\arctan\frac{1}{8}.$$

The first equality in this formula was given by HUTTON, in 1776, and EULER, in 1779, while the second form is attributed to G. VON VEGA (1756–1802) in 1794.

Beginning with $u = \frac{120}{119}$, so $v = -\frac{1}{239}$, hence

$$\pi/4 = \arctan\frac{120}{119} - \arctan\frac{1}{119} = 4\arctan\frac{1}{5} - \arctan\frac{1}{239} \qquad (*)$$

which is J. MACHIN's (1680–1752) formula.

The series in the above formulas converge much more rapidly, and allowed the computation of π to many decimals: J. MACHIN (1706) with 100 decimals, F. DE LAGNY (1719) with 127 decimals, and G. VON VEGA (1789) with 136 decimals.

The same method was used by the prodigy calculator Z. DAHSE, author of an extended table of primes. In 1844, in two months, DAHSE computed π to 200 decimals. The astronomer T. CLAUSEN went to 248 decimals in 1847. Later efforts in the last century were, among others, by W. RUTHERFORD, in 1853, up to 400 decimals, RICHTER, in 1855, up to 500 decimals, and W. SHANKS, in 1873, up to 707 decimals (but "only" 527 decimals were correct, as found out later).

With the advent of computers, π was calculated in 1962 by D. SHANKS and J. W. WRENCH up to 100 000 decimals using the formula

$$\frac{\pi}{4} = 6\arctan\frac{1}{8} + 2\arctan\frac{1}{57} + \arctan\frac{1}{239}.$$

The calculations were extended by J. GUILLOUD and M. BOUYER who computed π to one million decimals in 1974, using GAUSS' formula

$$\frac{\pi}{64} = \frac{3}{4}\arctan\frac{1}{18} + \frac{1}{2}\arctan\frac{1}{57} - \frac{5}{16}\arctan\frac{1}{239}.$$

For the history of calculation of π up to 1960, see the paper of WRENCH (1960) and, more particularly, to the special issue on π of the magazine *Petit Archiméde* (1980) which is quite well documented and interesting to read.

An explosive progress in the calculation of decimals of π came about with the implementation of the old arithmetic-geometric mean of GAUSS as proposed by the brothers BORWEIN. These matters are discussed in their very stimulating book *Pi and the AGM* (1987); see also the article of BAILEY, BORWEIN, BORWEIN, and PLOUFFE (1997) . At present, 50 billion decimals of π have been calculated. This record was attained by Y. KAHADA and D. TAKAHASHI in July 1997.

3.

14159	26535	89793	23846	26433	83279	50288	41971	69399	37510
58209	74944	59230	78164	06286	20899	86280	34825	34211	70679
82148	08651	32823	06647	09384	46095	50582	23172	53594	08128
48111	74502	84102	70193	85211	05559	64462	29489	54930	38196
44288	10975	66593	34461	28475	64823	37867	83165	27120	19091
45648	56692	34603	48610	45432	66482	13393	60726	02491	41273
72458	70066	06315	58817	48815	20920	96282	92540	91715	36436
78925	90360	01133	05305	48820	46652	13841	46951	94151	16094
33057	27036	57595	91953	09218	61173	81932	61179	31051	18548
07446	23799	62749	56735	18857	52724	89122	79381	83011	94912
98336	73362	44065	66430	86021	39494	63952	24737	19070	21798
60943	70277	05392	17176	29317	67523	84674	81846	76694	05132
00056	81271	45263	56082	77857	71342	75778	96091	73637	17872
14684	40901	22495	34301	46549	58537	10507	92279	68925	89235
42019	95611	21290	21960	86403	44181	59813	62977	47713	09960
51870	72113	49999	99837	29780	49951	05973	17328	16096	31859
50244	59455	34690	83026	42522	30825	33446	85035	26193	11881
71010	00313	78387	52886	58753	32083	81420	61717	76691	47303
59825	34904	28755	46873	11595	62863	88235	37875	93751	95778
18577	80532	17122	68066	13001	92787	66111	95909	21642	...

Pour en savoir plus... tournez cette carte

One wonders what, if any, information of interest can be obtained from such extensive calculations? A place in *The Guiness Book of Records*? Is there anything to do with atomic energy that would be revealed by special patterns of the decimal digits of π? As a matter or fact, in June 1949, J. VON NEUMANN suggested using ENIAC to determine many decimals for π and e and study the distribution of decimal digits statistically. It has been observed that any digit, or for that matter, any short sequence of digits like 40533, occurs among the decimals of π. For example the sequence 0123456789 occurs beginning at the 17 387 594 880th digit after the decimal point. Of course, this is an experimental observation and does not constitute or lead to a proof of the "normality" of π.

As seen above, the computations of π have used formulas like the ones marked with (*). In this respect, D. GRAVÉ from Saint-Petersburg asked which of the formulas

$$m \arctan \frac{1}{x} + n \arctan \frac{1}{y} = k \frac{\pi}{4}$$

are true, where m, n, k, x, y are nonzero integers, and k, x, $y \geq 1$.

This problem, related to the speedy calculation of π, provided a link with the equation $X^4 - 1 = Y^n$, studied by S. SELBERG.

In 1897, STØRMER solved GRAVÉ's problem; he gave a simpler proof in 1899. Writing $(a_1 + ib_1) \cdots (a_n + ib_n) = re^{i\phi}$ (a_j, b_j, ϕ real and r real positive), then $\phi = \arctan \frac{b_1}{a_1} + \cdots + \arctan \frac{b_n}{a_n}$. Thus, $\arctan \frac{b_1}{a_1} + \cdots + \arctan \frac{b_n}{a_n} = s\pi$ (where s is an integer) if and only if $(a_1 + ib_1) \cdots (a_n + ib_n)$ is a real number. In particular, $m \arctan \frac{1}{x} + n \arctan \frac{1}{y} = k\frac{\pi}{4}$ if and only if $(1 - i)^k (x + i)^m (y + i)^n$ is real. This leads to the solution in integers of the equations $1 + X^2 = Y^n$ or $1 + X^2 = 2Y^n$ (where $n \geq 3$, n odd). As we have seen in **(3.3)**, the first of these equations has only trivial solutions. As for the second, STØRMER showed:

(3.7) If n is not a power of 2 and $n > 1$, then the only solutions of $1 + X^2 = 2Y^n$ are $x = \pm 1$.

We note, incidentally, that the equation $1 + X^2 = 2Y^4$ had been solved by LAGRANGE (1777) by the method of descent, the only solutions in natural numbers being $x = 239$, $y = 13$ (and the trivial solutions $x = 1$, $y = 1$).

Using **(3.7)**, STØRMER showed that the only solutions for GRAVÉ's problem were the expressions already indicated with $(*)$. So, somehow, just by trial, the significant formulas involving two arctangents had already been found!

After this aside, we return to SELBERG's result.

(3.8) If $n \geq 2$, the equation $X^4 - Y^n = 1$ has only trivial solutions.

PROOF. This is easy to see when n is even, so we assume n to be odd. If y is odd, then $y^n = x^4 - 1 = (x^2 + 1)(x^2 - 1)$, hence there exist natural numbers a, b such that

$$x^2 + 1 = a^n$$
$$x^2 - 1 = b^n.$$

By subtracting, we easily reach a contradiction. If y is even, then

$$x^2 + 1 = 2a^n$$
$$x^2 - 1 = 2^{n-1}b^n$$

(with a, b integers). By STØRMER's result, $x^2 = 1$, $y = 0$, which is a trivial solution. $\qquad\square$

The next developments in the study of the equation $X^2 - Y^n = 1$ were fruitless attempts, false routes, and wanderings. It is, however, interesting to report on these partial results, since some of the methods were used again for other exponents.

In 1934, NAGELL showed that if $X^2 - Y^q = 1$ (with q prime, $q > 3$) has a non-trivial solution, then $q \equiv 1 \pmod 8$. Moreover, there are at most finitely many solutions, as follows from a general theorem of THUE, which will be discussed in the last section of this paper. In 1940/1941, OBLÁTH, inspired by theorems of WIEFERICH and MIRIMANOFF on Fermat's equation $X^p + Y^p = Z^p$, showed that if $X^2 - Y^q = 1$ has a non-trivial solution, then

$$2^{q-1} \equiv 1 \pmod{q^2} \quad \text{and} \quad 3^{q-1} \equiv 1 \pmod{q^2}.$$

As is known (and as we shall discuss later in this chapter), the above congruences are very rarely satisfied.

INKERI and HYYRÖ (1961) showed that if $x^2 - y^q = 1$, then $q^2 \mid x$, $q^3 \mid y + 1$; moreover, they improved on estimates of OBLÁTH (1941, 1954), showing that

$$x > 2^{q(q-2)} > 10^{3 \times 10^9} \quad \text{and} \quad y > 4^{q-2} > 10^{6 \times 10^5}.$$

Finally, as was becoming believable, it was shown by CHAO KO in two papers of 1960 and 1964 that $X^2 - Y^n = 1$ has only trivial solutions.

The proof was subsequently very much simplified by CHEIN (1976). It is based on elementary, though non-trivial, results. The first one concerns the equation $X^2 - DY^2 = 1$, with $D > 0$, D not a square.

In 1657, FERMAT stated in a letter to FRÉNICLE that this equation has infinitely many solutions in integers but, as usual, he gave no proof. For a history of this important equation, consult DICKSON's, *History of the Theory of Numbers, Vol. II* (1920) and HEATH's *Diophantus of Alexandria* (1885).

EULER contributed to the theory of this equation but he is also responsible for ascribing it to PELL, when it should rightfully be called Fermat's equation—another Fermat's equation!

LAGRANGE used the theory of continued fractions to prove that the equation has indeed infinitely many solutions in integers. He also applied his methods to prove his famous theorem that the real roots of quadratic equations have periodic regular continued fraction developments, and conversely.

Here is a brief summary of some of the more relevant properties of the solutions of $X^2 - DY^2 = 1$ ($D > 0$, D not a square):

(a) Besides the trivial solutions $x = \pm 1$, $y = 0$, it has an infinity of solutions; moreover, there exists a solution (x_1, y_1) with $y_1 > 0$, y_1 minimum possible.

(b) For every integer $n \neq 0$ let x_n, y_n be positive integers defined by $x_n + y_n\sqrt{D} = (x_1 + y_1\sqrt{D})^n$; then $x_n^2 - Dy_n^2 = 1$.

(c) Conversely, if x, y are positive integers such that $x^2 - Dy^2 = 1$, then there exists an integer $n \neq 0$ such that $x = x_n$, $y = y_n$.

(d) If D is square-free, then the solutions correspond to the units $x + y\sqrt{D}$ of $\mathbb{Q}(\sqrt{D})$ having norm equal to 1. (If $D \equiv 1 \pmod 4$, the units $\frac{x+y\sqrt{D}}{2}$ with $x \equiv y \equiv 1 \pmod 2$ having norm 1 correspond to the solutions of the equation $X^2 - DY^2 = 4$).

The unit $x_1 + y_1\sqrt{D}$ is called the *fundamental unit* of $\mathbb{Q}(\sqrt{D})$.

STØRMER showed in 1897 (and more simply in 1908) the following interesting lemma:

If (x_n, y_n) is a solution of $X^2 - DY^2 = 1$, with $n > 1$, then there exists some prime dividing y_n but not dividing D.

STØRMER also proved a similar result for the equation $X^2 - DY^2 = -1$; this was a bit easier.

Based on STØRMER's result, NAGELL showed in 1921, and again in 1924, the following divisibility criterion:

If $x^2 - y^q = 1$ (q prime, $q > 3$), then $2 \mid y$ and $q \mid x$. (As we shall see, this was later generalized by CASSELS).

CHAO KO's proof of the following result appeared in two installments (1960, 1964) and has now been replaced by the elegant proof by CHEIN (1976):

(3.9) The equation $X^2 - Y^q = 1$ has no solution in non-zero integers.

CHEIN's two-page proof appeared in the *American Mathematical Monthly*.

4 Divisibility properties

The guiding idea for the propositions in this section is to assume that there exist non-zero integers x, y such that $x^p - y^q = 1$ and to derive divisibility conditions which must be satisfied by x, y, p, q. These conditions should be so restrictive that they would preclude the existence of solutions. For example, GERONO showed in 1870/71:

(4.1) If q is a prime number, if $q^m - y^n = 1$ with $m, n \geq 2$, then $q = 3$, $y = 2$. Similarly, if p is a prime number, and if $x^m - p^n = 1$ with $m, n \geq 2$, then $p = 2$, $x = 3$.

This fact was proved again and again, for example by CATALAN (1885), CARMICHAEL (1909), CASSELS (1953), and ROTKIEWICZ (1960).

OBLÁTH indicated in 1941 a slight extension concerning the types of prime factors of x, y, assuming that $x^m - y^n = 1$ (with $m, n \geq 2$). See also HAMPEL (1960).

By far the most important result on divisibility conditions of hypothetical solutions is due to CASSELS (1960). It is very easy to state, and at first sight it is hard to anticipate that it plays such an important role in the study of Catalan's equation.

(4.2) If p, q are odd primes, $x, y \geq 2$, and $x^p - y^q = 1$, then $p \mid y$, $q \mid x$.

The proof uses EULER's Lemmas 1 and 2 and delicate estimates, but it remains strictly elementary and no appeal is made to high-powered theorems. In a short space we cannot give any intelligible sketch of the proof.

An immediate corollary is the solution by MĄKOWSKI (1962) of problem (III) of the Introduction; this was also done independently by HYYRÖ (1963).

(4.3) Three consecutive integers cannot be proper powers.

PROOF.   Assuming the contrary, we have $x^l - y^p = 1$, $y^p - z^q = 1$, where x, y, z are natural numbers and the exponents l, p, q may be assumed primes without loss of generality. By Cassels' theorem, $p \mid x, p \mid z$, so $p \mid x^l - z^q = 2$. Thus $x^l - y^2 = 1$, which, by LEBESGUE's result **(3.3)**, is impossible.   □

We now make an aside to apply Cassels' theorem to divisibility properties of the numbers of FERMAT and FERENTINOU-NICOLACOPOULOU.

The *nth Fermat number* is $F_n = 2^{2^n} + 1$ $(n \geq 0)$, thus $F_0 = 3$, $F_1 = 5$, $F_2 = 17$, $F_3 = 257$, $F_4 = 65537$, while F_5 has about 10 digits, etc.

FERMAT expressed the belief, and proposed as a problem (letter of October 18, 1640) to prove that all Fermat numbers are primes, this being true for F_n when $n \leq 4$: for F_5 and larger Fermat numbers, FERMAT was unable to perform explicit calculations, due to a lack of extended tables of primes.

However, EULER showed:

If p is a prime, and $p \mid F_n$, then $p = 2^{n+2}k+1$ (for some integer k).

Thus, with this criterion, for $n = 5$ it was enough to test the primes congruent to 1 modulo 128. In this way, EULER deduced in 1732 that

$$F_5 = 641 \times 6700417.$$

Thus, we see that FERMAT was wrong! But this was not only an accident. Indeed, up to now, all other Fermat numbers investigated are composite—in fact, they are square-free.

There is an interesting connection discovered by ROTKIEWICZ (1965) and WARREN and BRAY (1967) with the so-called Fermat quotient with base 2:

$$q_p(2) = \frac{2^{p-1} - 1}{p}.$$

Namely: if $p \mid F_n$, then $p^2 \mid F_n$ if and only if $(2^{p-1}-1)/p \equiv 0 \pmod{p}$, that is, $2^{p-1} \equiv 1 \pmod{p^2}$.

The latter congruence is very rare, as we have already mentioned. For $p < 4 \times 10^{12}$, we have $2^{p-1} \not\equiv 1 \pmod{p^2}$ except when $p = 1093$, 3511.

SCHINZEL and SIERPIŃSKI conjectured in 1958 that there exist infinitely many square-free Fermat numbers. This is much weaker than the conjecture of EISENSTEIN (1844) that there exist infinitely many prime Fermat numbers. If Schinzel's conjecture is true, since distinct Fermat numbers are relatively prime, the existence of infinitely many primes p such that $2^{p-1} \not\equiv 1 \pmod{p^2}$ follows. And, in turn, by a famous theorem of WIEFERICH (see my book, 1979), there would exist infinitely many primes p such that the *first case* of Fermat's Last Theorem holds for the exponent p, that is:

"there do not exist integers x, y, z, not multiples of p, such that $x^p + y^p = z^p$".

Even though Fermat's Last Theorem has been proved by WILES in 1994, the above connection with a special case of Fermat's theorem is still intriguing.

Deep waters!

Now we introduce the numbers of FERENTINOU-NICOLACOPOULOU (1963). If $a \geq 2$, $n \geq 0$, let

$$F_{a,n} = a^{a^n} + 1.$$

The following result is an easy corollary of Cassels' theorem (RIBENBOIM (1979b)):

(4.4) $F_{a,n}$ is not a proper power.

PROOF. If $F_{a,n}$ is a proper power, we may write $a^{a^n} + 1 = m^p$ for some prime p. If q is a prime dividing a, let $a^n = qa'$, hence $m^p - (a^{a'})^q = 1$, and so $q \mid m$, which is impossible.   □

In particular, F_n cannot be a proper power; this special case needs only Lebesgue's theorem.

Another consequence (RIBENBOIM (1979b)) is the following fact which is a slight improvement over **(4.1)**:

(4.5) If $x^p - y^q = 1$ with $p, q > 3$, then x, y have at least two odd prime factors. We now focus on the sharpening of Cassels' theorem by HYYRÖ AND INKERI

If p, q are primes and $x^p - y^q = 1$, from the previous results we must have $p, q > 3$ and $p \mid y$, $q \mid x$. Then EULER's Lemma 1 becomes:

$$x - 1 = p^{q-1}a^q$$

$$\frac{x^p - 1}{x - 1} = pu^p \quad \text{with} \quad p \nmid u, y = pau,$$

$$y + 1 = q^{p-1}b^p$$

$$\frac{y^q + 1}{y + 1} = qv^p \quad \text{with} \quad q \nmid v, x = qbv.$$

HYYRÖ showed in 1964:

(4.6) With the above notations:

$$a = qa_0 - 1, \quad b = pb_0 + 1 \quad \text{(with } a_0 \geq 1, b_0 \geq 1\text{)}$$
$$x \equiv 1 - p^{q-1} \pmod{q^2}, \quad y \equiv -1 + q^{p-1} \pmod{p^2};$$

thus $q^2 \mid x$ if and only if $p^{q-1} \equiv \pmod{q^2}$, and $p^2 \mid y$ if and only if $q^{p-1} \equiv 1 \pmod{p^2}$.

Thus, in particular, $a \geq q - 1$, $b \geq p + 1$. In view of later estimates of x, y, HYYRÖ showed also:

(4.7) If $m > 3$ is composite and $x^m - y^q = 1$ (where q is a prime, $q > 3$), and if p is any prime dividing m, then $p^{q-1} \equiv 1 \pmod{q^2}$ and also $q^2 \mid x$.

The next result of INKERI is quite interesting in that it establishes a connection with the class number of imaginary quadratic fields. It also uses an old result of GAUSS on the cyclotomic polynomial.

In *Disquisitiones Arithmeticae*, article 357 (1801), GAUSS showed:

If p is an odd prime, there exist polynomials $F, G \in \mathbb{Z}[X]$ such that

$$4\frac{X^p - 1}{X - 1} = F(X)^2 - (-1)^{\frac{p-1}{2}} pG(X)^2.$$

Incidentally, this proposition was used by GAUSS in the determination of the sign of the Gauss sum:

$$\tau = \sum_{j=1}^{p-1} \left(\frac{j}{p}\right) \zeta^j = \begin{cases} \sqrt{p} & \text{when } p \equiv 1 \pmod 4, \\ i\sqrt{p} & \text{when } p \equiv 3 \pmod 4. \end{cases}$$

It took four years for GAUSS to find the solution of this problem. It was not until 1805 when suddenly, "like lightning the solution appeared to him" (as he stated in a letter to his friend, the astronomer OLBERS).

If p is an odd prime, let $H(-p)$ denote the class number of the imaginary quadratic field $\mathbb{Q}(\sqrt{-p})$.

GUT showed in 1963 that $H(-p) < \frac{p}{4}$ and this inequality was used by INKERI. It basically reflects the fact that $H(-p)$ does not grow fast, as was shown before by SIEGEL (1936):

$$\log H(-p) \sim \log \sqrt{p}.$$

Here is INKERI's result (1964):

(4.8) With the same notations, if $x^p - y^q = 1$, then:

(a) If $p \equiv 3 \pmod 4$ and $q \nmid H(-p)$, then $q^2 \mid x$, $y \equiv -1 \pmod{q^{2p-1}}$, $p^{q-1} \equiv 1 \pmod{q^2}$.

(b) If $p \equiv q \equiv 3 \pmod 4$, $p > q > 3$ and $q \nmid H(-p)$, then $q^2 \mid x$, $p^2 \mid y$, $x \equiv 1 \pmod{p^{2q-1}}$, $y \equiv -1 \pmod{q^{2p-1}}$, $p^{q-1} \equiv 1 \pmod{q^2}$, and $q^{p-1} \equiv 1 \pmod{p^2}$.

INKERI used these divisibility conditions and congruences, together with RIESEL's tables (1964) for the residues of p^{q-1} modulo q^2, q^{p-1} modulo p^2, to show, for example: among the 946 pairs of primes (p, q) with $p \neq q$, $5 \leq p, q \leq 199$, there are 718 pairs for which the equation $X^p - Y^q = 1$ has only the trivial solution.

This work was continued in two papers of INKERI in 1990 and 1991 (one co-authored by AALTONEN). Among the many criteria in these papers, we single out the following:

(4.9) Let p, q be distinct odd primes and assume that there exist natural numbers x, y such that $x^p - y^q = 1$. Then

(i) If q does not divide h_p (the class number of the cyclotomic field of pth roots of 1), then $q^2 \mid x$ and $p^{q-1} \equiv 1 \pmod{q^2}$.

(ii) If p does not divide h_q (the class number of the cyclotomic field of qth roots of 1), then $p^2 \mid y$ and $q^{p-1} \equiv 1 \pmod{p^2}$.

This criterion is appropriate for computation and has been used to show that for many pairs of distinct odd primes (p, q) the equation $x^p - y^q = 1$ has no solution in positive integers. For example, in this way INKERI was able to show that $x^5 - y^7 = \pm 1$ has no solution in positive integers.

Another type of result, concerning solutions (x, y) of $x^m - y^n = 1$ for which $|x - y| = 1$, was obtained by HAMPEL in 1956. More generally, and with a very easy and elegant proof, ROTKIEWICZ showed in the same year:

(4.10) If $a \geq 1$ is an integer, if $\gcd(x, y) = 1$, $|x - y| = a$ and $x^m - y^n = a^n$, then $x = 3$, $y = 2$, $m = 2$, $n = 3$, $a = 1$.

The proof is based on a theorem of BANG (1886) and ZSIGMONDY (1892), see BIRKHOFF and VANDIVER (1904):

If $a > b \geq 1$, $\gcd(a, b) = 1$, for every $n > 1$ there exists a prime p such that $p \mid a^n - b^n$, but $p \nmid a^m - b^m$ for all m, $1 \leq m < n$ (except when $a = 2$, $b = 1$, $n = 6$, or $n = 2$, $a - b = 2$ and $a + b$ is a prime of 2).

5 Estimates

In this section, we shall indicate estimates for the size and number of solutions of Catalan's equation, assuming that non-trivial solutions exist.

First, given distinct integers $a, b \geq 2$ we look for solutions in natural numbers u, v of the equation $a^U - b^V = 1$.

Second, we consider fixed exponents $m, n \geq 2$ and examine the possible solutions of the equation $X^m - Y^n = 1$.

Finally, we shall consider the solutions in natural numbers of the exponential diophantine equation $X^U - Y^V = 1$.

A   The equation $a^U - b^V = 1$

LEVEQUE showed in 1952 the following result, which may also be obtained as an easy consequence of HAMPEL's result (see **(4.9)**):

(5.1) If $a, b \geq 2$, then $a^U - b^V = 1$ has at most one solution in natural numbers u, v, unless $a = 3$, $b = 2$ when there are two solutions $u = v = 1$ and $u = 2$, $v = 3$.

A somewhat interesting corollary concerns the sums of successive powers of integers:

$$S_1(n) = \sum_{j=1}^{n} j = \frac{n(n+1)}{2}$$

$$S_2(n) = \sum_{j=1}^{n} j^2 = \frac{n(n+1)(2n+l)}{6}$$

$$S_3(n) = \sum_{j=1}^{n} j^3 = \frac{n^2(n+1)^2}{4}, \quad \text{etc.} \ldots$$

More generally, $S_k(n) = \sum_{j=1}^{n} j^k$ is given by a polynomial expression of degree $k + 1$, with coefficients having denominator dividing $k + 1$ and expressible in terms of the Bernoulli numbers—which, incidentally, is irrelevant to the present purpose. As we saw above, $S_3(n) = [S_1(n)]^2$ for every $n \geq 1$.

The corollary to LEVEQUE's result is the following:

(5.2) If $t \geq 1$ and $u, v \geq 2$ are such that for every $n \geq 1$: $S_v(n) = [S_1(n)]^u$, then $v = 3$, $t = 1$, $u = 2$.

This holds just because the only solution in natural numbers of $2^V + 1 = (2^T - 1)^U$ is $t = 1$, $u = 2$, $v = 3$.

Let us note here that as early as 1908, THUE had obtained the following result:

If $E = \{p_1, \ldots, p_r\}$, where $r \geq 2$ and each p_i is a prime number, and if S is the set of natural numbers all of whose prime factors are in E, then for every $k \geq 2$ there exist at most finitely many integers $z, z' \in S$ such that $z - z' = k$. In particular, if $a, b \geq 2$, $k \geq 1$ the equation $a^U - b^V = k$ has at most finitely many solutions in integers u, v. This was obtained again by PÓLYA in 1918. In 1931, PILLAI indicated a quantitative form of this theorem, giving an upper bound for the number of solutions in natural numbers of the inequalities $0 < a^u - b^v \leq k$ (where $\frac{\log b}{\log a}$ is irrational). Later, in 1936, HERSCHFELD showed that $2^U - 3^V = 1$ has at most one solution for each sufficiently large k; PILLAI extended this result in 1936 for any bases $a, b \geq 2$.

CASSELS indicated in 1953 an algorithm to compute the solution of $a^U - b^V = 1$, if one exists, giving in this way a new proof of LEVEQUE's result.

(5.3) Let $a, b \geq 2$, and let A (respectively B) be the product of the distinct odd primes dividing a (respectively b). If $u, v \geq 2$ are such that $a^u - b^v = 1$, then:

(a) either $a = 3$, $b = 2$, $u = 2$, $v = 3$, or

(b) u, v are the smallest natural numbers such that $a^u \equiv 1 \pmod{B}$ and $b^v \equiv -1 \pmod{A}$.

Thus, we need only test these values u, v as possible solutions.

B The equation $X^m - Y^n = 1$

Our aim is to make statements about the number and size of solutions of this equation. The proof that this equation has only finitely many solutions may be achieved in the following ways:

(a) by showing that the existence of infinitely many solutions leads to a contradiction;

(b) by determining explicitly an integer $N \geq 1$ such that the number of solutions is at most equal to N;

(c) by determining explicitly some integer $C \geq 1$ such that every solution (x, y) must satisfy $x \leq C$, $y \leq C$. By trying all possible natural numbers up to C, it is possible to identify all the solutions.

In case (a) there is no indication of how many, or how large the solutions are.

In case (b), there is no indication of how large the solutions are; thus, even if $N - 1$ solutions are already known, nothing may be inferred about whether any other solution exists, or how large it may be.

Finally, case (c) is the most satisfactory. Yet, if the constant C provided by the method of proof is much too large—as is often the case—it is impossible to identify all the solutions.

Our first result is an easy consequence of a powerful and classical theorem that goes back to SIEGEL (1929) and is based on ideas of THUE about diophantine approximation.

It is convenient to use the following more explicit form of Siegel's theorem, indicated by INKERI and HYYRÖ (1964b) (see also a relevant paper by LEVEQUE (1964)):

Let $m, n \geq 2$ with $\max\{m, n\} \geq 3$. Let $f(X) \in \mathbb{Z}[X]$ have degree n and assume that all its zeroes are simple. If a is a non-zero integer, then the equation $f(X) = aY^m$ has at most finitely many solutions.

In particular:

(5.4) For every natural number k, the equation $X^m - Y^n = k$ has at most finitely many solutions.

This result may also be proved as a consequence of the following interesting theorem of MAHLER (1953):

If a, b are non-zero integers, $x, y \geq 1$, $\gcd(x, y) = 1$, $m \geq 2$, $n \geq 3$, then the greatest prime factor of the number $ax^m - by^n$ tends to infinity, as $\max\{x, y\}$ tends to infinity.

In particular, if x, y are sufficiently large, then $x^m - y^n$ cannot be equal to k.

A more elementary proof of a special case of **(5.3)**, due to HYYRÖ (1964), results from an application of a theorem of DAVENPORT and ROTH (1955) on diophantine approximation. Without giving any more indication of the method used, we state HYYRÖ's result.

(5.5) The number of solutions of $X^m - Y^n = 1$ is at most $\exp\{631m^2n^2\}$.

This upper bound is quite large, especially in light of the conjecture that there are no solutions!

Moreover, HYYRÖ showed:

(5.6) If p, q are primes, $x, y \geq 2$, and $x^p - y^q = 1$, then $x, y > 10^{11}$.

So, the solutions cannot be too small. Moreover, if one of the exponents is composite, then:

(5.7) If $x^m - y^n = 1$ and m is composite, then $x > 10^{84}$, while if n is composite, then $y > 10^{84}$.

And it is even worse (or better?) when m, n are both composite:

(5.8) If $x^m - y^n = 1$ and m, n are composite, then x^m, $y^n > 10^{10^9}$.

HYYRÖ has also indicated an algorithm to find the solutions (if any) of $X^p - Y^q = 1$, where p, q are primes. It involves regular continued fraction developments. If α is a positive real number, we define successively the integers c_0, c_1, c_2, ..., and the positive real numbers α_1, α_2, ... by the relations

$$\alpha = c_0 + \frac{1}{\alpha_1} \quad \text{where } c_0 = [\alpha], \text{ so } \alpha_1 > 1,$$

$$\alpha_1 = c_1 + \frac{1}{\alpha_2} \quad \text{where } c_1 = [\alpha_1], \text{ so } \alpha_2 > 1,$$

$$\alpha_2 = c_2 + \frac{1}{\alpha_3} \quad \text{where } c_2 = [\alpha_2], \text{ so } \alpha_3 > 1,$$

etc. Thus

$$\alpha = c_0 + \cfrac{1}{c_1 + \cfrac{1}{c_2 + \cfrac{1}{c_3 + \cdots + \cfrac{1}{c_n} + \cdots}}}$$

and we write

$$\alpha = [c_0, c_1, c_2, \ldots, c_n, \ldots].$$

The above fraction is called the *regular continued fraction of* α. We define also

$$A_0 = c_0 \quad A_1 = c_0 c_1 + 1$$
$$B_0 = 1 \quad B_1 = c_1,$$

and for $2 \leq i \leq n$,

$$A_i = c_i A_{i-1} + A_{i-2},$$
$$B_i = c_i B_{i-1} + B_{i-2}.$$

In particular, $B_0 \leq B_1 < B_2 < B_3 < \cdots$.

The fractions A_i/B_i are called the *convergents* of α, and $A_i/B_i = [c_0, c_1, \ldots, c_i]$ for every $i \geq 0$

We recall some basic properties:

(a) for every $i \geq 0$ we have $\gcd(A_i, B_i) = 1$,

(b) $A_i/B_i \leq \alpha$ if and only if i is even.

LAGRANGE showed in 1798:

(c) For every $i \geq 0$,

$$\frac{1}{B_i(B_i + B_{i+1})} < \left| \alpha - \frac{A_i}{B_i} \right| < \frac{1}{B_i^2}.$$

(d) If a, b are non-zero integers, $b \geq 1$, $\gcd(a, b) = 1$, and $\left| \alpha - \frac{a}{b} \right| < \frac{1}{2b^2}$, then there exists $i \geq 0$ such that $a = A_i$, $b = B_i$.

HYYRÖ's result is the following:

(5.9) Let p, q be distinct odd primes. If there exist integers $x, y \geq 2$ such that $x^p - y^q = 1$, they may be found by the following algorithm. Let

$$\alpha = \frac{q^{\frac{p-1}{p}}}{p^{\frac{q-1}{q}}};$$

consider its regular continued fraction development:

$$\alpha = [c_0, c_1, c_2, \ldots].$$

Let A_i/B_i be the convergents. Then any solution is of the form

$$x = p^{q-1} A_i^q + (-1)^i, \quad y = q^{p-1} B_i^p - (-1)^i,$$

where $i \geq 0$ is any index such that:

(i) $A_i > 1$, $B_i > 1$,

(ii) $A_i \equiv (-1)^{i+1} \pmod{q}$, $B_i \equiv (-1)^i \pmod{p}$,

(iii) $A_i \equiv (-1)^i \frac{q^{p-1}-1}{p} \pmod{p}$, $B_i \equiv (-1)^{i+1} \frac{p^{q-1}-1}{q} \pmod{q}$,

(iv) $c_{i+1} \geq (-1)^{i+1} A_i^{r-2}$, $c_{i+1} \geq (-1)^i B_i^{r-2}$, where $r = \min\{p, q\}$.

This algorithm will not determine if one non-trivial solution exists. But, if one exists, it will eventually find it.

In his paper, HYYRÖ also obtained other results on Catalan's equation as a consequence of his study of the exponential-diophantine equation $X^n - d^U Y^n = \pm 1$, where $n \geq 5$, $d \geq 2$ are given integers.

Having shown that this equation has at most one solution in integers u, x, y, with $0 \leq u < n$, $x \geq 2$, $y \geq 1$, he could prove:

(5.10) If p, q are odd primes, where $m > 2$, $e \geq q$ and p^e divides m, then $X^m - Y^q = \pm 1$ have no solutions in integers $x \geq 2$, $y \geq 1$.

(5.11) If $n \geq 5$, $a \geq 2$ are integers, then the exponential-diophantine equations $a^U - Y^n = \pm 1$ have at most finitely many solutions in integers.

(5.12) If $a \geq 2$, then the exponential-diophantine equations $a^U - Y^V = \pm 1$ have at most $(a+1)^\nu$ solutions in integers, where ν is the number of distinct prime factors of a.

C   The equation $X^U - Y^V = 1$

All the preceding is still not sufficient to conclude that the exponential-diophantine equation with four unknowns $X^U - Y^V = 1$ has at most finitely many solutions. In fact, this is true and was first shown by TIJDEMAN in 1976 using BAKER's estimate for linear forms in logarithms.

BAKER applied his estimates to give effective bounds for solutions of various types of diophantine equations (see, for example, his book, 1975): these results represented a definite improvement over the previous qualitative statements obtained by THUE, SIEGEL, and ROTH.

(5.13) If $m, n \geq 3$, $x, y \geq 1$, and $x^m - y^n = 1$, then

$$\max\{x, y\} < \exp\exp\{(5n)^{10}m^{10m}\}$$

and

$$\max\{x, y\} < \exp\exp\{(5m)^{10}n^{10n}\}.$$

TIJDEMAN showed:

(5.14) There exists a number $C > 0$ which may be effectively computed, such that if x, y, m, n are natural numbers, $m, n \geq 2$, and $x^m - y^n = 1$, then $\max\{x, y, m, n\} < C$.

TIJDEMAN's result may be viewed as "settling" the problem. Indeed, with this theorem the problem is shown to be decidable. It is now "only" a question of trying all the 4-tuples of natural numbers less than C. Here "only" still means too much because, as we shall indicate, the smallest value of C thus far obtained with the present method is much too large.

In 1996, LANGEVIN computed explicitly that

$$C < \exp\exp\exp\exp 730.$$

In recent years there has been considerable computational activity to enlarge the set of pairs of prime exponents (p, q) for which $x^p - y^q = 1$ is shown to be impossible when $|x|$, $|y| > 1$. These computations appeal to refined criteria.

A leader in these efforts is MIGNOTTE who has been publishing an extensive series of papers. We may single out his latest (still unpublished) paper which gives the state of the art. First we describe some of the newer criteria which are of two kinds.

On the one hand, there are more specific bounds for linear forms in logarithms. These are difficult to obtain and are technically involved. The reader may consult the papers by BENNETT, BLASS, GLASS, MERONK, and STEINER (1997) and by LAURENT, MIGNOTTE, and NESTERENKO (1995). These results are useful in finding upper bounds for the exponents.

After TIJDEMAN's Theorem, LANGEVIN (1976) showed that if $x^m - y^n = 1$ with m, n, x, $y > 1$, then $\max\{m, n\} < 10^{110}$.

Today it is known (O'NEIL (1995)):

(5.15) If $x^p - y^q = 1$, with p, q primes, x, $y > 1$, then $\max\{p, q\} < 3.18 \times 10^{17}$ and p or q is at most 2.60×10^{12}.

On the other hand, lower bounds for the possible exponents have been derived using criteria which are improvements of the groundbreaking results of INKERI already described in **(4.8)** and **(4.9)**; as stated in these references, if $p \equiv 1 \pmod 4$ the latter criterion required to ascertain whether q divides the class number h_p of the cyclotomic field $\mathbb{Q}(\zeta_p)$. It should be noted that up to now, h_p has been computed only for $p < 71$. The class number is the product of the factors $h_p = h_p^- h_p^+$ where h_p^+ is the class number of the real subfield $\mathbb{Q}(\zeta_p + \zeta_p^{-1})$—and this factor is the one that is so difficult to compute.

Some of the computational work was alleviated with the following criterion of MIGNOTTE and ROY (1993).

(5.16) Let $p - 1 = 2^d s$ where s is an odd integer. Let K' be the subfield of $\mathbb{Q}(\zeta_p)$ having degree 2^d, and let h' be its class number. If $x^p - y^q = 1$ with p, q prime, $|x|, |y| > 1$, and if $p \equiv 1 \pmod 4$, then q divides h' or $p^{q-1} \equiv 1 \pmod{q^2}$.

In 1994, SCHWARZ gave a criterion involving only the easier-to-compute factor h_p^- thereby implying a substantial extension of the computations:

(5.17) With the preceding notations, either $p^{q-1} \equiv 1 \pmod{q^2}$ or q divides the first factor h_p^- of the class number of K'.

In 1999, BUGEAUD and HANROT derived further necessary requirements for the largest exponent:

(5.18) If $p < q$, with the above notations q divides the class number h_p of $\mathbb{Q}(\zeta_p)$.

The most remarkable improvement has just been made by AILESCU (1999):

(5.19) With the previous notations, both $p^{q-1} \equiv 1 \pmod{q^2}$ and $q^{p-1} \equiv 1 \pmod{p^2}$.

The amount of calculations has been drastically reduced. It should be pointed out that the smallest pairs of primes satisfying the above two congruences are $(p, q) = (83, 4871)$ and $(911, 318017)$. These pairs have been excluded by other considerations.

WIth these criteria, MIGNOTTE and ROY (1999) have shown:

(5.20) If $x^p - y^q = 1$ with p, q primes, $x, y > 1$, then p and q are greater than 10^7.

Another interesting fact has just been shown by MIGNOTTE (2000):

(5.21) If $x^m - y^n = \pm 1$, with $2 \le m < n$, $x, y > 1$, then m is a prime number and n admits at most two non-trivial factors.

All this having been siad, it is still true that the amount of calculation needed remains ... (Is there a word to dwarf "astronomical"? Invent one).... "Catalanic!"

In conclusion, up to now it is still not known whether 8 and 9 are the only consecutive powers.

6   Final comments and applications

A simple corollary of the preceding result **(5.3)** is the following: Let $m, n \geq 2$ be distinct integers, let $z_1 < z_2 < z_3 < \cdots$ be the sequence of natural numbers which are either an mth power or an nth power of a natural number. Then $\lim_{i \to \infty} (z_{i+1} - z_i) = \infty$.

For $n = 2$ the above result is due to LANDAU and OSTROWSKI (1920) and for arbitrary m, n it was explicitly stated by INKERI and HYYRÖ in 1964.

The main *conjecture* referring to powers, which may be attributed to PILLAI (1936) or LANDAU, is the following:

If $k \geq 2$, there exist at most finitely many quadruples of natural numbers x, y, m, n, with $m \geq 2$, $n \geq 2$, such that $x^m - y^n = k$.

This conjecture may be equivalently stated as follows:

If $z_1 < z_2 < z_3 < \cdots$ is the sequence of powers of natural numbers, then $\lim_{i \to \infty} (z_{i+1} - z_i) = \infty$.

Now we consider the sequence of powers of given distinct numbers $a, b \geq 2$ or, more generally, the sequence S of natural numbers all of whose prime factors are in $E = \{p_1, \ldots, p_r\}$, $r \geq 2$, each p_i a prime number. We write

$$S : z_1 < z_2 < z_3 < \cdots,$$

so S is the sequence (d) of the Introduction.

In 1897, STØRMER showed that

$$\liminf_{i \to \infty} (z_{i+1} - z_i) \geq 2,$$

in other words, the equation $X - Y = 1$ has only finitely many solutions for $x, y \in S$.

STØRMER has indicated a constructive method to find all the solutions; see also the paper by LEHMER (1964). THUE's result of 1918, indicated after **(5.1)**, which is however not constructive, is the following:

$$\lim_{i \to \infty} (z_{i+1} - z_i) = \infty.$$

In 1965, ERDÖS proved that for every $\epsilon > 0$ there exists an i_0 such that if $i \geq i_0$, then

$$\frac{z_{i+1} - z_i}{z_i} > \frac{1}{z_i^\epsilon}.$$

In 1973 and 1974, TIJDEMAN showed that this result is, in a sense, the best possible. Indeed, there exist constants C, C', which may be effectively computed and depend only on the sequence S, and there exists an i_0 such that if $i \geq i_0$, then

$$\frac{1}{(\log z_i)^{C'}} \geq \frac{z_{i+1} - z_i}{z_i} \geq \frac{1}{(\log z_i)^C}.$$

Other results along the same lines, representing the present state of knowledge, are the following:

Let E be the set of primes less than N, and S the sequence of numbers all or whose prime factors are less than N.

Let $\tau > 0$. Then there exists an effectively computable constant C, depending only on N and τ, such that if $m \geq 2$, $n \geq 2$, $x \geq 2$, $y \geq 2$, if $\gcd(ax^m, k) \leq \tau$, if $|a|, |b|, |k| \in S$, and if

$$ax^m - by^n = k,$$

then $\max\{|a|, |b|, |k|, m, n, x, y\} < C$.

Similarly:

Let $\tau > 0$, $m \geq 2$. Then there exists an effectively computable constant C, depending only on N, τ and m, such that if $n \geq 2$, $x \geq 2$, $y \geq 2$, $mn \geq 5$, if $\gcd(ax^m, k) \leq \tau$, if $|a|, |b|, |k| \in S$ and $ax^m - by^n = k$, then $\max\{|a|, |b|, |k|, n, x, y\} < C$.

For more results along this line, see the important monograph of SHOREY and TIJDEMAN (1986).

References

1288–1344 Levi ben Gerson. See Dickson, L. E., *History of the Theory of Numbers*, Vol II, p. 731. Carnegie Institution, Washington, 1920. Reprinted by Chelsea Publ. Co., New York, 1971.

1640 P. Fermat. Lettre à Mersenne (Mai, 1640). In *Oeuvres, Vol. II*, 194–195. Gauthier-Villars, Paris, 1894.

1657 P. Fermat. Lettre à Frénicle (Février, 1657). In *Oeuvres, Vol. II*, 333–335. Gauthier-Villars, Paris, 1894.

1657 Frénicle de Bessy. Solutio duorum problematum circa numeros cubos et quadratos. Bibliothèque Nationale de Paris.

1732 L. Euler. Observationes de theoremate quodam Fermatiano aliisque ad numeros primos spectantibus. *Comm. Acad. Sci.*

Petrop., 6, 1732/3 (1738):103–107. Reprinted in *Opera Omnia*, Ser. I, Vol. II. Comm. Arithm., I, 1–5. B. G. Teubner, Leipzig, 1915.

1737 L. Euler. De variis modis circuli quadraturam numeros primos exprimendi. *Comm. Acad. Sci. Petrop.*, 9, 1737 (1744):222–236. Reprinted in *Opera Omnia*, Ser. I. Vol. XIV. Comm. Arithm., I, 245–259. B. G. Teubner, Leipzig. 1924.

1738 L. Euler. Theorematum quorundam arithmeticorum demonstrationes. *Comm. Acad. Sci. Petrop.*, 10, 1738 (1747): 125–146. Reprinted in *Opera Omnia*, Ser. I, Vol. II, Comm. Arithm., I, 38–58. B. G. Teubner, Leipzig. 1915.

1755 L. Euler. *Institutiones Calculi Differentialis*. Partis Posterioris (Caput V). Imp. Acad. Sci., St. Petersburg. Reprinted in *Opera Omnia*, Ser. I, Vol. X, 321–328. B. G. Teubner, Leipzig, 1913.

1777 J. L. Lagrange. Sur quelques problèmes de l'analyse de Diophante. *Nouveaux Mém. Acad. Sci. Belles Lettres, Berlin.* Reprinted in *Oeuvres*, Vol. IV, publiées par les soins de M. J.-A. Serret, 377–398, Gauthier-Villars, Paris. 1869.

1783 L. Euler. Variae observationnes circa angulos in progressione geometrica progredientes. *Opuscula Analytica, I*, 1783:345–352. Reprinted in *Opera Omnia*, Ser. I, Vol. XV. 498–508. B. G. Teubner, Leipzig, 1927.

1798 J. L. Lagrange. Addition aux "Eléments d'Algèbre" d'Euler—Analyse Indéterminée. Reprinted in *Oeuvres*, Vol. VII, publiées par les soins de M. J.-A. Serret, 3–180. Gauthier-Villars, Paris, 1877.

1801 C. F. Gauss. *Disquisitiones Arithmeticae*. G. Fleischer, Leipzig. Translated by A. A. Clarke, Yale Univ. Press, New Haven, 1966.

1830 A. M. Legendre. *Théorie des Nombres, Vol. II, (3 édition)*. Firmin Didot, Paris. Reprinted by A. Blanchard, Paris, 1955.

1844 E. Catalan. Note extraite d'une lettre addressée á l'éditeur. *J. reine u. angew. Math.*, 27:192.

1844 Z. Dase. Der Kreis-Umfang für den Durchmesser 1 auf 200 Decimalstellen berechnet. *J. reine u. angew. Math.*, 27:198.

1844 F. G. Eisenstein. Aufgaben und Lehrsätze. *J. reine*

u. angew. Math., 27:86–88. Reprinted in *Mathematische Werke*, Vol. I, 111-113, Chelsea, New York. 1975.

1850 V. A. Lebesgue. Sur l'impossibilité en nombres entiers de l'équation $x^m = y^2 + 1$. *Nouv. Ann. de Math.*, 9:178–181.

1870 G. C. Gérono. Note sur la résolution en nombres entiers et positifs de l'équation $x^m = y^n + 1$. *Nouv. Ann. de Math. (2)*, 9:469–471, and 10:204–206 (1871).

1885 E. Catalan. Quelques théorémes empiriques (Mélanges Mathématiques, XV). *Mém. Soc. Royale Sci. de Liége, Sér. 2*, 12:42–43.

1885 T. L. Heath. *Diophantus of Alexandria. A Study in the History of Greek Algebra.* Cambridge Univ. Press, Cambridge. Reprinted by Dover. New York. 1964.

1886 A. S. Bang. Taltheoretiske Untersogelser. *Tidskrift Math., Ser. 5*, 4:70–80 and 130–137.

1892 K. Zsigmondy. Zur Theorie der Potenzreste. *Monatsh. f. Math.*, 3:265–284.

1897 C. Størmer. Quelques théorèmes sur l'équation de Pell $x^2 - Dy^2 = \pm 1$ et leurs applications. *Christiania Videnskabens Selskabs Skrifter, Math. Nat. Kl.*, 1897, No. 2, 48 pages.

1899 C. Størmer. Solution complète en nombres entiers de l'équation $m \arctan \frac{1}{x} + n \arctan \frac{1}{y} = k\frac{\pi}{4}$. *Bull. Soc. Math. France*, 27:160–170.

1904 G. D. Birkhoff and H. S. Vandiver. On the integral divisors of $a^n - b^n$. *Ann. Math. (2)*, 5:173–180.

1908 A. Thue. Om en general i store hele tal ulösbar ligning. *Christiania Videnskabens Selskabs Skrifter, Math. Nat. Kl.*, 1908, No. 7, 15 pages. Reprinted in *Selected Matheinealcal Papers*, 219-231. Universitetsforlaget, Oslo, 1982.

1908 C. Størmer. Solution d'un problème curicux qu'on rencontre dans la théorie élémentaire des logarithmes. *Nyt Tidskrift f. Mat. (Copenhagen) B*, 19:1–7.

1909 R. D. Carmichael. Problem 155 (proposed and solved by R. D. Carmichael). *Amer. Math. Monthly*, 16:38–39.

1918 G. Pólya. Zur arithmetischen Untersuchung der Polynome. *Math. Z.*, 1:143–148.

1919 P. Bachmann. *Das Fermatproblem in seiner bisherigen Entwicklung.* W. De Gruyter, Berlin. Reprinted by Springer-Verlag, Berlin, 1976.

1920 L. E. Dickson. *History of the Theory of Numbers, Vol. II.* Carnegie Institution, Washington. Reprinted by Chelsea Publ. Co. New York, 1971.

1920 E. Landau and A. Ostrowski. On the diophantine equation $ay^2 + by + c = dx^n$. *Proc. London Math. Soc., (2)*, 19: 276–280.

1921a T. Nagell. Des équations indéterminées $x^2 + x + 1 = y^n$ et $x^2 + x + 1 = 3y^n$. *Norsk Mat. Forenings Skrifter, Ser. I*, 1921, No. 2, 14 pages.

1921b T. Nagell. Sur l'équation indéterminée $\frac{x^n-1}{x-1} = y^2$. *Norsk Mat. Forenings Skrifter, Ser. I*, 1921, No. 3, 17 pages.

1923 P. Franklin. Problem 2927. *Amer. Math. Monthly*, 30:81.

1924 T. Nagell. Über die rationale Punkte auf einigen kubischen Kurven. *Tôhoku Math. J.*, 24:48–53.

1929 C. L. Siegel. Über einige Anwendungen diophantischer Approximation. *Abhandl. Preuss. Akad. d. Wiss., No. I.* Reprinted in *Gesammelte Abhandlungen, Vol. 1*, 209–266. Springer-Verlag, Berlin, 1966.

1931 S. S. Pillai. On the inequality "$0 < a^x - b^y \leq n$". *J. Indian Math. Soc.*, 19:1–11.

1932 S. Selberg. Sur l'impossibilité de l'équation indéterminée $z^p + 1 = y^2$. *Norsk Mat. Tidsskrift*, 14:79–80.

1934 T. Nagell. Sur une équation diophantienne à deux indéterminées. *Det Kongel. Norske Vidensk. Selskab Forhandliger, Trondhejm*, 1934, No. 38, 136–139.

1936 A. Herschfeld. The equation $2^x - 3^y = d$. *Bull. Amer. Math. Soc.*, 42:231–234.

1936 S. S. Pillai. On $a^x - b^y = c$. *J. Indian Math. Soc., (New Series)*, 2:119–122.

1936 C. L. Siegel. Über die Classenzahl quadratischer Zahl-körper. *Acta Arith.*, 1:83–86. Reprinted in *Gesammelte Abhandlungen, Vol. I*, 406–409. Springer-Verlag, Berlin, 1966.

1940 R. Obláth. Az $x^2 - 1$ Számokól. (On the numbers $x^2 - 1$). *Mat. és Fiz. Lapok*, 47:58–77.

1941 R. Obláth. Sobre ecuaciones diofánticas imposibles de la forma $x^m + 1 = y^n$. (On impossible Diophantine equations of the form $x^m + 1 = y^n$). *Rev. Mat. Hisp.-Amer. IV*, 1: 122–140.

1942 W. Ljunggren. Einige Bemerkungen über die Darstellung ganzer Zahlen durch binäre kubische Formen mit positiver Diskriminante. *Acta Math.*, 75:1–21.

1943 J. E. Hofmann. Neues über Fermats zahlentheoretische Herausforderungen von 1657. *Abhandl. d. Preussischen Akad. d. Wiss.*, 1943, No. 9, 52 pages.

1943 W. Ljunggren. New propositions about the indeterminate equation $\frac{x^n-1}{x-1} = y^q$. *Norsk Mat. Tidsskr.*, 25:17–20.

1952 W. J. LeVeque. On the equation $a^x - b^y = 1$. *Amer. J. of Math.*, 74:325–331.

1953 J. W. S. Cassels. On the equation $a^x - b^y = 1$. *Amer. J. of Math.*, 75:159–162.

1953 K. Mahler. On the greatest prime factor of $ax^m + by^n$. *Nieuw Arch. Wisk. (3)*, 1:113–122.

1954 R. Obláth. Über die Gleichung $x^m + 1 = y^n$. *Ann. Polon. Math.*, 1:73–76.

1955 H. Davenport and K. F. Roth. Rational approximations to algebraic numbers. *Mathematika*, 2:160–167.

1956 R. Hampel. On the solution in natural numbers of the equation $x^m - y^n = 1$. *Ann. Polon. Math.*, 3:1–4.

1956 W. J. LeVeque. *Topics in Number Theory. Vol. II.* Addison-Wesley, Reading, Mass.

1956 A. Rotkiewicz. Sur l'équation $x^z - y^t = a^t$ où $|x - y| = a$. *Ann. Polon. Math.*, 3:7–8.

1958 A. Schinzel and W. Sierpiński. Sur certaines hypothèses concernant les nombres premiers. Remarques. *Acta Arith.*, 4:185–208 and 5:259 (1959).

1960 J. W. S. Cassels. On the equation $a^x - b^y = 1$. II. *Proc. Cambridge Phil. Soc.*, 56:97–103.

1960 R. Hampel. O zagadnieniu Catalana (On the problem of Catalan). *Roczniki Polskiego Towarzystwa Matematycznego, Ser. I, Prace Matematyczne*, 4:11–19.

1960 Chao Ko. On the Diophantine equation $x^2 = y^n + 1$. *Acta Sci. Natur. Univ. Szechuan*, 2:57–64.

1960 A. Rotkiewicz. Sur le problème de Catalan. *Elem. d. Math.*, 15:121–124.

1960 J. W. Wrench. The evolution of extended decimal approximations to π. *Math. Teacher*, 53:644–650.

1961 K. Inkeri and S. Hyyrö. On the congruence $3^{p-1} \equiv 1 \pmod{p^2}$ and the diophantine equation $x^2 - 1 = y^p$. *Ann. Univ. Turku. Ser. AI*, 1961, No. 50, 2 pages.

1962 A. Mąkowski. Three consecutive integers cannot be powers. *Colloq. Math.*, 9:297.

1963 J. Ferentinou-Nicolacopoulou. Une propriété des diviseurs du nombre $r^{r^m} + 1$. Applications au dernier théorème de Fermat. *Bull. Soc. Math. Grèce, Sér. 4*, (1):121–126.

1963 M. Gut. Abschätzungen für die Klassenzahlen der quadratischen Körper. *Acta Arith.*, 8:113–122.

1963 S. Hyyrö. On the Catalan problem (in Finnish). *Arkhimedes*, 1963, No. 1, 53–54. See Math. Reviews, 28, 1964, #62.

1964 S. Hyyrö. Über die Gleichung $ax^n - by^n = c$ und das Catalansche Problem. *Annales Acad. Sci. Fennicae, Ser. AI*, 1964(355):50 pages.

1964a K. Inkeri. On Catalan's problem. *Acta Arith.*, 9:285–290.

1964b K. Inkeri and S. Hyyrö. Über die Anzahl der Lösungen einiger diophantischer Gleichungen. *Ann. Univ. Turku. Ser. AI*, 1964, No. 78, 7 pages.

1964 Chao Ko. On the Diophantine equation $x^2 = y^n + 1$. *Scientia Sinica (Notes)*, 14:457–460.

1964 W. J. LeVeque. On the equation $y^m = f(x)$. *Acta Arith.*, 9:209–219.

1964 D. H. Lehmer. On a problem of Størmer. *Illinois J. Math.*, 8:57–79.

1964 H. Riesel. Note on the congruence $a^{p-1} \equiv 1 \pmod{p^2}$. *Math. of Comp.*, 18:149–150.

1965 P. Erdös. Some recent advances and current problems in number theory. In *Lectures on Modern Mathematics, Vol. III*, edited by T. L. Saaty, 169–244. Wiley, New York.

1965 A. Rotkiewicz. Sur les nombres de Mersenne dépourvus de diviseurs carrés et sur les nombres naturels n tels que $n^2 \mid 2^n - 2$. *Matematicky Vesnik, Beograd, (2)*, 17:78–80.

1967 L. J. Warren and H. Bray. On the square-freeness of Fermat and Mersenne numbers. *Pacific J. Math.*, 22:563–564.

1973 R. Tijdeman. On integers with many small prime factors. *Compositio Math.*, 26:319–330.

1974 R. Tijdeman. On the maximal distance between integers composed of small primes. *Compositio Math.*, 28:159–162.

1975 A. Baker. *Transcendental Number Theory.* Cambridge Univ. Press, London.

1976 E. Z. Chein. A note on the equation $x^2 = y^n + 1$. *Proc. Amer. Math. Soc.*, 56:83–84.

1976 M. Langevin. Quelques applications des nouveaux résultats de van der Poorten. *Sém. Delange-Pisot-Poitou, 17e année,* 1976, No. G12, 1–11.

1976 R. Tijdeman. On the equation of Catalan. *Acta Arith.*, 29: 197–209.

1979a P. Ribenboim. *13 Lectures on Fermat's Last Theorem.* Springer-Verlag, New York. Second edition with a new Epilogue, 1995.

1979b P. Ribenboim. On the square factors of the numbers of Fermat and Ferentinou-Nicolacopoulou. *Bull. Soc. Math. Grèce (N.S.)*, 20:81–92.

1980 _____. *Numéro Spécial π, Supplément au "Petit Archimède", Nos. 64-65.* 289 pages. 61 Rue St. Fuscien. 80000, Amiens (France).

1986 T. N. Shorey and R. Tijdeman. *Exponential Diophantine Equations.* Cambridge University Press, Cambridge.

1990 K. Inkeri. On Catalan's conjecture. *J. Nb. Th.*, 34:142–152.

1991 M. Aaltonen and K. Inkeri. Catalan's equation $x^p - y^q = 1$ and related congruences. *Math. of Comp.*, 56:359–370.

1993 M. Mignotte. Un critère élémentaire pour l'équation de Catalan. *C. R. Math. Rep. Acad. Sci. Canada*, 15:199–200.

1994 P. Ribenboim. *Catalan's Conjecture.* Academic Press, Boston.

1995 M. Laurent, M. Mignotte, and Y. Nesterenko. Formes linéaires en deux logarithmes et déterminants d'interpolation. *J. Nb. Th.*, 55:285–321.

1995a M. Mignotte. A criterion on Catalan's equation. *J. Nb. Th.*, 52:280–283.

1995b M. Mignotte and Y. Roy. Catalan's equation has no new solution with either exponent less than 10651. *Experiment. Math.*, 4:259–268.

1995 O'Neil. Improved upper bounds on the exponents in Catalan's equation. Manuscript.

1995 W. Schwarz. A note on Catalan's equation. *Acta Arith.*, 72:277–279.

1996 F. Jongmans. *Eugène Catalan*. Soc. Belge Prof. Math. Expr. Française, Mons.

1996 M. Mignotte. Sur l'équation $x^p - y^q = 1$ lorsque $p \equiv 5$ mod 8. *C. R. Math. Rep. Acad. Sci. Canada*, 18:228–232.

1997 D. H. Bailey, J. M. Borwein, P. B. Borwein, and S. Plouffe. The quest for pi. *Math. Intelligencer*, 19(1):50–57.

1997 C. D. Bennett, J. Blass, A. M. W. Glass, D. B. Meronk, and R. P. Steiner. Linear forms in the logarithms of three positive rational numbers. *J. Théor. Nombres Bordeaux*, 9: 97–136.

1997 M. Mignotte and Y. Roy. Minorations pour l'équation de Catalan. *C. R. Acad. Sci. Paris, Sér. I*, 324:377–380.

1999 Maurice Mignotte. Une remarque sur l'équation de Catalan. In *Number Theory in Progress, Vol. 1 (Zakopane-Kościelisko, 1997)*, 337–340. de Gruyter, Berlin.

1999 P. Mihăilescu. A class number free criterion for Catalan's conjecture. Manuscript, Zurich.

2000 Maurice Mignotte. Catalan's equation just before 2000. To appear in the Proceedings of the Symposium in memory of Kustaa Inkeri.

8

1093

1093! If you are wondering what this paper is all about, I hasten to say that it is not about the quality of wines from the year 1093. Indeed, no records exist for such a remote time. It was not until 1855 that a select group of Bordeaux wine authorities ranked the finest vineyards in their region, distinguishing among others the outstanding Châteaux Lafitte, Margaux, Latour, and Haut-Brion as Premiers Crus in Médoc, and Chateau Yquem in Sauternes as Premier Grand Crus. Not to mention all the marvelous wines of Bourgogne ... (see FADIMAN (1981)). This paper is not about wines, however.

In fact, the idea for this lecture came to me after a discussion with F. LE LIONNAIS, in Paris. He is a science writer, now in his eighties, with an acute curiosity. Just after the War, in 1946, he edited a book *Les Grands Courants de la Pensée Mathématique*, containing contributions by eminent mathematicians like ANDRÉ WEIL and several other members of the Bourbaki group. The article "L'Avenir des Mathématiques" by WEIL is worth reading, given the perspective of the more than fifty years that have elapsed. This book has been translated into English and it is readily available. I became aware of it shortly after its appearance and was always eager to meet LE LIONNAIS.

So, it was with great pleasure that in 1976 I was introduced to him, at a seminar on the history of mathematics. During the conversation

I learned that he was preparing a book on distinguished or important numbers like 2, 7, π, e, etc.... He asked me if I knew of any numbers with interesting properties that could be included in his book.* After searching, I decided on "1093" about which he had as yet heard nothing.

My intention is to tell you why I consider 1093 an interesting number. (Later I was glad to see that Le Lionnais included 1093 in his book on remarkable numbers.) Of course, one could say that *every* natural number is remarkable. If not, there is a smallest number N which is not remarkable—and having this property N is certainly remarkable ....

Well ... 1093 is the smallest prime p satisfying the congruence

$$2^{p-1} \equiv 1 \pmod{p^2}. \tag{1}$$

Thus,

$$2^{1092} \equiv 1 \pmod{1093^2}.$$

This was discovered by MEISSNER (1913) by actually doing the calculation. According to Fermat's Little Theorem,

$$2^{p-1} \equiv 1 \pmod{p}, \tag{2}$$

but in general there is no reason to expect the stronger congruence (1).

In Volume III of LANDAU (1927), you find gives the following proof of (1) for $p = 1093$:

$$3^7 = 2187 = 2p + 1,$$

so squaring

$$3^{14} \equiv 4p + 1 \pmod{p^2}. \tag{3}$$

On the other hand,

$$2^{14} = 16384 = 15p - 11, \quad \text{so}$$
$$2^{28} = -330p + 121 \pmod{p^2},$$

hence

$$3^2 \times 2^{28} \equiv -1876p - 4 \pmod{p^2}$$

---

*In the meantime, this book has been published: F. Le Lionnais *Les Nombres Remarquables*, Hermann Editeurs, Paris, 1983.

and dividing by 4,

$$3^2 \times 2^{26} \equiv -469p - 1 \pmod{p^2}.$$

Raising now to the 7th power:

$$3^{14} \times 2^{182} \equiv -4p - 1 \pmod{p^2}$$

and taking (3) into account,

$$2^{182} \equiv -1 \pmod{p^2}.$$

Finally, raising to the 6th power,

$$2^{1092} \equiv 1 \pmod{p^2}.$$

Of course, since the congruence (1) does not hold in general, any proof which is valid for $p = 1093$ has to be ad-hoc.

BEEGER (1922) found that 3511 satisfies the congruence (1). Imagine, he did his calculation in 1921, before the age of computers. What tenacity! Even more so, since it was not known if any other example $p > 1093$ would exist.

The search was continued up to 6×10^9 by D. H. LEHMER, then by W. KELLER, D. CLARK, and lately by R. E. CRANDALL, K. DILCHER, and C. POMERANCE up to 4×10^{12} (see CRANDALL, DILCHER, and POMERANCE (1997)). No other prime satisfying the congruence (1) was found.

All this is fine, but why is anyone interested in the congruence (1)?

It was ABEL, apparently, who asked in the third volume of Crelle's Journal (1828) whether it is possible to have

$$a^{p-1} \equiv 1 \pmod{p^m} \tag{4}$$

with $m \geq 2$ and p a prime not dividing a.

JACOBI, who was extremely proficient in numerical computations, gave various examples of (4):

$$3^{10} \equiv 1 \pmod{11^2},$$
$$7^4 \equiv 1 \pmod{5^2},$$
$$31^6 \equiv 1 \pmod{7^2},$$

and, with $m = 3$,

$$19^6 \equiv 1 \pmod{7^3}.$$

But the problem is not to find examples, but to answer the question:

Are there infinitely many primes p such that $2^{p-1} \equiv 1 \pmod{p^2}$?

To study this question I rephrase it in terms of the Fermat quotient. If $a \geq 2$, and p is a prime not dividing a, then

$$q_p(a) = \frac{a^{p-1} - 1}{p} \tag{5}$$

is an integer (by Fermat's Little Theorem) called the *Fermat quotient*, with base a and exponent p. Thus, $q_p(a) \equiv 0 \pmod{p}$ if and only if $a^{p-1} \equiv 1 \pmod{p^2}$.

So, this leads to the residue modulo p of $q_p(a)$ and one is immediately struck by the many interesting results linking the Fermat quotient to various interesting arithmetical quantities. I want to illustrate this with some examples.

The residue modulo p of $q_p(a)$ behaves like a logarithm. This fact was noted by EISENSTEIN in 1850:

$$q_p(ab) \equiv q_p(a) + q_p(b) \pmod{p}. \tag{6}$$

A   Determination of the residue of $q_p(a)$

The first recorded result is due to SYLVESTER (1861a), who proved the nice congruence:

$$q_p(2) \equiv \frac{1}{2}\left(1 + \frac{1}{2} + \frac{1}{3} + \cdots + \frac{1}{\frac{p-1}{2}}\right)$$
$$\equiv 1 + \frac{1}{3} + \cdots + \frac{1}{p-1} \pmod{p}. \tag{7}$$

And more generally, for any base a:

$$q_p(a) = \sum_{j=1}^{p-1} \frac{a_j}{j} \pmod{p} \tag{8}$$

where $0 \leq a_j \leq p - 1$ and $pa_j + j \equiv 0 \pmod{a}$.

In 1910, MIRIMANOFF showed: If $p = 2^r \pm 1$ is a prime, then $q_p(2) \equiv \mp 1/r \not\equiv 0 \pmod{p}$.

More recently, JOHNSON (1977) obtained a practical means for determining the Fermat quotient. If r is the smallest integer such that $a^r \equiv \pm 1 \pmod{p}$, letting $a^r \equiv \pm 1 + tp$ it follows that

$$q_p(a) \equiv \mp \frac{t}{r} \pmod{p}. \tag{9}$$

B   Identities and congruences for the Fermat quotient

The first paper of substance on Fermat quotients is LERCH (1905), and today it is almost forgotten. I am pleased to be able to bring his nice results to your attention.

First he showed

$$\sum_{j=1}^{p-1} q_p(j) \equiv W(p) \pmod{p} \tag{10}$$

where $W(p)$ denotes the Wilson quotient. It is defined in a manner similar to the one used for the Fermat quotient. Namely, Wilson's theorem says that

$$(p-1)! \equiv -1 \pmod{p}, \tag{11}$$

hence the quotient

$$W(p) = \frac{(p-1)! + 1}{p} \tag{12}$$

is an integer, called the *Wilson quotient* of p. Before I return to the Fermat quotients, let me just say that the problem of determining the residue of $W(p)$ modulo p is just as interesting as the corresponding problem for the Fermat quotient. In particular, if $W(p) \equiv 0 \pmod{p}$, then p is called a *Wilson prime*. For example, $p = 5, 13$ are easily seen to be Wilson primes. The search for new Wilson primes uncovered only one more (GOLDBERG): $p = 563$, up to 4×10^{12} (CRANDALL, DILCHER, and POMERANCE (1997)). The question whether there are infinitely many Wilson primes appears to be very difficult. For example, VANDIVER said in 1955:

> This question seems to be of such a character that if I should come to life any time after my death and some mathematician were to tell me it had been definitely settled, I think I would immediately drop dead again.

Let me return to the determination of the residue of Wilson's quotient.

LERCH showed that

$$W(p) = B_{2(p-1)} - B_{p-1} \pmod{p}. \tag{13}$$

Here B_n denotes the nth Bernoulli number, these being generated by the function

$$\frac{x}{e^x - 1} = \sum_{n=0}^{\infty} B_n \frac{x^n}{n!}. \tag{14}$$

Thus, $B_0 = 1$, $B_1 = -1/2$, and $B_n = 0$ for every odd $n > 1$. The Bernoulli numbers satisfy the recurrence relation:

$$\binom{n+1}{1} B_n + \binom{n+1}{2} B_{n-1} + \binom{n+1}{3} B_{n-2} + \cdots + \binom{n+1}{n} B_1 + 1 = 0. \tag{15}$$

This may be written symbolically as

$$(B + 1)^{n+1} - B^{n+1} = 0. \tag{16}$$

(Treat B as an indeterminate and, after computing the polynomial in the left-hand side, replace B^k by B_k.)

An important property of the Bernoulli numbers was discovered by EULER and connects these numbers to the Riemann zeta function ζ:

$$B_{2n} = (-1)^{n-1} \frac{2(2n)!}{(2\pi)^{2n}} \zeta(2n) \quad \text{(for } n \geq 1) \tag{17}$$

where

$$\zeta(s) = \sum_{j=1}^{\infty} \frac{1}{j^s}, \quad s \text{ complex, } \operatorname{Re}(s) > 1. \tag{18}$$

Using the functional equation of Riemann zeta functions, it follows that

$$\zeta(1 - n) = -\frac{B_n}{n} \quad \text{(for } n \geq 2) \tag{19}$$

and also $\zeta(0) = -\frac{1}{2}$.

Returning to Lerch's formula (10), the point is the following: Since the Fermat quotient is somehow hard to compute, it is more natural to relate their sum, over all the residue classes, to quantities defined by p.

This has also been done for weighted sums. In a letter of 1909 to HENSEL, FRIEDMANN and TAMARKINE proved: If $1 \leq n \leq p - 1$, then

$$\sum_{j=1}^{p-1} j^n q_p(j) \equiv (-1)^{[n/2]} \frac{B_n}{n} \pmod{p}. \tag{20}$$

From (19), it follows that if $2 \leq n \leq p - 1$, then

$$\sum_{j=1}^{p-1} j^n q_p(j) \equiv (-1)^{[(n-2)/2]} \zeta(1-n) \pmod{p} \tag{21}$$

and also

$$\sum_{j=1}^{p-1} j q_p(j) \equiv -\frac{1}{2} = \zeta(0) \pmod{p}.$$

LERCH also connected the Fermat quotient with the *Legendre quotient*

$$\lambda_p(j) = \frac{j^{\frac{p-1}{2}} - \left(\frac{j}{p}\right)}{p}. \tag{22}$$

Recall that if $p \nmid j$ and $\left(\frac{j}{p}\right)$ is the Legendre symbol, then

$$\left(\frac{j}{p}\right) \equiv j^{\frac{p-1}{2}} \pmod{p}, \tag{23}$$

so $\lambda_p(j)$ is an integer.

LERCH proved

$$q_p(j) \equiv 2 \left(\frac{j}{p}\right) \lambda_p(j) \pmod{p}. \tag{24}$$

Another nice relation involves the distribution of quadratic residues, and the class number of quadratic fields.

Let $H(a)$ denote the class number of the quadratic field $\mathbb{Q}(\sqrt{a})$ (where a is a square-free integral). DIRICHLET proved the famous formula, when $p \equiv 3 \pmod{4}$, $p \neq 3$:

$$H(-p) = -\frac{1}{p} \sum_{j=1}^{p-1} \left(\frac{j}{p}\right) j = \frac{\rho - \rho'}{2 - \left(\frac{2}{p}\right)} \tag{25}$$

where

ρ = number of quadratic residues between 0 and $p/2$,

ρ' = number of non-quadratic residues between 0 and $p/2$.

It is known that $\rho > \rho'$ when $p \equiv 3 \pmod{4}$. This is a difficult theorem. The only proofs now known require analysis, in spite of its purely arithmetical nature.

FRIEDMANN and TAMARKINE noted that

$$\rho - \rho' \equiv \left[2 - \left(\frac{2}{p}\right)\right] 2B_{\frac{p+1}{2}} \pmod{p} \tag{26}$$

which amounts to

$$H(-p) \equiv 2B_{\frac{p+1}{2}} \pmod{p}. \tag{27}$$

LERCH showed

$$\sum \left(\frac{j}{p}\right) jq_p(j) \equiv \begin{cases} 0 & \text{when } p \equiv 1 \pmod{4}, \\ H(-p) & \text{when } p \equiv 3 \pmod{4}. \end{cases} \tag{28}$$

Let me now turn to a theorem which has sparked interest in Fermat quotients and opened up new areas of research.

The following results have historical importance. They have not completely lost their interest with the proof of Fermat's Last Theorem by WILES, since the methods are applicable to similar diophantine equations, for example, to Catalan's equation

$$X^m - Y^n = 1. \tag{29}$$

In 1909, WIEFERICH proved: Suppose that there exist integers x, y, z, not multiples of the odd prime p such that $x^p + y^p + z^p = 0$ (that is, the *first* case of Fermat's Last Theorem fails for p). Then

$$2^{p-1} \equiv 1 \pmod{p^2}.$$

Hence, by what I said in the beginning of this chapter, the first case of Fermat's Last Theorem ($= FLT$) is true for every prime exponent $p < 4 \times 10^{12}$, except, possibly, for $p = 1093$, and 3511.

The proof of WIEFERICH was very difficult and technical. It was based on the following deep result of KUMMER: If the first case of FLT fails for the exponent p, with $x^p + y^p + z^p = 0$, and p not dividing xyz, then

$$\left[\frac{d^{2s} \log(x + e^v y)}{dv^{2s}}\right] \times B_{2s} \equiv 0 \pmod{p} \tag{30}$$

for $2s = 2, 4, \ldots, p - 3$ (and similar congruences for the pairs (y, x), (y, z), (z, y), (x, z), (z, x)).

Moreover, WIEFERICH used complicated congruences that are satisfied by the Bernoulli numbers. In 1912, FURTWÄNGLER found another proof of Wieferich's theorem, using class field theory—a further confirmation of the power of class field theory. The theorem of Wieferich was the first of a series of criteria involving Fermat quotients.

MIRIMANOFF showed in 1910:

If the first case of *FLT* fails for the exponent p, then $q_p(3) \equiv 0 \pmod{p}$.

Since $p = 1093, 3511$ do not satisfy the above congruence, then this establishes the first case for the range of primes less than 4×10^{12}.

This work was continued by FROBENIUS, VANDIVER, POLLACZEK, ROSSER, and, more recently, by GRANVILLE and MONAGAN. With this method, it was proved that if the first case fails for p, then

$$q_p(\ell) \equiv 0 \pmod{p} \quad \text{for } \ell \text{ prime, } \ell \leq 89. \tag{31}$$

A corollary worth mentioning is the following one, due to SPUNAR:
Let p be an odd prime satisfying the following property (P89):

There exists k not a multiple of p such that $kp = a \pm b$ where all prime factors of a, b are at most equal to 89.

Then the first case of *FLT* holds for the exponent p.

The proof is actually so simple that I will give it now.

If the first case fails for the exponent p, then for every prime ℓ, $\ell \leq 89$, it follows that $\ell^{p-1} \equiv 1 \pmod{p^2}$, by formula (31). So $a^{p-1} \equiv 1 \pmod{p^2}$ and $b^{p-1} \equiv 1 \pmod{p^2}$. Hence, $a^p \equiv a \pmod{p^2}$ and $b^p \equiv b \pmod{p^2}$. But $a = \mp b + kp$, so $a^p \equiv \mp b^p \pmod{p^2}$. From $kp = a \pm b \equiv a^p \pm b^p \equiv 0 \pmod{p^2}$ it follows that p divides k, contrary to the hypothesis.

In this connection, there is the following open problem:

Are there infinitely many primes p with property (P89)?

PUCCIONI showed in 1968: If the above set is finite, then for every prime ℓ, $\ell \leq 89$, $\ell \not\equiv 1 \pmod{8}$, the set $M_\ell - \{p \mid \ell^{p-1} \equiv 1 \pmod{p^2}\}$ is infinite.

Unfortunately, this theorem may mean nothing since both sets in question, sparse as they seem to be, may turn out to be infinite.

A corollary of SPUNAR's result, already given by MIRIMANOFF, is the following one:

The first case of *FLT* holds for the prime exponents p of the form $p = 2^m \pm 1$.

It is an easy exercise to show:

If $2^m + 1$ is a prime number, then $m = 2^n$ ($n \geq 0$). The numbers

$$F_n = 2^{2^n} + 1$$

were first considered by FERMAT, and are now called Fermat numbers.

Similarly, if $2^m - 1$ is a prime number, then $m = p$ is a prime. The numbers

$$M_p = 2^p - 1 \quad (p \text{ prime})$$

are called the Mersenne numbers.

FERMAT believed that all Fermat numbers are primes. Indeed,

$$F_1 = 5, \quad F_2 = 17, \quad F_2 = 257, \quad F_4 = 65537.$$

F_5 is much larger, with about 10 digits.

EULER proved in 1747 the following criterion: If p divides F_n ($n \geq 2$), then $p = 2^{n+1}k + 1$ (with $k \geq 1$).

He applied this criterion to F_5 and found that 641 divides F_5, so therefore F_5 is not a prime number.

This can also be seen in the following way:

$$641 = 2^4 + 5^4 = 5 \times 2^7 + 1,$$
$$2^{32} = 2^4 \times 2^{28} = (641 - 5^4) \times 2^{28}$$
$$= 641 \times 2^{28} - (5 \times 2^7)^4$$
$$= 641 \times 2^{28} - (641 - 1)^4 \equiv -1 \pmod{641}.$$

So, 641 divides $2^{32} - 1 = F_5$.

The study of the Fermat numbers and Mersenne numbers led to the discovery of the first primality tests for large numbers.

The first criterion was devised by LUCAS in the form of a converse of Fermat's Little Theorem:

Let $n \geq 3$ be odd and assume that there exists a, $1 < a < n$, such that

$$a^{p-1} \equiv 1 \pmod{n},$$

and if q is any prime dividing $n - 1$, then

$$a^{\frac{n-1}{q}} \not\equiv -1 \pmod{n}.$$

Then n is a prime.

Based on such criterion PEPIN showed:

The Fermat number F_n is a prime if and only if

$$3^{\frac{F_n-1}{2}} \equiv -1 \ (\text{mod } F_n).$$

There has been an extensive search for Fermat primes. According to my information, which may already be outdated, the largest Fermat number tested is F_{303088} (composite). It usually takes long calculations to decide if a Fermat number is composite.

Contrary to FERMAT's belief, the only known Fermat primes are the ones he already knew!

The following conjectures are weaker than FERMAT's original assertion:

(a) EISENSTEIN (1844): there are infinitely many Fermat primes;

(b) SCHINZEL (1963): there exist infinitely many Fermat numbers which are square-free (i.e., products of distinct primes).

For the Mersenne numbers, EULER gave the first test for factors: If p is a prime, $p > 3$, $p \equiv 3 \ (\text{mod } 4)$, then $2p + 1$ divides Mp if and only if $2p + 1$ is a prime.

In this way, EULER concluded that 23 divides M_{11}; ...; 503 divides M_{251}, etc. ...

The problem becomes the determination of primes p such that $2p+1$ is again a prime. Such primes are justly called *Sophie Germain primes*. They were first considered around 1820 when she proved the following beautiful theorem, which was of a totally new nature:

If p and $2p + 1$ are primes, then the first case of *FLT* is true for the exponent p.

An open question is: Are there infinitely many Sophie Germain primes?

Compare this question with the following one (the twin prime problem):

Are there infinitely many primes p such that $p+2$ is also a prime?

In both cases there are linear polynomials $2X + 1$, $X + 2$ respectively, and the question is whether they infinitely often assume prime values at primes.

Let me now describe a very effective primality test for Fermat and Mersenne numbers, devised by LUCAS in 1878. It uses second order linear recurrences, more specifically, Fibonacci and Lucas numbers.

In the thirteenth century, FIBONACCI considered the sequence of numbers $F_0 = 0$, $F_1 = 1$, $F_2 = 1$, $F_3 = 2$, $F_4 = 3$, $F_5 = 5$, $F_6 = 8$,

..., and, more generally,

$$F_n = F_{n-1} + F_{n-2}.$$

(I hope no confusion with the Fermat numbers will arise.)

The Fibonacci numbers have a wealth of arithmetic properties. Books have been written about them, and a quarterly journal has these numbers as its main topic.

A companion sequence is the one of Lucas numbers: $L_0 = 2$, $L_1 = 1$, $L_2 = 3$, $L_3 = 4$, $L_4 = 7$, $L_5 = 11$, ..., and $L_n = L_{n-1} + L_{n-2}$, for $n \geq 2$.

More generally, given the numbers U_0, U_1, and α, $\beta \in \mathbb{Q}$, let

$$U_n = \alpha U_{n-1} - \beta U_{n-2}. \tag{32}$$

The equation $X_2 - \alpha X + \beta = 0$ has roots

$$a = \frac{\alpha + \sqrt{\alpha^2 - 4\beta}}{2}, \qquad b = \frac{\alpha - \sqrt{\alpha^2 - 4\beta}}{2},$$

so

$$\alpha = a + b, \beta = ab$$

and

$$U_n = (a + b)U_{n-1} - abU_{n-2}. \tag{33}$$

For Fibonacci and Lucas numbers, $\alpha = 1$, $\beta = -1$, so

$$a = \frac{1 + \sqrt{5}}{2}\text{(the ``Golden ratio''),} \qquad b = \frac{1 - \sqrt{5}}{2}.$$

Binet's formula gives

$$U_n = \frac{a^n - b^n}{a - b}. \tag{34}$$

The sequence companion is

$$V_n = a^n + b^n. \tag{35}$$

Putting

$$W_n = \frac{V_{2^{n-1}}}{Q^{2^{n-2}}} \qquad \text{(for } n \geq 2\text{),} \tag{36}$$

so,

$$W_1 = \frac{\alpha^2 - 2\beta}{\beta}, \qquad W_{n+1} = W_n^2 - 2. \tag{37}$$

With appropriate choices of α, β, LUCAS obtained the useful "testing sequence":

If $p \equiv 1 \pmod 4$, let $W_2 = -4$, $W_{n+1} = W_n^2 - 2$, so the sequence is $-4, 14, 194, \ldots$.

His criterion is:

$M_p = 2^p - 1$ is a prime if and only if M_p divides W_p.

If $p \equiv 3 \pmod 4$, $p > 3$, let $W_2 = -3$, $W_{n+1} = W_n^2 - 2$, so the sequence is $-3, 7, 47, \ldots$.

His criterion in this case is again: $M_p = 2^p - 1$ is a prime if and only if M_p divides W_p.

This is the method which is at present used in testing the primality of Mersenne numbers.

In 1944, MERSENNE knew that for $p = 2, 3, 5, 7, 13, 17, 19, 31$, M_p is a Mersenne prime.

In 1878, LUCAS showed that if $p = 61, 89, 107, 127$, then M_p is also a Mersenne prime.

With the advent of computers, we now know 37 Mersenne primes, the largest ones being $M_{3021377}$ with 909526 digits, and $M_{2976221}$ which has 895932 digits.

SCHINZEL conjectured the following:

There exist infinitely many square-free Mersenne numbers. To date, no Fermat or Mersenne number with a square factor has ever been found.

In 1965, ROTKIEWICZ took up the above conjecture and showed:

If Schinzel's conjecture on Mersenne numbers is true, there exist infinitely many primes p such that

$$2^{p-1} \not\equiv 1 \pmod{p^2}.$$

By the way, ROTKIEWICZ made use of the following interesting, (and many times rediscovered) theorem of ZSIGMONDY (1892):

If $n \neq 6$, $n \geq 3$, $a \geq 2$, then there exists a prime p such that the order of a modulo p is equal to n. Equivalently, there exists a prime p such that p divides $a^n - 1$, but p does not divide $a^m - 1$ for $m < n$.

This theorem was discovered by ZSIGMONDY (earlier by BANG for $a = 2$), BIRKHOFF and VANDIVER, DICKSON, CARMICHAEL, KANOLD, ARTIN, HERING, LÜNEBURG, POMERANCE, and ... who else? I would like to know.

From the Rotkiewicz theorem it follows that there is rather surprising, and I dare say, deep, connection between such dissimilar topics as Fermat's Last Theorem, the congruence $2^{p-1} \equiv 1 \pmod{p^2}$, and the factorization of Mersenne numbers. But, I have been digressing from the main question.

There is a heuristic reason to believe that there exist infinitely many primes p such that $2^{p-1} \equiv 1 \pmod{p^2}$. The argument is as follows. Since nothing to the contrary is known, it may be assumed (heuristically) that for each prime p the probability that $\frac{2^{p-1}-1}{p} \equiv 0 \pmod{p}$ is just $\frac{1}{p}$ since there are p residue classes modulo p. If x is any large positive real number, then the number of primes $p \leq x$ with $2^{p-1} \equiv 1 \pmod{p^2}$ should be

$$\sum_{p \leq x} \frac{1}{p} = \log \log x + \text{error term.}$$

So there would exist infinitely many p satisfying the above congruence. However, this argument cannot be made rigorous. From the calculations, apart from two exceptions, $2^{p-1} \not\equiv 1 \pmod{p^2}$, so it should be expected that there are infinitely many primes p satisfying $2^{p-1} \not\equiv 1 \pmod{p^2}$. This has not yet been proved, but it follows from the important and interesting (ABC) conjecture of MASSER and OESTERLÉ:

For each $\epsilon > 0$ there exists a real number $K(\epsilon) > 0$ such that for any positive integers A, B, and C with $\gcd(A, B, C) = 1$, and $A + B = C$, then

$$C \leq K(\epsilon) r^{1+\epsilon}$$

where r (the *radical* of ABC) is the product of the distinct prime factors of ABC.

For example, if $\epsilon = \frac{1}{2}$, and if $A = 2^m$, $B = 3^n$ (with m and n large), and $C = A^m + B^n$, then from $C < K(\frac{1}{2}) r^{3/2}$ and $r = 6 \prod_{p | C} p$ it follows that C must have large radical.

SILVERMAN (1988) proved:

If the (ABC) conjecture is assumed true, then there exists infinitely many primes p such that $2^{p-1} \not\equiv 1 \pmod{p^2}$.

It would be of the greatest importance to prove this conjecture.

The number 1093 is indeed interesting after all $\ldots$.

References

1828 N. H. Abel. Aufgabe von Herrn N. H. Abel zu Christiania (in Norwegian). *J. reine u. angew. Math.*, 3:212.

1828 C. G. J. Jacobi. Beantwortung der aufgabe S. 212 dieses Bandes: "Kann $\alpha^{m-1} - 1$ wenn μ eine Primzahl und α eine ganze Zahl und kliener als μ und größer als 1 ist, durch $\mu\mu$ theilbar sein?". *J. reine u. angew. Math.*, 3:301–303.

1844 F. G. Eisenstein. Aufgaben. *J. reine u. angew. Math.*, 27:87. (Reprinted in *Mathematische Werke*, Vol. 1. No. 3, Chelsea, New York. 1975).

1850 F. G. Eisenstein. Eine neue Gattung zahlentheoretischer Funktionen, welche von zwei Elementen abhangen und durch gewisse lineare Funktionalgleichungen definiert werden. *ber. über verhandl. der königl. Preuß. Akad. d. Wiss. zu Berlin*, 36–42. Reprinted in *Mathematische Werke*, Vol. 2. 705–712, Chelsea, New York, 1975.

1861a J. J. Sylvester. Note relative aux communications faites dans les séances du 28 Janvier et 4 Février 1861. *C. R. Acad. Sci. Paris*, 52:307–308. Reprinted in *Math. Papers*, Vol. 2: 234–235; and *Corrigenda*, 241, Cambridge University Press, 1908.

1861b J. J. Sylvester. Sur une propriété des nombres premiers qui se rattache au théorème de Fermat. *C. R. Acad. Sci. Paris*, 52:161–163. Reprinted in *Math. Papers*, Vol. 2: 229–231, Cambridge University Press, 1908.

1876 E. Lucas. Sur la recherche des grands nombres premiers. *Congrès de l'Assoc. Française pour l'Avancement des Sciences*, Clermont-Ferrand 5:61–68.

1877 T. Pepin. Sur la formule $2^{2^n} + 1$. *C. R. Acad. Sci. Paris*, 85:329–331.

1878 E. Lucas. Théorie des fonctions numériques simplement périodiques. *Amer. J. of Math.*, 1:184–240 and 289–321.

1905 M. Lerch. Zur Theorie der Fermatschen Quotienten $a^{p-1} - 1/p \equiv q(a)$. *Math. Annalen*, 60:471–490.

1909 A. Friedmann and J. Tamarkine. Quelques formules concernant la théorie de la fonction $\{x\}$ et des nombres de Bernoulli. *J. reine u. angew. Math.*, 135:146–156.

1910 D. Mirimanoff. Sur le dernier théorème de Fermat. *C. R. Acad. Sci. Paris*, 150:204–206.

1913 W. Meissner. Uber die Teilbarkeit von $2^n - 2$ durch das Quadrat der Primzahl $p = 1093$. *Sitzungsber. Akad. d. Wiss., Berlin*, 51:663–667.

1914 H. S. Vandiver. Extension of the critereon of Wieferich and Mirimanoff in connection with Fermat's last theorem. *J. reine u. angew. Math.*, 144:314–318.

1922 N. G. W. H. Beeger. On a new case of the congruence $2^{p-1} \equiv 1 \pmod{p^2}$. *Messenger of Math.*, 51:149–150.

1927 E. Landau. *Vorlesungen über Zahlentheorie, Vol. 3.* S. Hirzel, Leipzig. Reprinted by Chelsea, New York, 1969.

1946 F. Le Lionnais. *Les Grands Courants de la Pensée Mathématique.* A. Blanchard, Paris.

1953 K. Goldberg. A table of Wilson quotients and the third Wilson prime. *J. London Math. Soc.*, 28:252–256.

1955 H. S. Vandiver. Divisibility problems in number theory. *Scripta Math.*, 21:15–19.

1963 A. Schinzel. Remarque au travail de W. Sierpiński sur les nombres $a^{2^n} + 1$. *Colloq. Math.*, 10:137–138.

1963 W. Sierpiński. Sur les nombres composés de la forme $a^{2^n} + 1$. *Colloq. Math.*, 10:133–135.

1965 A. Rotkiewicz. Sur les nombres de Mersenne dépourvus de diviseurs carrés et sur les nombres naturels n tels que $n^2 \mid 2^n - 2$. *Matematicky Vesnik, Beograd, (2)*, 17:78–80.

1968 S. Puccioni. Un teorema per una resoluzione parziale del famoso teorema di Fermat. *Archimede*, 20:219–220.

1969 R. K. Guy. The primes 1093 and 3511. *Math. Student*, 35:204–206 (1969).

1977 W. Johnson. On the non-vanishing of Fermat quotients (mod p). *J. reine u. angew. Math.*, 292:196–200.

1981 C. Fadiman and S. Aaron. *The Joys of Wine.* Galahad Books, New York.

1988 J. H. Silverman. Wieferich's criterion and the abc-conjecture. *J. Nb. Th.*, 30:226–237.

1996 P. Ribenboim. *The New Book of Prime Number Records.* Springer-Verlag, New York.

1997 R. E. Crandall, K. Dilcher, and C. Pomerance. A search for Wieferich and Wilson primes. *Math. of Comp.*, 66:433–449.

9

Powerless Facing Powers

I gave this lecture many times in many countries. Can you guess who came to listen to it?

Political scientists! Third-world countries facing the big powers? And powerless Paulo would tell how to resist or to become one of them ....

No, I am just a mathematician not knowing how to solve many problems involving powers of integers or the so-called powerful numbers.

My intention is to present several problems of this kind, in a few cases, to advance conjectures of what should be true.

The following notations will be used. If S is a finite set, $\#S$ denotes the number of elements of S. If S is a set of positive integers, and $x \geq 1$, let $S(x) = \{s \in S \mid s \leq x\}$.

The integers of the form a^n, where $|a| > 1$, $n > 1$, are said to be powers. Thus, 1 is not a power.

1   Powerful numbers

The first paper about powerful numbers was by ERDÖS (1935); however, the name "powerful number" was coined later by GOLOMB (1970).

Let $k \geq 2$. The natural number $n \geq 1$ is said to be a k-*powerful number* when the following property is satisfied: if a prime p divides n, then p^k also divides n.

In other words, the k-powerful numbers are exactly the integers which may be written in the form $a_0^k a_1^{k+1} \cdots a_{k-1}^{2k-1}$ (where a_0, ..., a_{k-1} are positive integers that may not be coprime). A 2-powerful number is simply called a *powerful* number. In particular, powerful numbers are those of the form $a_0^2 a_1^3$, with $a_0, a_1 \geq 1$. We note that 1 is a powerful number. I shall denote by W_k the set of k-powerful numbers.

The main problems about powerful numbers are of the following kinds:

1. Distribution of powerful numbers.
2. Additive problems.
3. Difference problems.

A.   Distribution of powerful numbers

The aim is to estimate the number of elements in the set

$$W_k(x) = \{n \in W_k \mid 1 \leq n \leq x\}, \tag{1}$$

where $x \geq 1$, $k \geq 2$.

Already in 1935, ERDÖS and SZEKERES gave the first result about $W_2(x)$:

$$\#W_2(x) = \frac{\zeta(\frac{3}{2})}{\zeta(3)} x^{1/2} + O(x^{1/3}) \quad \text{as } x \to \infty, \tag{2}$$

where $\zeta(s)$ is the Riemann zeta function; see also BATEMAN (1954) and GOLOMB (1970).

To describe the more recent results, I introduce the zeta function associated to the sequence of k-powerful numbers. Let

$$j_k(n) = \begin{cases} 1 & \text{if } n \text{ is } k\text{-powerful,} \\ 0 & \text{otherwise.} \end{cases}$$

The series $\sum_{n=1}^{\infty} \frac{j_k(n)}{n^s}$ is convergent for $\text{Re}(s) > \frac{1}{k}$ and defines a function $F_k(s)$. This function admits the following Euler product representation

$$F_k(s) = \prod_p \left(1 + \frac{\frac{1}{p^{ks}}}{1 - \frac{1}{p^s}} \right) = \prod_p \left(1 + \frac{1}{p^{(k-1)s}(p^s - 1)} \right), \quad (3)$$

which is valid for $\operatorname{Re}(s) > \frac{1}{k}$.

With well-known methods, IVIĆ and SHIU showed in 1982:

(1.1) $\#W_k(x) = \gamma_{0,k} x^{\frac{1}{k}} + \gamma_{1,k} x^{\frac{1}{k+1}} + \cdots + \gamma_{k-1,k} x^{\frac{1}{2k-1}} + \Delta_k(x),$
where $\gamma_{i,k}$ is the residue at $\frac{1}{k+i}$ of $\frac{F_k(s)}{s}$.

Explicitly,

$$\gamma_{i,k} = C_{k+i,k} \frac{\Phi_k(\frac{1}{k+i})}{\zeta(\frac{2k+2}{k+i})}, \quad (4)$$

where

$$C_{k+i,k} = \prod_{\substack{j=k \\ j \neq k+i}}^{2k-1} \zeta\left(\frac{j}{k+i}\right), \quad (5)$$

$\Phi_2(s) = 1$, and if $k > 2$, then $\Phi_k(s)$ has a Dirichlet series with abscissa of absolute convergence $\frac{1}{2k+3}$, and $\Delta_k(x)$ is the error term.

ERDÖS and SZEKERES had already considered this error term and showed that

$$\Delta_k(x) = O(x^{\frac{1}{k+1}}) \quad \text{as } x \to \infty. \quad (6)$$

Better estimates of the error have since been obtained. Let

$$\rho_k = \inf\{\rho > 0 \mid \Delta_k(x) = O(x^\rho)\}.$$

BATEMAN and GROSSWALD showed in 1958 that $\rho_2 \leq \frac{1}{6}$ and $\rho_3 \leq \frac{7}{46}$.
Sharper results are due to IVIĆ and SHIU:

$$\rho_2 \leq 0.128 < \frac{1}{6}, \quad \rho_3 \leq 0.128 < \frac{7}{46}, \quad \rho_4 \leq 0.1189,$$
$$\rho_5 \leq \frac{1}{10}, \qquad \rho_6 \leq \frac{1}{12}, \qquad \rho_7 \leq \frac{1}{14}, \quad \text{etc.}$$

I refer also to the work of KRÄTZEL (1972) on this matter.

It is conjectured that, for every $k \geq 3$,

$$\Delta_k(x) = O(x^{\frac{1}{2k}}) \quad \text{for } x \to \infty. \quad (7)$$

More specifically, taking $k = 2$:

$$\#W_2(x) = \frac{\zeta(\frac{3}{2})}{\zeta(3)}x^{\frac{1}{2}} + \frac{\zeta(\frac{2}{3})}{\zeta(2)}x^{\frac{1}{3}} + \Delta_2(x),$$ (8)

with $\Delta_2(x) = O(x^{\frac{1}{6}})$, as $x \to \infty$.

B.   Additive problems

If $h \geq 2$, $k \geq 2$, I shall use the following notation:

$$\sum hW_k = \{\sum_{i=1}^{h} n_i \mid \text{each } n_i \in W_k \cup \{0\}\},$$

$$\sum hW_k(x) = \{n \in \sum hW_k \mid n \leq x\} \quad \text{(for } x \geq 1\text{)}.$$

The additive problems concern the comparison of the sets $\sum hW_k$ with the set of natural numbers, the distribution of the sets $\sum hW_k$, and similar questions.

The distribution of $\sum 2W_2$ was treated by ERDÖS in 1975:

(1.2)

$$\#\sum 2W_2(x) = o\left(\frac{x}{(\log x)^\alpha}\right) \quad \text{(as } x \to \infty\text{), where } 0 < \alpha < \tfrac{1}{2}.$$

In particular, $\#\sum 2W_2(x) = o(x)$, so there exist infinitely many natural numbers which are not the sum of two powerful numbers.

ODONI showed in 1981 that there is no constant $C > 0$ such that

$$\#\sum 2W_2(x) \sim \frac{Cx}{(\log x)^{1/2}} \quad \text{(as } x \to \infty\text{)}.$$

The following result was conjectured by ERDÖS and IVIĆ in the 1970's and proved by HEATH-BROWN (1988):

(1.3) There is an effectively computable number n_0, such that every $n \geq n_0$ is the sum of at most three powerful numbers.

The only known exceptions up to 32000 are 7, 15, 23, 87, 111, and 119. MOLLIN and WALSH conjectured in 1986 that there are no other exceptions.

The following problem concerning 3-powerful numbers remains open:

Do there exist infinitely many natural numbers which are not sums of three 3-powerful numbers? Probably, yes.

C.   Difference problems

The problems of this kind are the following.

Problem D1. Given $k \geq 2$, determine which numbers N are of the form $N = n_1 - n_2$, where n_1, $n_2 \in W_k$. Such an expression of N is called a *representation as a difference of k-powerful numbers*, or simply a *k-powerful representation*. When $k = 2$, I simply say a powerful representation. If $\gcd(n_1, n_2) = 1$, the representation is called *primitive*; if n_1 or n_2 is a power, or 1, the representation is called *degenerate*.

Problem D2. Given $k \geq 2$, $N \geq 1$, determine the set, or just the number of representations (primitive or not, degenerate or not) of N as a difference of k-powerful numbers.

In the same vein is the following problem:

Problem D3. Given integers $N_1, N_2 \geq 1$, determine if there exist k-powerful numbers n_1, n_2, n_3 such that

$$n_2 - n_1 = N_1 \quad \text{and} \quad n_3 - n_2 = N_2.$$

In such a case, study the possible triples of such numbers.

One may also think of similar problems with several differences N_1, $N_2, \ldots, N_r \geq 1$ given in advance, but as I shall indicate, problem D3 in its simplest formulation is unsolved and certainly very difficult.

I begin by discussing problems D1 and D2. The first remark, due to MAHLER, also shows that these questions are in close relationship to the equations $X^2 - DY^2 = C$.

Thus, MAHLER said: since the equation $X^2 - 8Y^2 = 1$ has infinitely many solutions in integers (x, y), and since the number $8y^2$ is powerful, then 1 admits infinitely many degenerate (primitive) powerful representations.

In 1976, WALKER showed that 1 also has an infinite number of non-degenerate powerful (primitive) representations.

In 1981, SENTANCE showed that 2 has infinitely many primitive degenerate powerful representations, the smallest ones being:

$$2 = 27 - 25 = 70227 - 70225 = 189750627 - 189750625.$$

More recently, putting together the results in various papers, published independently and almost simultaneously by MCDANIEL, MOLLIN and WALSH, and VANDEN EYNDEN, it has been established that:

(1.4) Every natural number has infinitely many primitive degenerate powerful representations and also infinitely many primitive non-degenerate powerful representations.

Moreover, there is an algorithm to determine such representations. For a survey of the above results, see also MOLLIN (1987).

It has been asked by ERDÖS whether consecutive powerful numbers may be obtained other than as solutions of appropriate equations $EX^2 - DY^2 = 1$.

Concerning the distribution of pairs of consecutive powerful numbers, there are several conjectures by ERDÖS (1976).

First* Erdös conjecture:
$\#\{n \mid n \text{ and } n+1 \text{ are powerful}, n \leq x\} < (\log x)^c$, *where* $c > 0$ *is a constant.*

It is not even yet proved that $c'x^{\frac{1}{3}}$ is an upper bound (with a constant $c' > 0$).

Second Erdös conjecture:
There do not exist two consecutive 3-powerful numbers.

It is interesting to note that the only known examples of consecutive integers, such that one is 2-powerful and the other is 3-powerful, are $(8, 9)$ and $(12167, 12168)$.

A related conjecture is the following:

Third Erdös conjecture:
Let $a_1 < a_2 < a_3 < \cdots$ *be the sequence of 3-powerful numbers. There exist constants* $c > 0$, $c' > 0$, *such that for every sufficiently large* m,

$$a_{m+1} - a_m > cm^{c'}.$$

In particular,
$$\lim_{m \to \infty} (a_{m+1} - a_m) = \infty.$$

Fourth Erdös conjecture:
There are infinitely many 3-powerful numbers which are sums of two 3-powerful numbers.

Now I consider problem D3 in its simplest form, which concerns three consecutive powerful numbers.

---

*No one can state which was ERDÖS' first-conjecture—I would not be surprised if it was his first meaningful sentence, as a child....

With his awesome insight, ERDÖS conjectured:

Fifth Erdös conjecture:
There do not exist three consecutive powerful numbers.

This goes beyond the fact, proved by MAKOWSKI (1962) and independently by HYYRÖ (1963), that there do not exist three consecutive powers.

Of these, only the fourth conjecture has been established. In 1995, NITAJ proved that there are infinitely many 3-powerful numbers which are sums $x + y$, where x is a cube and y is a 3-powerful number. In 1998, COHN proved, more specifically, that there are infinitely many 3-powerful numbers which are sums $x + y$, where x and y are both 3-powerful numbers which are not cubes.

Later in this chapter I shall say more about the second, third, and fifth conjectures. These are difficult problems and calculations could only be of use to find three consecutive powerful numbers—if they exist. But when should the calculations be stopped, since no bounds are available?

It is very unexpected and intriguing that the existence of three consecutive powerful numbers has a relation to Fermat's Last Theorem. I shall discuss this later in this chapter.

2   Powers

I shall discuss whether a sum of two or more powers may be a power; if so, how often. A more demanding problem requires that the exponents in these powers be the same.

A.   Pythagorean triples and Fermat's problem

It is well-known that there are infinitely many primitive Pythagorean triples of integers (x, y, z) with $0 < x, y, z$, $\gcd(x, y) = 1$, y even, and $x^2 + y^2 = z^2$. All these triples are parameterized as follows:

$$x = a^2 - b^2$$
$$y = 2ab$$
$$z = a^2 + b^2$$

where $1 \leq b < a$, with $\gcd(a, b) = 1$.

In this respect, the following problem remains open: are there infinitely many Pythagorean triples (x, y, z) such that x and y are primes? This question has been treated assuming the truth of the conjecture of BOUNIAKOWSKI (1857), which is very strong.

An irreducible polynomial $f \in \mathbb{Z}[X]$ is said to be *strongly primitive* when there is no prime p such that p divides $f(k)$ for every integer k. In particular, the greatest common divisor of the coefficients of f is equal to 1.

The conjecture of BOUNIAKOWSKY is the following:

If $f \in \mathbb{Z}[X]$ is any irreducible strongly primitive polynomial, then there exist infinitely many integers n such that $|f(n)|$ is a prime.

Note that if $f(X)$ has degree 1, then $f(X) = aX + b$ with $\gcd(a, b) = 1$, and the above conjecture is true—it is the theorem of DIRICHLET on primes in arithmetical progressions.

In 1958, SCHINZEL and SIERPIŃSKI reformulated this and other conjectures and derived many consequences of the above conjecture. In particular, they showed:

(2.1) Assume that the conjecture of BOUNIAKOWSKY is true. Let a, b, c, and d be integers with $a > 0$, $d > 0$, $b^2 - 4ac \neq 0$. Assume that there exist integers x_0, y_0 such that $ax_0^2 + bx_0 + c = dy_0$. Then there exist infinitely many pairs (p, q) of prime numbers such that $ap^2 + bp + c = dq$.

Now it is easy to show:

(2.2) Every positive rational number $a/b \neq 1$ ($a > 0$, $b > 0$, $\gcd(a, b) = 1$) may be written in infinitely many ways in the form $\frac{a}{b} = \frac{p^2 - 1}{q - 1}$, where p and q are prime numbers.

PROOF.   Indeed, the equation $bX^2 - (b - a) = aY$ has the solution $(x_0, y_0) = (1, 1)$. Note that if $b > 0$, $a > 0$, then $4b(b - a) \neq 0$. By (2.1), there exist infinitely many pairs (p, q) of prime numbers such that $bp^2 - (b - a) = aq$, hence

$$\frac{a}{b} = \frac{p^2 - 1}{q - 1}. \qquad \square$$

Applying (2.2) with the rational number 2 gives:

(2.3) If the conjecture of BOUNIAKOWSKY is true, there exist infinitely many Pythagorean triples (a, b, c) where a and c are prime numbers.

PROOF. By (2.2), there exist infinitely many pairs (p, q) with p, q primes, such that $2 = \frac{p^2-1}{q-1}$. Then $p^2 = 2q - 1$. Hence, $p^2 + (q-1)^2 = q^2$, so $(p, q - 1, q)$ is a Pythagorean triple.   □

Of course, what is difficult is to prove the conjecture of BOUNIAKOWSKY. For the consequences of this conjecture see also my book RIBENBOIM (1996).

Now I turn my attention to Fermat's Last Theorem. If $n > 2$, WILES proved in 1995 that if $a^n + b^n = c^n$, then $abc = 0$. This was the long sought solution of Fermat's problem. Among the numerous partial results that were obtained before the complete proof by WILES, I want to mention just two, which are connected with the present discussion.

It has been traditional to say that *the first case of Fermat's Last Theorem is true for the prime exponent p* when there do not exist integers a, b, and c not multiples of p, such that $a^p + b^p = c^p$.

In 1909, WIEFERICH proved

(2.4) If p is an odd prime such that

$$2^{p-1} \not\equiv 1 \pmod{p^2}, \tag{1}$$

then the first case of Fermat's Last Theorem is true for p.

As I mentioned in my book *13 Lectures on Fermat's Last Theorem* (1979, second edition 1995), the first case of Fermat's Last Theorem is true for p if there exists a prime $l \leq 89$ such that

$$l^{p-1} \not\equiv 1 \pmod{p^2} \tag{2}$$

(see, in particular, GRANVILLE (1988)).

In 1985, ADLEMAN, HEATH-BROWN, and FOUVRY proved:

(2.5) There exist infinitely many prime exponents p for which the first case of Fermat's Last Theorem is true.

However, the method of proof did not allow the determination of any of these prime exponents p explicitly.

B.   Variants of Fermat's problem

It is easy to formulate variants of Fermat's problem.

a   The twisted Fermat's problem

Let A, B, $C > 0$, $\gcd(A, B, C) = 1$ be given integers; let $n \geq 3$. The problem is to determine all solutions in integers of the equation

$$AX^n + BY^n = CZ^n. \tag{3}$$

For $n > 3$ the curve with the above equation has genus greater than 1, so by the powerful theorem of FALTINGS (1983) (proof of Mordell's conjecture), there exist only finitely many solutions, i.e., triples (x, y, z) of pairwise relatively prime integers, which satisfy the given equations. Often, such equations may have easily-detected trivial solutions which are finite in number.

For each $N > 1$, let $S(N)$ denote the set of all exponents $n \leq N$ for which the equation (3) have only the trivial solutions.

It has been shown (see GRANVILLE (1985b), HEATH-BROWN (1985)) that:

(2.6) With the above notations,

$$\lim_{N \to \infty} \frac{\#S(N)}{N} = 1.$$

In words, for "almost all" exponents n, the twisted Fermat equations (for each triple (A, B, C)) has only trivial solutions.

Nevertheless, there is presently no criterion to tell if for arbitrary A, B, C, and n, the twisted Fermat equation has only trivial solutions. Also, there is no theorem giving an upper bound for the size of the integers x, y, z that might be solutions of the twisted Fermat equation.

b   Homework

In a recent paper (1999) that I titled *Homework* (and that aimed at making my colleagues work hard, now that I am retired), I stated the following conjecture:

(2.7) Let $d \geq 1$. Then there exists a natural number $n_0(d)$ such that if K is any number field of degree at most d and if $n \geq n_0(d)$, then the equation $X^n + Y^n = Z^n$ has only trivial solutions in K.

Here a trivial solution in K is any triple (x, y, z) with x, y, $z \in K$ and $xyz = 0$, and, if K contains a primitive sixth root of 1, any triple $(a, a\zeta^2, a\zeta)$ or their permutations, where a is any non-zero element of K. (In the case when such a triple is a solution, then $n \equiv \pm 1 \pmod 6$.)

For $d = 1$, taking $n_0(1) = 3$, the conjecture is no more than Fermat's Last Theorem, which was proved recently. It is also conjectured that $n_0(2) = 5$. In this respect I note: There are infinitely many quadratic fields $\mathbb{Q}(\sqrt{D})$ (D not a square) such that $X^3 + Y^3 = Z^3$ has nontrivial solutions in $\mathbb{Q}(\sqrt{D})$.

The quartic equation $X^4 + Y^4 = Z^4$ has nontrivial solutions in $\mathbb{Q}(\sqrt{D})$ if and only if $D = -7$; for more on these results, consult RIBENBOIM (1979). For $p = 5, 7, 11$, the equation $X^p + Y^p = Z^p$ has only trivial solutions in any quadratic field (see GROSS and RÖHRLICH (1978)). No more is known when $p > 11$.

Here is a different, but related, problem:

Let $n \geq 3$. How large can d be so that there exist only finitely many fields K of degree at most d such that $X^n + Y^n = Z^n$ has a nontrivial solution in K?

C.   The conjecture of Euler

EULER proved that a (non-zero) cube is not the sum of two cubes (different from zero).

In 1769, EULER conjectured, for each $k > 3$: *A non-zero kth power is not equal to the sum of $k - 1$ non-zero kth powers.*

However, a counterexample was given by LANDER and PARKIN in 1966, for $k = 5$:

$$144^5 = 27^5 + 84^5 + 110^5 + 133^5.$$

This was found by computer search, and, as far as I know, is the only example for 5th powers.

In 1988, ELKIES gave a parametrized infinite family of triples of coprime 4th powers whose sum is a 4th power. The smallest example was

$$20615673^4 = 2682440^4 + 15365639^4 + 18796760^4.$$

These examples were obtained using the arithmetic theory of elliptic curves.

It is conceivable that for every $k > 5$ there are also counterexamples to the conjecture of Euler.

I wish to formulate a problem. Let $k \geq 3$ and define $v(k)$ to be the minimum of the integers $m > 1$ such that there exists a kth power which is the sum of m natural numbers which are kth powers or 1. By Fermat's Last Theorem, $v(k) > 2$. The problem is the determination of $v(k)$. Since $3^3 + 4^3 + 5^3 = 6^3$, then $v(3) = 3$.

From ELKIES' example, $v(4) = 3$. From LANDER and PARKIN's example, $v(5) \leq 4$. It is not known if there exists a 5th power which is the sum of three 5th powers.

Clearly, $v(k) \leq 2^k$ since 2^k is the sum of 2^k integers all equal to 1. There is no experimental supporting evidence to suggest any conjecture about $v(k)$. Equivalently, nothing is known about the existence of rational points in hypersurfaces $\sum_{i=1}^{n} x_i^k = 1$. This is another instance supporting the title chosen for this lecture.

D.   The equation $AX^l + BY^m = CZ^n$

Let A, B, C be non-zero coprime integers, and let l, m, $n \geq 2$. According to the exponents, the equation

$$AX^l + BY^m = CZ^n \qquad (4)$$

exhibits a very different behavior. There are three possibilities:

$$\frac{1}{l} + \frac{1}{m} + \frac{1}{n} \quad \begin{cases} < 1 & \text{hyperbolic case,} \\ = 1 & \text{Euclidean case,} \\ > 1 & \text{spherical case.} \end{cases}$$

a   The hyperbolic case

This case was studied by DARMON and GRANVILLE in 1995. Using Faltings Theorem, they showed

(2.8) If $\frac{1}{l} + \frac{1}{m} + \frac{1}{n} < 1$, the equation (4) has only finitely many solutions in non-zero coprime integers (x, y, z).

The case when $A = B = C = 1$ was the object of more scrutiny. Only ten solutions are known (in the hyperbolic case):

$$1^l + 2^3 = 3^2,$$
$$2^5 + 7^2 = 3^4,$$
$$7^3 + 13^2 = 2^9,$$

$$2^7 + 17^3 = 71^2,$$
$$3^5 + 11^4 = 122^2,$$
$$17^7 + 76271^3 = 21063928^2,$$
$$1414^3 + 2213459^2 = 65^7,$$
$$9262^3 + 15312283^2 = 113^7,$$
$$43^8 + 96222^3 = 30042907^2,$$
$$33^8 + 1549034^2 = 15613^3.$$

These relations appear in the paper of BEUKERS (1988). It may be observed that in each case, one of the exponents is equal to 2. Must it be so?

The theorem (2.8) does not indicate when the equation has only trivial solutions. In this direction, I shall give a density theorem.

Let $k \geq 1$ and let S be a set of k-tuples of natural numbers. For each $N \geq 1$ let

$$S(N) = \{(a_1, \ldots, a_k) \in S \mid 1 \leq a_1, \ldots, a_k \leq N\}.$$

Thus, $S(N)$ has at most N^k elements.

The number

$$\underline{\delta}(S) = \liminf \frac{\#S(N)}{N^k} \tag{5}$$

is the *lower asymptotic density* of S; the number

$$\bar{\delta}(S) = \limsup \frac{\#S(N)}{N^k} \tag{6}$$

is the *upper asymptotic density* of S. If the upper and lower asymptotic densities coincide, they are simply denoted by $\delta(S)$ and this number is called the *asymptotic density* of S.

Let $S = \{(l, m, n) \mid 2 \leq l, m, n$ and the equation (4) has only trivial solutions}. Together with POWELL, I proved in 1985:

$$\underline{\delta}(S) = \liminf \frac{\#S(N)}{N^3} > 1 - \frac{8}{7} \times \frac{27}{26} \times \frac{1}{\zeta(3)} > 0$$

where $\zeta(3) = \sum_{n=1}^{\infty} \frac{1}{n^3}$ (value at 3 of the zeta function).

This is, of course, a weak result even though its proof uses in an essential way the strong theorem of Faltings.

In 1993, I proved other density results. For definiteness, I shall consider specifically the equation

$$X^l + Y^m = Z^n \tag{7}$$

with l, m, $n \geq 2$.

For each $n \geq 2$ let $D_n = \{(l, m) \mid \text{equation (7) has only trivial}$ solutions$\}$. Then (see RIBENBOIM (1993)):

(2.9) (a) The set $\{n \mid D_n \neq \varnothing\}$ has density 1. (b) The set $\{n \mid \delta(D_n) = 0\}$ has density 0.

In words, for almost all n, there is an nth power which is a sum of two powers, and for almost all n there is a positive proportion of (l, m) such that an lth power plus an mth power is not an nth power. These statements give an indication of the little that is known.

Recent work with the method developed by WILES has led to the following results (RIBET 19?? and DARMON and MEREL (1997)) for very special equations.

(2.10) (a) If $n \geq 3$, then $X^n + Y^n = 2Z^n$ has only trivial solutions (in integers with absolute value at most 1).

(b) If n is odd and $n \geq 3$, then $X^n + Y^n = Z^2$ has only trivial solutions.

(c) If $n \geq 3$, the equation $X^n + Y^n = Z^3$ has only trivial solutions.

These statements were proved using the important, and now celebrated, Theorem of Wiles (1995, see also TAYLOR (1995)):

The conjecture of SHIMURA and TANIYAMA is true for semi-stable elliptic curves defined over $\mathbb{Q}$, that is, every such curve is a modular elliptic curve.

I shall say no more about it, but the reader may wish to consult the expository article by KRAUS (1999).

b   The Euclidean case

If

$$\frac{1}{l} + \frac{1}{m} + \frac{1}{n} = 1,$$

there are only the following possibilities (up to permutation):

$$(l, m, n) \in \{(2, 3, 6), (2, 4, 4), (3, 3, 3)\}.$$

In this case, equation (4) may be dealt with by the theory of elliptic curves. This will not be discussed here.

c The spherical case

If
$$\frac{1}{l} + \frac{1}{m} + \frac{1}{n} > 1,$$
then up to permutation,
$$(l, m, n) \in \{(2, 2, n) \mid n \geq 2\} \cup \{(2, 3, 3), (2, 3, 4), (2, 3, 5)\}.$$

In 1998, BEUKERS published a theorem about equation (4) in the spherical case. To fix the terminology, the homogeneous polynomials $f, g, h \in \mathbb{Z}[X, Y]$ furnish a parametric family of solutions of (4) when
$$Af^l + Bg^m = Ch^n.$$
Thus, for all pairs of integers (s, t),
$$Af(s, t)^l + Bg(s, t)^m = Ch(s, t)^n,$$
so $(f(s, t), g(s, t), h(s, t))$ are solutions, for all choices of s, t. BEUKERS proved:

(2.11) The set of solutions of (4) in the spherical case consists of finitely many families of parametrized solutions. If the equation has one nontrivial solution, then it has infinitely many solutions.

The result for the Pythagorean equation was already described and has been known for a very long time.

ZAGIER determined the solutions for the equations $X^3 + Y^3 = Z^2$, $X^4 + Y^3 = Z^2$, and $X^4 + Y^2 = Z^3$, and these are included in BEUKERS' paper.

The equation $X^3 + Y^3 = Z^2$ has the following three families of parametrized solutions:

$$\begin{cases} x = s^4 + 6s^2t^2 - 3t^4, \\ y = -s^4 + 6s^2t^2 + 3t^4, \\ z = 6st(s^4 + 3t^4); \end{cases}$$

$$\begin{cases} x = (1/4)(s^4 + 6s^2t^2 - 3t^4), \\ y = (1/4)(-s^4 + 6s^2t^2 + 3t^4), \\ z = (3/4)st(s^4 + 3t^4); \end{cases}$$

$$\begin{cases} x = s^4 + 8st^3, \\ y = -4s^3t + 4t^4, \\ z = s^6 - 20s^3t^3 - 8t^6. \end{cases}$$

The equation $X^4 + Y^3 = Z^2$ has the following six families of parametrized solutions:

$$\begin{cases} x = (s^2 - 3t^2)(s^4 + 18s^2t^2 + 9t^4), \\ y = -(s^4 + 2s^2t^2 + 9t^4)(s^4 - 30s^2t^2 + 9t^4), \\ z = 4st(s^2 + 3t^2)(s^4 - 6s^2t^2 + 81t^4)(3s^4 - 2s^2t^2 + 3t^4); \end{cases}$$

$$\begin{cases} x = 6st(s^4 + 12t^4), \\ y = s^8 - 168s^4t^4 + 144t^8, \\ z = (s^4 - 12t^4)(s^8 + 408s^4t^4 + 144t^8); \end{cases}$$

$$\begin{cases} x = 6st(3s^4 + 4t^4), \\ y = 9s^8 - 168s^4t^4 + 16t^8, \\ z = (3s^4 - 4t^4)(9s^8 + 408s^4t^4 + 16t^8); \end{cases}$$

$$\begin{cases} x = s^6 + 40s^3t^3 - 32t^6, \\ y = -8st(s^3 - 16t^3)(s^3 + 2t^3), \\ z = s^{12} - 176s^9t^3 - 5632s^3t^9 - 1024t^{12}; \end{cases}$$

$$\begin{cases} x = -5s^6 + 6s^5t + 15s^4t^2 - 60s^3t^3 + 45s^2t^4 - 18st^5 + 9t^6, \\ y = 6s^8 - 56ts^7 + 112t^2s^6 - 168t^3s^5 + 252t^4s^4 - 168t^5s^3 + 72t^7s - 18t^8, \\ z = -29s^{12} + 156ts^{11} - 726t^2s^{10} + 2420t^3s^9 - 4059t^4s^8 + 3960t^5s^7 \\ \qquad - 2772t^6s^6 + 2376t^7s^5 - 3267t^8s^4 + 3564t^9s^3 - 1782t^{10}s^2 \\ \qquad + 324t^{11}s + 27t^{12}; \end{cases}$$

$$\begin{cases} x = s^6 + 6s^5t - 15s^4t^2 + 20s^3t^3 + 15s^2t^4 + 30st^5 - 17t^6, \\ y = 2s^8 - 8ts^7 - 56t^3s^5 - 28t^4s^4 + 168t^5s^3 - 112t^6s^2 + 88t^7s + 42t^8, \\ z = -3s^{12} + 12ts^{11} - 66t^2s^{10} - 44t^3s^9 + 99t^4s^8 + 792t^5s^7 \\ \qquad - 924t^6s^6 + 2376t^7s^5 - 1485t^8s^4 - 1188t^9s^3 + 2046t^{10}s^2 - 156t^{11}s \\ \qquad + 397t^{12}; \end{cases}$$

The equation $X^4 + Y^2 = Z^3$ has the following four families of parametrized solutions:

$$\begin{cases} x = (s^2 + 3t^2)(s^4 - 18s^2t^2 + 9t^4), \\ y = 4st(s^2 - 3t^2)(s^4 + 6s^2t^2 + 81t^4)(3s^4 + 2s^2t^2 + 3t^4), \\ z = (s^4 - 2s^2t^2 + 9t^4)(s^4 + 30s^2t^2 + 9t^4); \end{cases}$$

$$\begin{cases} x = 6st(s^4 - 12t^4), \\ y = (s^4 + 12t^4)(s^8 - 408s^4t^4 + 144t^8), \\ z = s^8 + 168s^4t^4 + 144t^8; \end{cases}$$

$$\begin{cases} x = 6st(3s^4 - 4t^4), \\ y = (3s^4 + 4t^4)(9s^8 - 408s^4t^4 + 16t^8), \\ z = 9s^8 + 168s^4t^4 + 16t^8; \end{cases}$$

$$\begin{cases} x = (3/2)st(s^4 - 3t^4), \\ y = (1/8)(s^4 + 3t^4)(s^8 - 102s^4t^4 + 9t^8), \\ z = (1/4)(s^8 + 42s^4t^4 + 9t^8). \end{cases}$$

E.   Powers as values of polynomials

The question treated now is the following: How often does a polynomial $f \in \mathbb{Z}[X]$ have values which are powers? Of course, the question is only interesting when f itself is not the power of another polynomial.

The following important and useful theorem was proved by SCHINZEL and TIJDEMAN in 1976 and holds for polynomials with rational coefficients:

(2.12) Let $f \in \mathbb{Q}[X]$ and assume that f has at least 3 simple roots (respectively, 2 simple roots). Then there exists an effectively computable constant $C > 0$ (depending on f) such that if x, y, h are integers with $y \geq 2$, $h \geq 2$ (respectively, $h \geq 3$) and $f(x) = y^h$, then $|x|$, y, $h \leq C$.

So, f may assume only finitely many values which are powers. The proof of this result required the theory of linear forms of logarithms, as had been developed by BAKER.

3   Exponential congruences

A.   The Wieferich congruence

Motivated by a criterion for the first case of Fermat's Last Theorem, I consider the following *Wieferich congruence*:

$$a^{p-1} \equiv 1 \pmod{p^2} \tag{1}$$

where p is an odd prime, and $2 \le a$, $p \nmid a$.

Due to Fermat's Little Theorem, $q_p(a) = \frac{a^{p-1}-1}{p}$ is an integer called the *Fermat quotient of p in base a*. Thus (1) holds exactly when

$$q_p(a) \equiv 0 \pmod{p}. \tag{2}$$

The Fermat quotient satisfies the following property, which was first observed by EISENSTEIN:

$$q_p(ab) \equiv q_p(a) + q_p(b) \pmod{p}. \tag{3}$$

As noted by actual computation, only rarely is $q_p(a) \equiv 0 \pmod{p}$. Thus, for $a = 2$, and $p < 4 \times 10^{12}$, if $q_p(2) \equiv 0 \pmod{p}$, then $p = 1093$ or 3511.

More generally, I consider also the congruences

$$a^{p-1} \equiv 1 \pmod{p^k}, \tag{4}$$

where $k \ge 1$, p is an odd prime, $a \ge 2$, and $p \nmid a$.

Let $l \ge 2$, l prime, $k \ge 1$, and let

$$W_l^{(k)} = \{p \text{ odd prime} \mid l^{p-1} \equiv 1 \pmod{p^k}\},$$
$$W_l^{(k)'} = \{p \text{ odd prime} \mid l^{p-1} \not\equiv 1 \pmod{p^k}\}.$$

Thus, $W_l^{(1)}$ is the set of all primes $p \ne l$. Clearly,

$$W_l^{(1)} \supseteq W_l^{(2)} \supseteq \cdots \supseteq W_l^{(k)} \supseteq \cdots.$$

Heuristically, $W_l^{(2)}$ is an infinite set, while $W_l^{(k)}$ is finite, for all $k \ge 3$. This is seen in the following way: not knowing what should be $q_p(l)$-modulo p, and supposing that the Fermat quotient may assume with the same probability each value, then if x is any positive real number,

$$\#\{p \le x \mid p \in W_l^{(2)}\} = \sum_{p \le x} \frac{1}{p} = \log \log x + O(1);$$

thus, $W_l^{(2)}$ should be infinite. For $k \geq 3$,

$$\#\{p \leq x \mid p \in W_l^{(k)}\} = \sum_{p \leq x,\, p \nmid a} \frac{1}{p^{k-1}} < \zeta(k-1) < \infty.$$

Even though the above arguments are heuristically acceptable, they are not fully justified. For $k \geq 2$ it is not known whether $W_l^{(k)}$ of $W_l^{(k)'}$ is finite or infinite.

I note the following interesting result by POWELL (1982):

(3.1) Let l be any prime. Then the set

$$S = \bigcup_{k \text{ odd}} \left(W_l^{(k)} \setminus W_l^{(k+1)} \right)$$

is infinite.

PROOF. The prime q belongs to S if and only if the q-adic value $v_q(l^{q-1}-1)$ is odd. Assume that $\{q_1, \ldots, q_n\}$ (with $n \geq 0$) is the set of odd primes in S. Let $s = 1$ when $n = 0$, or $s = \prod_{i=1}^{n}(q_i - 1)^2$ when $n \geq 1$.

Since $q_i - 1$ divides $4s$, then q_i divides $l^{4s} - 1$ (for each $i = 1, \ldots, n$). It will be shown that $l^{4s} + 1$ is a square or a double square.

Let p be any odd prime dividing $l^{4s} + 1$, so $p \mid l^{8s} - 1$ and $p \nmid l^{4s} - 1$. Hence, $p \neq q_i$ for each $i = 1, \ldots, n$. Let r be the order of l modulo p. So, $p - 1 = rk$ (implying $p \nmid r$, $p \nmid k$), and also $8s = rhp^f$ with $f \geq 0$, $p \nmid h$. Since s is a square, f is even.

Note that

$$\frac{l^{p-1} - 1}{l^r - 1} = l^{r(k-1)} + l^{r(k-2)} + \cdots + l^r + 1$$

$$\equiv k \pmod{p},$$

and similarly

$$\frac{l^{rh} - 1}{l^r - 1} \equiv h \pmod{p}.$$

So, $v_p(l^{p-1} - 1) = v_p(l^r - 1) = v_p(l^{rh} - 1)$. Since $p \neq q_1, \ldots q_n$, then $d = v_p(l^{p-1} - 1)$ is even. Now, $v_p(l^{8s} - 1) = d + f$ is also even, noting that $8s = rhf$. Since p is an arbitrary odd prime divisor of $l^{4s} + 1$, then $l^{4s} + 1 = c^2$ or $2c^2$ (for some $c \geq 1$). But, as is well-known, FERMAT had shown that the equations $X^4 + Y^4 = Z^2$ or $X^4 + Y^4 =$

$2Z^2$ have only trivial solutions (x, y, z) with $|x|, |y|, |z| \leq 1$. So this is a contradiction, proving that the set S is indeed infinite.   □

In 1985, GRANVILLE showed:

(3.2) Let $l \geq 2$ be a prime. If $W_l^{(3)}$ is finite, then there exist infinitely many primes p such that $p \equiv 1 \pmod 4$ and $l^{p-1} \not\equiv 1 \pmod{p^2}$. In particular, if $W_l^{(3)}$ is finite, then $W_l^{(2)'}$ is infinite.

From POWELL's result, $W_l^{(2)'}$ is infinite; here is also asserted that $W_l^{(2)'}$ contains infinitely many primes $p \equiv 1 \pmod 4$.

It is also interesting to consider the following question. Given an odd prime p, estimate the number of elements in the set

$$B(p) = \{a \mid 2 \leq a < p \text{ such that } a^{p-1} \equiv 1 \pmod{p^2}\}$$

or of the subset

$$B'(p) = \{q \text{ prime} \mid 2 \leq q < p, \text{ such that } q^{p-1} \equiv 1 \pmod{p^2}\}.$$

In 1966, KRUYSWIJK showed:

(3.3) There is a constant $C > 0$ such that for every p

$$\#B(p) < p^{\frac{1}{2} + \frac{C}{\log\log p}}.$$

The result is better for $B'(p)$. GRANVILLE showed in 1987:

(3.4) Let $u \geq 1$ be an integer and let p be a prime such that $p > u^{2u}$. Then

$$\#\{q \text{ prime} \mid 2 \leq q < p^{\frac{1}{u}} \text{ and } q^{p-1} \equiv 1 \pmod{p^2}\} < up^{1/2u}.$$

In particular, for every prime p,

$$\#\{q \text{ prime} \mid 2 \leq q < p^{1/2} \text{ and } q^{p-1} \equiv 1 \pmod{p^2}\} < p^{1/2}.$$

B. Primitive factors

The following theorem was proved by BANG in 1886 (for $a = 2$) and extended by ZSIGMONDY in 1892:

(3.5) Let $a \geq 2$. For every $n \geq 2$ (with the exceptions indicated below) there exists a prime p which is a primitive factor of $a^n \mp 1$, that is, p divides $a^n \mp 1$, but p does not divide $a^m \mp 1$ for all $m < n$. The only exceptions to the above are:

(i) $2^6 - 1$, $2^3 + 1$
(ii) $(2^k - 1)^2 - 1$

A detailed proof may be found, for example, in my book *Fermat's Last Theorem for Amateurs* (1999).

Let $a^n \mp 1 = AB$ with $\gcd(A, B) = 1$ and $p \mid A$ if and only if p is a primitive factor. Then A is called the *primitive part* of $a^n \mp 1$, and I shall use the notation $A = (a^n \mp 1)^*$.

The above theorem may be applied to the Mersenne numbers $M_q = 2^q - 1$ (q prime) as well as to the Fermat numbers $F_n = 2^{2^n} + 1$. So it makes sense to consider their primitive parts M_q^*, respectively F_n^*.

For each prime $L \geq 2$ let $\mathcal{N}_L = \{p$ prime $\mid$ there exists $c \geq 1$, $p \nmid c$, such that $pc = a \pm b$, where each prime factor of ab is at most equal to $L\}$. It is not known if the sets $\mathcal{N}_L$ are finite or infinite.

To further analyze the situation, I introduce other sets of primes. If $k \geq 1$ and l is any prime, let

$$\mathcal{N}_l^{(k)} = \{p \text{ prime} \mid \text{there exists } s \geq 1 \text{ such that } p^k \text{ divides } l^s + l,$$
$$\text{but } p^{k+1} \text{ does not divide } l^s + l\}.$$

For example, if $l \leq L$, then $\mathcal{N}_l^{(1)} \subseteq \mathcal{N}_L$. So, in order to show that $\mathcal{N}_L$ is infinite, it suffices to find a prime $l \leq L$ such that $\mathcal{N}_l^{(1)}$ is infinite. In other words, consider the sequence of integers $\{l+1, l^2+1, l^3+1, \ldots\}$. By (3.5), there are infinitely many primes p dividing some number of the sequence because (with the only exception $l = 2$, $s = 3$) each number $l^s + 1$ has a primitive prime factor. Are there still infinitely many such primes belonging to $\mathcal{N}_l^{(1)}$?

This is true if there would exist infinitely many primitive prime factors whose squares are not factors. A hard question to settle, but once again full of important consequences.

A result of 1968 by PUCCIONI has been improved as follows (see my own paper (1998)):

(3.6) For every $k \geq 1$ and prime $l \geq 2$:

1. $\mathcal{N}_l^{(k)} \cap \mathcal{W}_l^{(k+1)} = \begin{cases} \varnothing & \text{if } l \not\equiv 1 \pmod{2^{k+1}}, \\ \{2\} & \text{if } l \equiv 1 \pmod{2^{k+1}}, \end{cases}$

2. $\mathcal{N}_l^{(k)} \cup \mathcal{W}_l^{(k+2)}$ is an infinite set.

PROOF.   (1) First, it will be shown by induction on k that $\mathcal{N}_l^{(k)} \cap \mathcal{W}_l^{(k+1)} \subseteq \{2\}$.

If $k = 1$ and p is an odd prime such that $p \in \mathcal{N}_l^{(1)} \cap \mathcal{W}_l^{(2)}$, then $l^{p-1} \equiv 1 \pmod{p^2}$ and there exist $s \geq 1$, $c \geq 1$, such that $p \nmid c$, $l^s + 1 = pc$; since $l^p \equiv 1 \pmod{p^2}$, then $l^s \equiv l^{ps} = (pc - 1)^p \equiv -1 \pmod{p^2}$, so $p^2 \mid l^s + 1$, which is absurd.

Assume the statement true for $k \geq 1$. First note that $\mathcal{N}_l^{(k)} \cap \mathcal{W}_l^{(k+2)} \subseteq \mathcal{N}_l^{(k)} \cap \mathcal{W}_l^{(k+1)} \subseteq \{2\}$. It suffices to show that $(\mathcal{N}_l^{(k+1)} \setminus \mathcal{N}_l^{(k)}) \cap \mathcal{W}_l^{(k+1)} = \varnothing$. Let p be a prime in this set, so $l^{p-1} \equiv 1 \pmod{p^{k+2}}$ and there exists $s \geq 1$, $c \geq 1$, such that $p \nmid c$, $l^s + 1 = p^{k+1}c$; since $l^p \equiv l \pmod{p^{k+2}}$, then $l^s \equiv l^{ps} \equiv (p^{k+1}c - 1)^p \pmod{p^k + 2}$. If $p \neq 2$, then $l^s \equiv -1 \pmod{p^{k+2}}$, which is absurd. If $p = 2$, then $l^s \equiv 1 \pmod{2^{k+2}}$ and $2^{k+1}c \equiv l^s + 1 \equiv 2 \pmod{2^{k+2}}$, hence $k + 1 = 1$ and $k = 0$, which is absurd.

This shows that $\mathcal{N}_l^{(k)} \cap \mathcal{W}_l^{(k+1)} \subseteq \{2\}$.

Finally, if $2 \in \mathcal{N}_l^{(k)} \cap \mathcal{W}_l^{(k+1)}$, then $l \equiv 1 \pmod{2^{k+1}}$.

Conversely, if $l \equiv 1 \pmod{2^{k+1}}$, then $2 \in \mathcal{W}_l^{(k+1)}$ and $l + 1 \equiv 2 \pmod{2^{k+1}}$, so $s \in \mathcal{N}_l^{(1)} \subseteq \mathcal{N}_l^{(k)}$.

(2) In this proof, (2.12) will be used. For the polynomial $f(X) = 2X^{k+1} - 1$, let C be the corresponding effectively computable constant.

If $\mathcal{N}_l^{(k)} \cup \mathcal{W}_l^{(k+2)}$ is assumed to be finite, let m be a prime number such that $m > C$ and $m > \max\{p \mid p \in \mathcal{N}_l^{(k)} \cup \mathcal{W}_l^{(k+2)}\}$. Let $P = \prod_{l \neq q \leq m} q$ (each factor q being a prime number). Hence, $\varphi(P) = \prod_{l \neq q \leq m}(q - 1)$, and so $\varphi(P)$ is even and greater than C.

It is clear that $l^{\varphi(P)} \equiv 1 \pmod{P}$. Also, if $q \neq 2$ and q divides $l^{\varphi(P)} + 1$, then $q > m$—otherwise, $l \neq q \leq m$, so q divides P, hence $l^{\varphi(P)} - 1$ and $q = 2$.

It is well-known, and easy to show, that if $n \geq 2$ and l is a prime, then $l^h + 1$ is not a power. For a proof, see my book *Catalan's Conjecture* (1994), page 201.

First case. There exists a prime q such that q^{k+2} divides $l^{\varphi(P)} + 1$. If $q = 2$, then l is odd and $l^{\varphi(P)} \equiv -1 \pmod 8$. But $l^2 \equiv 1 \pmod 8$ and $l^{\varphi(P)} \equiv 1 \pmod 8$, which is absurd.

So, $q \neq 2$, hence $q > m$, and therefore $q \nmid \varphi(P)$.

Let g be the order of l modulo q, hence g divides $q - 1$. But $q \mid l^{2\varphi(P)} - 1$, so $g \mid 2\varphi(P)$, and therefore $2\varphi(P) = gh$, with q not dividing h.

Since q^{k+2} divides $l^{gh} - 1 = (l^g - 1)(l^{g(h-1)} + l^{g(h-2)} + \cdots + l^g + 1)$, and $l^g \equiv 1 \pmod q$, then the second factor above is congruent to $h \not\equiv 0 \pmod q$. Therefore, q^{k+2} divides $l^g - 1$. So, $l^{q-1} \equiv 1 \pmod{q^{k+2}}$, that is, $q \in \mathcal{W}_l^{(k+2)}$ and hence $q < m$, which is a contradiction.

Second case. If q divides $l^{\varphi(P)} + 1$, then q^{k+2} does not divide $l^{\varphi(P)} + 1$.

Since $l^{\varphi(P)} + 1$ is not a $(k+1)$th power, there exists a prime q such that $q \mid l^{\varphi(P)} + 1$, but $q^{k+1} \nmid l^{\varphi(P)} + 1$. Hence, $q \in \mathcal{N}_l^{(k)}$ and $q \leq m$. This implies that $q = 2$, and so $l^{\varphi(P)} + 1 = 2^e t^{k+1}$, where $1 \leq e \leq k$ and t is odd. But l is odd and $\varphi(P)$ is even, so $l^{\varphi(P)} \equiv 1 \pmod 4$. Hence $e = 1$, that is, $l^{\varphi(P)} + 1 = 2t^{k+1}$.

Thus, the integers $t, l \neq 0, \varphi(P) \geq 1$ are solutions of the equation $2X^{k+1} - 1 = Y^Z$. Hence, $\varphi(P) \leq C$, which is an absurdity. $\square$

In particular, $\mathcal{N}_l^{(1)} \cup \mathcal{W}_l^{(3)}$ is an infinite set. It suffices to show that $\mathcal{W}_l^{(3)}$ is a finite set (for some prime l) to conclude that $\mathcal{N}_l^{(1)}$ is infinite. For example, if $l = 2$, no integer in $\mathcal{W}_l^{(3)}$ is known.

4   Dream mathematics

One day, mathematicians will become smarter and will be able to prove many statements that are today only conjectured to be true. For the moment, it is only possible to dream. But such dreams may be organized.

A.   The statements

To demonstrate my ignorance beyond any doubt, let me discuss binomials, Mersenne numbers, Fermat numbers, powerful numbers, square-free numbers, numbers with a square factor, prime numbers, and Wieferich congruences. Isn't that enough?

Nobody knows if the the statements listed below are true.

Notations

$$P = \text{prime}$$
$$C = \text{composite}$$
$$SF = \text{square-free}$$
$$S = \text{with a square factor (different than 1)}$$
$$W = \text{powerful}$$
$$\neg W = \text{not powerful}$$

A star refers to the primitive part.

Let $\alpha \in \{P, C, SF, S, W, \neg W\}$ and let $\epsilon \in \{\text{finite}, \infty\}$. Let

$$M_q = 2^q - 1 \text{ (for } q \text{ prime): Mersenne number,}$$
$$F_n = 2^{2^n} + 1 \text{ (for } n \geq 0\text{): Fermat number.}$$

B.   Statements

I begin by considering Mersenne numbers.

$$(M_{\alpha,\epsilon}) := \#\{q \mid M_q \text{ satisfies } \alpha\} = \epsilon,$$
$$(M^*_{\alpha,\epsilon}) := \#\{q \mid M^*_q \text{ satisfies } \alpha\} = \epsilon.$$

There are many obvious implications among these statements:

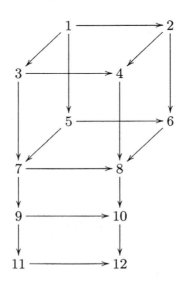

$$(1) = (M_{C,\,\text{finite}}) \qquad (2) = (M_{P,\infty})$$
$$(3) = (M^*_{C,\,\text{finite}}) \qquad (4) = (M^*_{P,\infty})$$
$$(5) = (M_{S,\,\text{finite}}) \qquad (6) = (M_{SF,\infty})$$
$$(7) = (M^*_{S,\,\text{finite}}) \qquad (8) = (M^*_{SF,\infty})$$
$$(9) = (M^*_{W,\,\text{finite}}) \qquad (10) = (M^*_{\neg W,\infty})$$
$$(11) = (M_{W,\,\text{finite}}) \qquad (12) = (M_{\neg W,\infty})$$

One may also consider the negations of these properties and for these the reverse implications are satisfied.

It is believed that $(M_{P,\infty})$ and $(M_{C,\infty})$ are both true. It is also a very deep problem to decide whether (7), (9), or even (11), is true.

Now I consider the analogous statements for Fermat numbers.

$$(F_{\alpha,\epsilon}) := \#\{n \mid F_n \text{ has property } \alpha\} = \epsilon,$$
$$(F^*_{\alpha,\epsilon}) := \#\{n \mid F^*_n \text{ has property } \alpha\} = \epsilon.$$

The same diagram of obvious implications hold for Fermat numbers by just replacing M by F.

There is no opinion as to whether $(F_{P,\infty})$ or even $(F_{\neg W,\infty})$ is true.

The next statements concern binomials $a^n \pm 1$ (where $a \geq 2$, $n \geq 1$).

It is easy to show that if $a^n - 1$ is a prime, then $a = 2$, and n is a prime. Also, if $a^n + 1$ is a prime, then $a = 2$ and n is a power of 2.

Consider the statements

$$(B(a,\pm)_{\alpha,\epsilon}) := \#\{n \mid a^n \pm 1 \text{ has property } \alpha\} = \epsilon,$$
$$(B(a,\pm)^*_{\alpha,\epsilon}) := \#\{n \mid (a^n \pm 1)^* \text{ has property } \alpha\} = \epsilon.$$

For $a = 2$, $(B(2,-)_{P,\epsilon}) = (M_{P,\epsilon})$ and $(B(2,-)_{C,\infty})$ is true. Also, $(B(2,+)_{P,\epsilon}) = (F_{P,\epsilon})$ and $(B(2,+)_{C,\infty})$ is true. The same obvious implications of the diagram (and reverse-implications) are satisfied by the properties (respectively, their negations) of the sequences of numbers $a^n \pm 1$.

Now I introduce statements concerning Wieferich congruences. Let $a \geq 2$ and

$$(W(a)_\epsilon) := \#\{p \text{ prime} \mid a^{p-1} \equiv 1 \pmod{p^2}\} = \epsilon,$$
$$(\neg W(a)_\epsilon) := \#\{p \text{ prime} \mid a^{p-1} \not\equiv 1 \pmod{p^2}\} = \epsilon.$$

Clearly, $(W(a)_{\text{finite}}) \to (\neg W(a)_\infty)$ and $(\neg W(a)_{\text{finite}}) \to (W(a)_\infty)$.

In Section 1 I indicated ERDÖS' conjecture about powerful numbers:

(E)   There do not exist three consecutive powerful numbers.

In the same order of ideas, consider the statement

(E_{finite})   There exist at most finitely many n such that $n - 1$, n, $n + 1$ are powerful.

Clearly, (E) implies (E_{finite}).

C.   Binomials and Wieferich congruences

I begin with the following useful result which was proposed in 1977 by POWELL as a problem (solution published by DE LEON in 1978):

(4.1) Let p be an odd prime, and $a \geq 2$, $m \geq 1$. If $a^m \equiv 1 \pmod{p}$, and $a^{m-1} \equiv 1 \pmod{p^2}$, then $a^m \equiv 1 \pmod{p^2}$.

PROOF.   Let $h = \operatorname{ord}(a \bmod p)$, so $h \mid m$, say $m = hk$. Also, $h \mid p-1$, so $p - 1 = hl$. Writing $a^h = 1 + cp$, then $a^{p-1} = (1 + cp)^l \equiv 1 + lcp \pmod{p^2}$. Thus, $p \mid lc$, so $p \mid c$, and hence $a^m = a^{hk} = (1+cp)^k \equiv 1 \pmod{p^2}$.   □

The following properties will be required:

$$(\mathcal{P}B(a, -)_{\alpha,\epsilon}) := \#\{p \text{ prime} \mid a^p - 1 \text{ satisfies property } \alpha\} = \epsilon,$$
$$(\mathcal{P}B(a, -)^*_{\alpha,\epsilon}) := \#\{p \text{ prime} \mid (a^p - 1)^* \text{ satisfies property } \alpha\} = \epsilon.$$

(4.2) For each $a \geq 2$ the following implications hold:

$$(W(a)_{\text{finite}}) \longrightarrow (\mathcal{P}B^*(a, -)_{SF,\infty}) \longrightarrow (B^*(a, -)_{SF,\infty})$$

$$(\mathcal{P}B^*(a, -)_{\neg W,\infty}) \longrightarrow (B^*(a, -)_{\neg W,\infty}) \longrightarrow (\neg W(a)_\infty)$$

$$(\mathcal{P}B(a, -)_{\neg W,\infty}) \longrightarrow (B(a, -)_{\neg W,\infty})$$

PROOF.   All implications but two are trivial.
$(W(a)_{\text{finite}}) \rightarrow (\mathcal{P}B^*(a, -)_{SF,\infty})$. Let

$$\{p \text{ prime} \mid a^{p-1} \equiv 1 \pmod{p^2}\} = \{p_1, \ldots, p_m\}$$

and

$$h_0 = \max\{\operatorname{ord}(a \bmod p_i) \mid i = 1, \ldots, m\},$$

and let $p > h_0$. Let $q = q(p)$ be any primitive prime factor of $a^p - 1$, so $\operatorname{ord}(a \bmod q) = p$. In particular, if $p \neq p'$, then $q(p) \neq q(p')$.

Moreover, since $p > h_0$, it follows that $q \neq p_1, \ldots, p_m$, and hence $a^{q-1} \equiv 1 \pmod{q^2}$. By (4.1), $q^2 \nmid a^p - 1$. This shows that $(a^p - 1)^*$ is square-free.

$(B^*(a, -)_{\neg W, \infty}) \rightarrow (\neg W(a)_\infty)$. Let n be such that $(a^n - 1)$ is not powerful, so there exists a prime p_n such that $p_n \mid (a^n - 1)^*$ but $p_n^2 \nmid (a^n - 1)^*$. Hence $p_n \mid a^n - 1$, but $p_n^2 \nmid a^n - 1$. By (4.1), $p_n^2 \nmid (a^{p_n-1} - 1)$. Note that $n = \operatorname{ord}(a \bmod p_n)$. So, if $n \neq m$, then $p_n \neq p_m$. This shows $(\neg W(a)_\infty)$. $\qquad \square$

In particular, taking $a = 2$, it follows that

$$(W(2)_{\text{finite}}) \rightarrow (M^*_{SF,\infty}) \rightarrow (M^*_{\neg W,\infty}) \rightarrow (\neg W(2)_\infty).$$

The following implication holds for special values of a:

(4.3) If a is even and $\sqrt{a-1}$ is powerful, then $(B(a, -)_{W,\text{finite}}) \rightarrow (B^*(a, -)_{\neg W, \infty})$.

PROOF. If $(B^*(a, -)_{\neg W, \infty})$ is false, there exists m_0 such that for every $m > m_0$ and for every primitive prime factor p_m of $a^m - 1$, $p_m^2 \mid a^m - 1$.

Choose a prime $q > m_0$; if $s \geq 1$ and l is a prime dividing $a^{q^s} - 1$, then there exists h, $0 \leq h \leq s$, such that l is a primitive prime divisor of $a^{q^h} - 1$. If $h = 0$, then $l^2 \mid a - 1$, by hypothesis. If $h \geq 1$, again $l^2 \mid a^{q^h} - 1$, because $q > m_0$. Hence $l^2 \mid a^{q^s} - 1$. This shows that $a^{q^s} - 1$ is a powerful number for every $s \geq 1$, contradicting the hypothesis. $\qquad \square$

I note also the following implication which will be useful very shortly:

(4.4) $(B(a^2, -)_{W,\text{finite}}) \rightarrow (\neg W(a)_\infty)$.

PROOF. Assume that $(\neg W(a)_\infty)$ is not true, so there exists p_0 such that if p is a prime, $p > p_0$, then $a^{p-1} \equiv 1 \pmod{p^2}$. Let $t = \prod_{p \leq p_0} p$, hence $\varphi(t) = \prod_{p \leq p_0}(p-1)$. For every $h \geq 1$ let $a_h = a^{ht\varphi(t)}$. Then $a_h - 1$ is a powerful number, as I proceed to show.

Note that $2 \nmid a_h - 1$, since a is even. If p is a prime such that $2 < p \leq p_0$, then $p(p-1)$ divides $t\varphi(t)$; from $a^{p-1} \equiv 1 \pmod{p}$, it follows that $a^{p(p-1)} \equiv 1 \pmod{p^2}$, therefore, $p^2 \mid a^{ht\varphi(t)} - 1 = a_h - 1$. Finally, if $p > p_0$ and $p \mid a_h - 1$, then by hypothesis $a^{p-1} \equiv 1 \pmod{p^2}$; hence by (4.1), $p^2 \mid a_h - 1$. Since h is arbitrary, this contradicts the hypothesis. $\square$

For the next result, the following notation will be used:

$$(QB(a,+)_{\alpha,\epsilon}) := \#\{2^n \mid a^{2^n} + 1 \text{ satisfies property } \alpha\} = \epsilon,$$
$$(QB^*(a,+)_{\alpha,\epsilon}) := \#\{2^n \mid (a^{2^n} + 1)^* \text{ satisfies property } \alpha\} = \epsilon.$$

The next proposition is the analog of (4.2):

(4.5) For each $a \geq 2$ the following implications hold:

$$
\begin{array}{ccc}
(W(a)_{\text{finite}}) \longrightarrow (QB^*(a,+)_{SF,\infty}) \longrightarrow (B^*(a,+)_{SF,\infty}) \\
\downarrow \qquad\qquad\qquad \downarrow \\
(QB^*(a,+)_{\neg W,\infty}) \longrightarrow (B^*(a,+)_{\neg W,\infty}) \longrightarrow (\neg W(a)_\infty) \\
\downarrow \qquad\qquad\qquad \downarrow \\
(QB(a,+)_{\neg W,\infty}) \longrightarrow (B(a,+)_{\neg W,\infty})
\end{array}
$$

PROOF.   Only two implications need a proof.

$(W(a)_{\text{finite}}) \to (QB^*(a,+)_{SF,\infty})$. Let $\{p \mid a^{p-1} \equiv 1 \pmod{p^2}\} = \{p_1, \ldots, p_m\}$. Let $h_0 = \max\{\text{ord}(a \bmod p_i) \mid i = 1, \ldots, m\}$. Let n be such that $2^n > h_0$ and let q be a prime dividing $(a^{2^n} + 1)^*$. So, $q \mid a^{2^n} + 1$, thus $q \mid a^{2^{n+1}} - 1$ and $\text{ord}(a \bmod q) = 2^{n+1} > h_0$; thus $q \neq p_i$ (for all i) and therefore $a^{q-1} \not\equiv 1 \pmod{q^2}$. Since $2^{n+1} \mid q - 1$ then $a^{2^{n+1}} \not\equiv 1 \pmod{q^2}$ and again $a^{2^n} \not\equiv 1 \pmod{q^2}$, showing that $(a^{2^n} + 1)^*$ is square-free.

$(B^*(a,+)_{\neg W,\infty}) \to (\neg W(a)_\infty)$. Let $n > 1$ be such that $(a^n + 1)^*$ is not powerful, so there exists a prime $p = p(a)$ such that p divides $(a^n + 1)^*$, but $p^2 \nmid (a^n + 1)^*$. It follows that $p \mid a^{2n} - 1$, but $p^2 \nmid a^n + 1$ since $(a^n + 1)^*$ and $\frac{a^n+1}{(a^n+1)^*}$ are coprime. If $p^2 \mid a^{2n} - 1$, then $p \mid a^n - 1$, so $p = 2$. But this implies that a is odd, so $p \mid a + 1$ and from the assumption that $n > 1$ it would follow that p would not be a divisor of $(a^n + 1)^*$. By (4.1), then, $a^{p-1} \not\equiv 1 \pmod{p^2}$.

To conclude, note that $2n = \text{ord}(a \bmod p_n)$, so if $n \neq n'$, then $p_n \neq p_{n'}$. This implies that $(\neg W(a)_\infty)$ is satisfied. $\square$

Taking $a = 2$, one obtains the implications involving Fermat numbers:

$$(W(2)_{\text{finite}}) \longrightarrow (F^*_{SF,\infty}) \longrightarrow (F^\alpha_{\neg W,\infty}) \longrightarrow (\neg W(2)).$$

There are, of course, many statements along these lines which may be proved in exercises by the reader. For example, consider the following ones due to ROTKIEWICZ (1965) and to WARREN and BRAY (1967):

1. Let p be a prime such that p^2 divides some Mersenne number. Then $2^{p-1} \equiv 1 \pmod{p^2}$; conversely, if p divides M_q and $2^{p-1} \equiv 1 \pmod{p^2}$, then p^2 divides M_q (this converse is just (4.1)).
2. The analogous statement holds for Fermat numbers.

D. Erdös conjecture and Wieferich congruence

I begin with an easy illustration of the connection between the Wieferich congruence and the Erdös conjecture.

(4.6) $(E_{\text{finite}}) \to (B(a^2, -)_{W, \text{finite}})$ (for any even a)

PROOF.   Indeed, if a is even and $a^{2k} - 1 = (a^k - 1)(a^k + 1)$ is powerful, the fact that $\gcd(a^k - 1, a^k + 1) = 1$ implies that $a^k - 1$, a^k, $a^k + 1$ are three consecutive powerful numbers. Thus (E_{finite}) implies $(B(a^2, -)_{W, \text{finite}})$.   □

From (4.4) and (4.6), it follows that $(E_{\text{finite}}) \to (\neg W(a)_\infty)$ for all a even. This remarkable implication was proved by GRANVILLE in 1986. In particular, $(\neg W(2)_\infty)$. In view of the theorem of WIEFERICH, (E_{finite}) implies the theorem of ADLEMAN, HEATH-BROWN, and FOUVRY (the first case of Fermat's Last Theorem is true for infinitely many prime exponents) already quoted in (2.5). Notwithstanding the fact that WILES proved Fermat's Last Theorem in all cases, the above connection with powerful numbers is very intriguing.

E. The dream in the dream

In your dreams, you have a marvelous dream and you wish it to be real. "It" is spelled ABC and it is the most tantalizing conjecture one may imagine (or dream). Yet, so simple to state!

MASON (1983, 1984) proved a theorem about polynomials that inspired MASSER in 1985 to formulate a conjecture, rephrased by OESTERLÉ in 1988 as follows:

(ABC) For every $\epsilon > 0$ there exists $K(\epsilon) > 0$ such that if A, B, C are positive integers with $\gcd(A, B, C) = 1$ and $A + B = C$, then

$$C < K(\epsilon)R^{1+\epsilon} \qquad (1)$$

where

$$R = \prod_{p|ABC} p.$$

In this respect, the following terminology is convenient. If $n \neq 0$, then $r = \prod_{p|n,\, p \text{ prime}} p$ is called the *radical* of n. So, R is the radical of ABC.

There is no attempt in the statement of the conjecture to give any indication of an effective lower bound for $K(\epsilon)$.

What is the gist of the conjecture? Taking, for example, $\epsilon = \frac{1}{2}$, $A = 2^m$ (m large), $B = 3^n$ (n large), if (1) holds, then $C < K(\frac{1}{2})6^{3/2} \prod_{p|C} p^{3/2}$. Since C is large, than C must have a large prime factor or many prime factors. At any rate, (ABC) expresses a deep connection between addition and multiplication.

A conjecture is interesting if it keeps being a conjecture for a long time, so it resists attempts to prove or to disprove it. Since $K(\epsilon)$ is not explicit, it is hard to see how to disprove the (ABC) conjecture. On the other hand, the (ABC) conjecture implies many other difficult conjectures. So it is both important and difficult to establish (ABC).

If you, a respected mathematician, would tell that you are studying (ABC), what will be the reaction? Perhaps derogatory. So, say instead that you are studying the (XYZ) conjecture. It is more mysterious.

Leaving fun aside, I will indicate what (ABC) implies, not everything however. Here is one striking implication (see OESTERLÉ (1988)).

(4.7) (ABC) $\rightarrow$ Fermat's Last Theorem is true for all sufficiently large exponents.

PROOF.   Assume that $n \geq 5$, a, b, c are positive integers with $\gcd(a, b, c) = 1$, $a < b < c$, and $a^n + b^n = c^n$. Let $\epsilon = \frac{1}{2}$, and let $K = K(\frac{1}{2})$ as per the (ABC) conjecture. So,

$$c^n < K(abc)^{3/2} < Kc^{9/2},$$

and hence $c^{n-\frac{9}{2}} < K$, and thus n is bounded, proving Fermat's Last Theorem is true for all sufficiently large exponents n.   □

Practically the same proof was used by GRANVILLE in 1997 to deal with the equation

$$AX^n + BY^n = CZ^n, \tag{2}$$

where A, B, C are non-zero coprime integers.

(4.8) (ABC) $\rightarrow$ For all sufficiently large n, the equation (2) has only trivial solutions (x, y, z), with $|x|, |y|, |z| \leq 1$.

I considered in Section D. the equation

$$AX^l + BY^m = CZ^n, \tag{3}$$

where $l, m, n \geq 2$ and A, B, C are non-zero coprime integers. It was indicated in (2.8) that if $\frac{1}{l} + \frac{1}{m} + \frac{1}{n} < 1$, then (3) has only finitely many solutions (x, y, z) with x, y, z coprime integers.
    I have shown (in 1999):

(4.9) (ABC) $\rightarrow$ There are only finitely many tuples (l, m, n) satisfying (3) for which the equation (2) has a nontrivial solution (x, y, z) in coprime integers, i.e., $|x|, |y|,$ or $|z| > 1$.

The proof of (4.9) requires the simpler (4.10) below. Let A, B, C be non-zero coprime integers, and let U be the set of 4-tuples (l, m, x, y) such that (1) $|x|, |y| > 1$, (2) $l, m \geq 2$, $\frac{1}{l} + \frac{1}{m} < 1$, and (3) $Ax^l + By^m = C$.

(4.10) (ABC) $\rightarrow$ U is a finite set.

Among the applications, one may consider differences of powers $(A = 1, B = -1)$. This includes Catalan's problem on consecutive powers (see Chapter 7 of this book). TIJDEMAN's celebrated theorem asserts that there exists an effectively computable bound $C > 0$ such that if x, y, m, n are integers with $x, y \neq 0$, $m, n \geq 2$, and $x^m - y^n = 1$, then $|x|, |y|, m, n < C$.

Let $z_1 < z_2 < z_3 < \cdots$ be the sequence of all integers which are powers, with arbitrary exponents (greater than 1). Tijdeman's theorem means that $\limsup(z_{i+1} - z_i) > 1$.

LANDAU conjectures that $\limsup(z_{i+1} - z_i) = \infty$. This has never been proved, however,

(4.11) (ABC) $\rightarrow$ Landau's conjecture is true.

ELKIES proved in 1991:

(4.12) (ABC) $\rightarrow$ Faltings theorem (i.e., Mordell's conjecture is true).

The proof is subtle.

Combining results (4.7) and (4.10), it follows from (ABC) that there exists at most finitely many 4-tuples (x, y, z, n) with $n \geq 3$, x, y, $z > 0$, $\gcd(x, y, z) = 1$, and $x^n + y^n = z^n$. Of course, no effective bound is provided. This is less than stating that (ABC) would imply Wiles' theorem (i.e., Fermat's Last Theorem is true).

The following result was proved by SILVERMAN in 1998; the simpler proof given here was kindly communicated to me by RAM MURTY.

(4.13) (ABC) $\rightarrow$ $(\neg W(a)_\infty)$ for every $a \geq 2$.

PROOF.   For every $n \geq 1$, let $a^n - 1 = u_n v_n$ where u_n is square-free, $\gcd(u_n, v_n) = 1$; so, v_n is powerful. Note that $\lim_{n \to \infty}(u_n v_n) = \infty$. Let $U = \{p \text{ prime } | \text{ there exists } n \text{ such that } p \mid u_n\}$. Since each u_n is square-free, then U is finite if and only if the set $\{u_n \mid n \geq 1\}$ is bounded.

Given $\epsilon = \frac{1}{2}$, let $K = K(\frac{1}{2})$ as mandated in the (ABC) conjecture. So,

$$u_n v_n < a^n < K(a u_n v_n^{1/2})^{3/2}$$

because v_n is powerful. So, $v_n^{1/4} < K a^{3/2} u_n^{1/2}$. If the set $\{u_n \mid n \geq 1\}$ is bounded, then so is the set $\{v_n \mid n \geq 1\}$, hence $\lim u_n v_n \to \infty$, a contradiction.

If $\{u_n \mid n \geq 1\}$ is unbounded, then U is infinite. If $p \in U$, then $p^2 \nmid (a^n - 1)$, so by (4.1), $a^{p-1} \not\equiv 1 \pmod{p^2}$, and hence $(\neg W(a)_\infty)$ is true. $\square$

Now I give another implication. With a similar argument, it is easy to show

(4.14) (ABC) $\rightarrow$ Let $a > b \geq 1$ with $\gcd(a, b) = 1$. Then the set $\{n \geq 1 \mid a^b \pm b^n$ is powerful$\}$ is finite.

PROOF.   For each n, let $a^n \pm b^n = u_n v_n$ where u_n is square-free, v_n is powerful, and $\gcd(u_n, v_n) = 1$. Let $\epsilon = \frac{1}{2}$, $K = K(\frac{1}{2})$, so by the (ABC) conjecture

$$u_n v_n < K(abu_n v_n^{1/2})^{3/2}.$$

Note that $a^n \pm b^n$ is powerful exactly when $u_n = 1$. In this case, $v_n^{1/4} < K(ab)^{3/2}$. Therefore, v_n is bounded and so is n. $\square$

As particular cases, with $a = 2$, $b = 1$, note

(4.15) (ABC) $\rightarrow$ $(M_{W,\text{finite}})$ and $(F_{W,\text{finite}})$, in words, there exist only finitely many powerful Mersenne and Fermat numbers.

The result (4.14) has been generalized by RIBENBOIM and WALSH (1999d). Let $R > 0$ be a square-free integer, let h, $k \geq 2$, and let A, B, E be non-zero integers, such that $\gcd(A, ER) = \gcd(B, ER) = 1$. For each $C \neq 0$ such that the radical of C divides R, consider the equation

$$AX^h + BY^k = EC. \qquad (4)$$

Let $S_C = \{(x, y) \mid x \geq 1, y \geq 1, \gcd(x, y) = 1 \text{ and } Ax^h + By^k = EC\}$. Let $S = \bigcup\{S_C \mid$ radical of C divides $R\}$. For each integer $n > 0$ denote by $w(n)$ the powerful part of n. So, $n = w(n)n'$ where n' is square-free and $\gcd(w(n), n') = 1$. With the above notations,

(4.16) (ABC) $\rightarrow$ For every $\epsilon > 0$ there exists only finitely many $(x, y) \in S$ such that $w(x) > x^\epsilon$ or $w(y) > y^\epsilon$. In particular, there exists only finitely many $(x, y) \in S$ such that x or y is powerful.

It is useful to observe that if $R = 1$ and $\max\{h, k\} \geq 3$, that is, $\frac{1}{h} + \frac{1}{k} < 1$, then by the well-known theorem of SIEGEL, there are only finitely many pairs (x, y) with x, $y \geq 1$, $\gcd(x, y) = 1$, such that $Ax^h + By^k = E$. If $h = k = 2$, this is the situation of the Pell equations and, as is well-known, the solutions of these equations are terms in certain binary linear recurring sequences.

In the same paper, RIBENBOIM and WALSH applied the above result to deal with powerful terms in binary linear recurring sequences. Let P, Q be non-zero coprime integers such that $P > 0$,

$D = P^2 - 4Q \neq 0$. The following two Lucas sequences are associated to the parameters (P, Q):

$$U_0 = 0, \quad U_1 = 1, \quad U_n = PU_{n-1} - QU_{n-2} \quad \text{(for } n \geq 2\text{)},$$

and

$$V_0 = 2, \quad V_1 = P, \quad V_n = PV_{n-1} - QV_{n-2} \quad \text{(for } n \geq 2\text{)}.$$

The additive relation

$$V_n^2 - DU_n^2 = 4Q^n \tag{5}$$

holds (for all $n \geq 0$). If $Q = \pm 1$ (for example, for the sequences of Fibonacci and Lucas numbers which have parameters $(1, -1)$) one has:

$$V_n^2 - DU_n^2 = 4(-1)^n. \tag{6}$$

The following result is well-known (see MOLLIN (1996)):

(4.17) (ABC) $\rightarrow$ There are only finitely many Fibonacci or Lucas numbers which are powerful.

An extension for all Lucas sequences with discriminant $D > 0$ requires the relation (5) which is dealt with in (4.16):

(4.18) (ABC) $\rightarrow$ If $D > 0$, for every $\epsilon > 0$ the sets $\{n \geq 1 \mid w(U_n) > U_n^\epsilon\}$ and $\{n \geq 1 \mid w(V_n) \geq V_n^\epsilon\}$ are finite. In particular, there are only finitely many $n \geq 1$ such that U_n or V_n is powerful.

Other types of binary linear recurring sequences were also considered in the same paper, with similar results.

A question which has been investigated is the differences between powers. For consecutive powers, see Chapter 7 of this book. MORDELL, HALL, and many others studied the differences between squares and cubes, that is, the equation

$$y^3 = x^2 + d$$

(where x, $y \geq 1$, d is any integer, not necessarily positive).

HALL conjectured:

(H) For every $\epsilon > 0$, there exists $K > 0$ (depending on ϵ) such that if $y^3 = x^2 + d$ with $x, y \geq 1$, $d \neq 0$, then $y < K|d|^{2+\epsilon}$.

It is reasonable to consider similar conjectures $(H_{m,n})$ for each pair (m, n) of positive integers such that $\frac{1}{m} + \frac{1}{n} < 1$:

$(H_{m,n})$ For every $\epsilon > 0$ such that $0 < \epsilon < \frac{1}{6}$, there exists $K > 0$ (depending on ϵ, m, n) such that if x, $y > 0$, $d \neq 0$ and $y^m = x^n + d$, then $y < K|d|^{t+\epsilon}$ where $t = n/(mn - m - n)$.

In RIBENBOIM (1999a) it is proven

(4.19) (ABC) $\rightarrow$ $(H_{m,n})$ holds for all pairs (m, n) as indicated above.

The conjectures $(H_{m,n})$ have interesting consequences indicated in the paper.

There are also strong conjectures about primes dividing values of polynomials and about powerful numbers which are values of polynomials.

First I state the conjecture of LANGEVIN (1993) :

(L) Let $f \in \mathbb{Z}[X]$ with degree $d \geq 2$ and having no multiple roots. For every $\epsilon > 0$ there exists $K = K(f, \epsilon) > 0$ such that if n is sufficiently large, then $R(f(n)) > Kn^{d-1-\epsilon}$ (where $R(f(n))$ is the radical of $f(n)$).

The conjecture of SCHINZEL (1976) is the following:

(ST) Let $f \in \mathbb{Q}[X]$ with at least three simple zeros. Then $\#\{n \geq 1 \mid f(n) \text{ is powerful}\} < \infty$.

This conjecture shoud be compared with the theorem quoted in (2.12).

It is very easy to show:

(4.20) (ST) $\rightarrow$ (E_{finite}).

PROOF.   Let $f(X) = X(X^2 - 1)$; so all the roots of f are simple. If $n-1$, n, $n+1$ are three powerful numbers, then $f(n) = (n-1)n(n+1)$ is powerful. Since $\#\{n : |f(n)| \text{ is powerful}\} < \infty$ by hypothesis, then (E_{finite}) holds. $\square$

WALSH proved in 1997 (to appear in 1999):

(4.21) (L) $\rightarrow$ (ST).

PROOF. (1) First, let $f \in \mathbb{Z}[X]$, with positive leading coefficient, $\deg(f) = d \geq 3$, and all the roots of f simple. Then there exists $C > 0$ such that for all n sufficiently large, $|f(n)| < C|n|^d$.

Let ϵ be such that $0 < \epsilon < \frac{1}{2}$ and let $K > 0$ be the constant indicated by the hypothesis (L) such that

$$R(f(n)) > K|n|^{d-1-\epsilon}$$

for all n sufficiently large.

If, moreover, $|f(n)|$ is powerful, then $R(f(n)) \leq |f(n)|^{1/2}$. Hence, $C|n|^d > K^2|n|^{2(d-1-\epsilon)}$, and therefore $C > K|n|^{d-2-2\epsilon}$. Since $d - 2 - 2\epsilon > 0$, it follows that $|n|$ remains bounded when $|f(n)|$ is powerful.

(2) Let $f \in \mathbb{Z}[X]$, with positive leading coefficient, $\deg(f) = d \geq 3$ and assume that f has at least three simple roots. The polynomial f may be written as a product of irreducible polynomials, which, by Gauss' Lemma, may be taken to be from $\mathbb{Z}[X]$. Moreover, since f has at least three simple roots, the above decomposition yields an expression $f = gh$, with g, $h \in \mathbb{Z}[X]$, $\deg(g) \geq 3$, the roots of g being the simple roots of f; moreover, g and h have positive leading coefficients and $\gcd(g, h) = 1$.

Hence, there exist polynomials g_1, $h_1 \in \mathbb{Z}[X]$ such that

$$g_1 g + h_1 h = 1.$$

If $|n|$ is sufficiently large, then $g(n)$, $g_1(n)$, $h(n)$, $h_1(n)$ are not equal to 0; as $g_1(n)g(n) + h_1(n)h(n) = 1$, it follows that $\gcd(g(n), h(n)) = 1$.

Now, if $|f(n)| = |g(n)||h(n)|$ is powerful, then also $|g(n)|$ is powerful, hence, by (1), $|n|$ is bounded.

(3) Let $f \in \mathbb{Q}[X]$ such that there exists $a^2 \in \mathbb{Z}$ and $a^2 f \in \mathbb{Z}[X]$. If f has positive leading coefficients and at least three simple roots, so does $a^2 f$. By (2), there are only finitely many $n \in \mathbb{Z}$ such that $a^2 f(n)$ is powerful, a fortiori the same holds for f.

(4) Assume that the leading coefficient a of f is negative. If the degree of f is even, let $f^-(X) = -f(X)$. If d is odd, let $f^-(X) = f(-X)$. So in both cases the leading coefficient of f^- is positive. By (3), $\{n \in \mathbb{Z} : |f^-(n)| \text{ is powerful}\}$ is finite. Therefore, $\{n : |f(n)| \text{ is powerful}\}$ is also finite. □

Langerin proved (1993):

(4.22) (ABC) $\rightarrow$ (L).

From the above results, it may be said, for example, that there are only finitely many integers n such that $n^3 + n + 1$ is powerful.

I illustrate the strength of the (ABC) conjecture with further results taken from my paper (1999). The first result concerns differences between 3-powerful numbers and powerful numbers. I will state it in a particular form, for simplicity.

Let $R \geq 1$ be a square-free integer, and let V_R be the set of all 3-powerful integers k such that there exists c, $1 \leq c < k$, with $\gcd(k, c) = 1$ and radical of c dividing R, such that $k + c$ or $k - c$ is powerful. Then,

(4.23) (ABC) $\to$ For every R as above, the set V_R is finite.

In particular, taking $R = 1$ there are only finitely many 3-powerful numbers k such that $k + 1$ or $k - 1$ is powerful. As was mentioned in subsection C., the only known examples are $2^3 + 1 = 3^2$ and $23^3 + 1 = 2^3 \times 3^2 \times 13^2$.

The next result concerns triples of powerful numbers which I state, for simplicity, in a particular case.

Let $R \geq 1$ be a square-free integer, let T_R be the set of all pairs (k, c) such that $1 \leq c < k$, $\gcd(k, c) = 1$, the radical of c divides R, and $k - c$, k, $k + c$ are powerful numbers. I proved (1999):

(4.24) (ABC) $\to T_R$ is a finite set for each square-free integer $R \geq 1$.

In particular, if $R = 1$, this shows yet again that (ABC) $\to (E_{\text{finite}})$; see GRANVILLE (1990).

It has now been amply illustrated how the (ABC) conjecture is interesting. An accessible paper on this conjecture is by NITAJ (1996).

And if it happens that you really reply that you are studying the (ABC) (and not the (XYZ)) conjecture, you know now what to explain.

References

1857 A. Bouniakowski. Nouveaux théorèmes relatifs à la distributíon des nombres premiers et á la décomposition desl entiers en facteur. *Mém. Acad. Sci. St. Petersbourg*, 6: 305–329.

1886 A. S. Bang. Taltheoretiske Untersogelser. *Tidskrift Math., Ser. 5*, 4:70–80 and 130–137.

1892 K. Zsigmondy. Zur Theorie der Potenzreste. *Monatsh. f. Math.*, 3:265–284.

1909 A. Wieferich. Zum letzten Fermatschen Theorem. *J. reine u. angew. Math.*, 136:293–302.

1935 P. Erdös and S. Szekeres. Uber die Anzahe der Abelsuhen Gruppen gegebner Ordnung und über ein verwandtes Zahlentheortisches. *Acta Sci. Math. Szeged*, 7:95–102.

1954 P. Bateman. Solution of problem 4459. *Amer. Math. Monthly*, 61:477–479.

1958 P. T. Bateman and E. Grosswald. On a theorem of Erdös and Szekeres. *Illinois J. Math.*, 2:88–98.

1958 A. Schinzel and W. Sierpiński. Sur certaines hypothèses concernant les nombres premiers. Remarques. *Acta Arith.*, 4:185–208 and 5:259 (1959).

1962 A. Makowski. Three consecutive integers cannot be powers. *Colloq. Math.*, 9:297.

1963 S. Hyyrö. On the Catalan problem (in Finnish). *Arkhimedes*, 1963, No. 1, 53–54. See Math. Reviews, 28, 1964, #62.

1965 A. Rotkiewicz. Sur les nombres de Mersenne dépourvus de diviseurs carrés et sur les nombres naturels n tels que $n^2 \mid 2^n - 2$. *Matematicky Vesnik, Beograd, (2)*, 17:78–80.

1966 D. Kruyswijk. On the congruence $u^{p-1} \equiv 1 \pmod{p^2}$. *Math. Centrum Amsterdam*, 7 pages. In Dutch.

1966 L. J. Lander and T. R. Parkin. Counterexamples to Euler's conjecture on sums of like powers. *Bull. Amer. Math. Soc.*, 72:1079.

1967 L. J. Warren and H. Bray. On the square-freeness of Fermat and Mersenne numbers. *Pacific J. Math.*, 22:563–564.

1968 S. Puccioni. Un teorema per una resoluzione parziale del famoso teorema di Fermat. *Archimede*, 20:219–220.

1970 S. W. Golomb. Powerful numbers. *Amer. Math. Monthly*, 77:848–852.

1972 E. Krätzel. Zahlen k-ter Art. *Amer. J. of Math.*, 94:309–328.

1975 P. Erdös. Problems, and results on consecutive integers. *Eureka*, 38:3–8.

1976 P. Erdös. Problems and results on consecutive integers. *Publ. Math. Debrecen*, 23:271–282.

1976 A. Schinzel and R. Tijdeman. On the equation $y^m = F(x)$. *Acta Arith.*, 31:199–204.

1976 D. T. Walker. Consecutive integer pairs of powerful numbers and related Diophantine equations. *Fibonacci Q.*, 11: 111–116.

1977 B. Powell. Problem E2631 (prime satisfying Mirimanoff's condition). *Amer. Math. Monthly*, 84:57.

1978 B. H. Gross and D. E. Röhrlich. Some results on the Mordell-Weil groups of the Jacobian of the Fermat curve. *Invent. Math.*, 44:210–224.

1978 M. J. De Leon. Solution of problem E2631. *Amer. Math. Monthly*, 85:279–280.

1979 P. Ribenboim. *13 Lectures on Fermat's Last Theorem*. Springer-Verlag, New York. Second edition with a new Epilogue, 1995.

1981 R. W. K. Odoni. On a problem of Erdoös on sums of two squareful numbers. *Acta Arith.*, 39:145–162.

1981 W. A. Sentance. Occurrences of consecutive odd powerful numbers. *Amer. Math. Monthly*, 88:272–274.

1982 A. Ivić and P. Shiu. The distribution of powerful integers. *Illinois J. Math.*, 26:576–590.

1982 B. Powell. Problem E2948 ($p|x^{p-1} - y^{p-1}$, $2 \nmid pe$, p prime occurs frequently). *Amer. Math. Monthly*, 89:334.

1983 G. Faltings. Endlichkeitssätze für abelsche Varietäten über Zahlkörpern. *Invent. Math.*, 73:349–366.

1983 R. C. Mason. Equations over function fields. In *Number Theory Voondwijkerhout*, Lecture Notes in Mathematics, 1068, 149–157. Springer-Verlag.

1984 R. C. Mason. Diophantine equations over function fields. In *London Math. Soc. Lecture Notes 96*. Cambridge University Press, Cambridge.

1985 L. M. Adleman and D. R. Heath-Brown. The first case of Fermat's last theorem. *Invent. Math.*, 79:409–416.

1985 E. Fouvry. Théorème de Brun-Titchmarsh: applications au théorème de Fermat. *Invent. Math.*, 79:383–407.

1985a A. Granville. Refining the conditions on the Fermat quotient. *Math. Proc. Cambridge Phil. Soc.*, 98:5–8.

1985b A. Granville. The set of exponents for which Fermat's last theorem is true has density one. *C. R. Math. Rep. Acad. Sci. Canada*, 7:55–60.

1985 D. R. Heath-Brown. Fermat's last theorem is true for "almost all" exponents. *Bull. London Math. Soc.*, 17:15–16.

1985 D. W. Masser. Open problems. In *Proceedings Symposium Analytic Number Theory*, edited by W. W. L. Chen, London. Imperial College.

1985 A. Nitaj. On a conjecture of Erdös on 3-powerful numbers. *Bull. London Math. Soc.*, 27:317–318.

1985 B. Powell and P. Ribenboim. Note on a paper by M. Filaseta regarding Fermat's last theorem. *Ann. Univ. Turkuensis*, 187:3–22.

1986 A. Granville. Powerful numbers and Fermat's last theorem. *C. R. Math. Rep. Acad. Sci. Canada*, 8:215–218.

1986a R. A. Mollin and P. G. Walsh. A note on powerful numbers, quadratic fields and the Pellian. *C. R. Math. Rep. Acad. Sci. Canada*, 8:109–114.

1986b R. A. Mollin and P. G. Walsh. On powerful numbers. *Intern. J. Math. and Math. Sci.*, 9:801–806.

1986 C. Eynden Vanden. Differences between squares and powerful numbers. *Fibonacci Q.*, 24:347–348.

1987 A. Granville. *Diophantine Equations with Variable Exponents with Special Reference to Fermat's Last Theorem.* PhD thesis, Queen's University.

1987 W. L. McDaniel. Representatations of every integer as the difference of nonsquare powerful numbers. *Port. Math.*, 44: 69–75.

1987 R. A. Mollin. The power of powerful numbers. *Intern. J. Math. and Math. Sci.*, 10:125–130.

1988 F. Beukers. The Diophantine equation $AX^p + BY^q = CZ^r$. *Duke Math. J.*, 91:61–88.

1988 N. D. Elkies. On $A^4 + B^4 + C^4 = D^4$. *Math. of Comp.*, 51: 825–835.

1988 A. Granville and M. B. Monagan. The first case of Fermat's last theorem is true for all prime exponents up to 714,591,116,091,389. *Trans. Amer. Math. Soc.*, 306: 329–359.

1988 D. R. Heath-Brown. Ternary quadratic forms and sums

of three square-full numbers. In *Sém. Th. Numbers Paris 1986–87*, edited by C. Goldstein. Birkhäuser, Boston.

1988 J. Oesterlé. Nouvelles approches du "théorème" de Fermat. Séminaire Bourbaki, 40ème anée, 1987/8, No. 694, *Astérisque*, 161–162, 165–186.

1988 P. Ribenboim. Remarks on exponential congruences and powerful numbers. *J. Nb. Th.*, 29:251–263.

1988 J. H. Silverman. Wieferich's criterion and the *abc*-conjecture. *J. Nb. Th.*, 30:226–237.

1990 A. Granville. Some conjectures related to Fermat's last theorem. In *Number Theory*, 177–192. W. de Gruyter, Berlin.

1990 K. A. Ribet. On modular representations of $\mathrm{Gal}(\overline{\mathbf{Q}}/\mathbf{Q})$ arising from modular forms. *Invent. Math.*, 100(2):431–476.

1991 W. D. Elkies. *ABC* implies Mordell. *Internat. Math. Res. Notices (Duke Math. J.)*, 7:99–109.

1993 M. Langevin. Cas d'égalité pour le théorème de Mason et applications de la conjecture (*abc*). *C. R. Acad. Sci. Paris, Sér. I*, 317(5):441–444.

1993 P. Ribenboim. Density results on families of Diophantine equations with finitely many solutions. *L'Enseign. Math.*, 39:3–23.

1994 P. Ribenboim. *Catalan's Conjecture.* Academic Press, Boston.

1995 H. Darmon and A. Granville. On the equations $Z^m = F(x, y)$ and $A^p + By^q = CZ^r$. *Bull. London Math. Soc.*, 27:513–544.

1995 A. Granville. On the number of solutions of the generalized Fermat equation. *Can. Math. Soc. Conference Proc.*, 15:197–207.

1995 R. Taylor and A. Wiles. Ring theoretic properties of certain Hecke algebras. *Annals of Math. (2)*, 141:553–572.

1995 A. Wiles. Modular elliptic curves and Fermat's last theorem. *Annals of Math. (2)*, 141:443–551.

1996 R. A. Mollin. Masser's conjecture used to prove results about powerful numbers. *J. Math. Sci.*, 7:29–32.

1996 A. Nitaj. La conjecture *abc*. *L'Enseign. Math.*, 42:3–24.

1996 P. Ribenboim. *The New Book of Prime Number Records.* Springer-Verlag, New York.

1997 H. Darmon and L. Merel. Winding quotients and some variations of Fermat's last theorem. *J. reine u. angew. Math.*, 490:81–100.

1997 K. A. Ribet. On the equation $a^p + 2^\alpha b^p + c^p = 0$. *Acta Arith.*, 79(1):7–16.

1997 P. G. Walsh. On the conjecture of Schinzel and Tijdeman. To appear in the Proceeding of the Zucopane Conference in honor of A. Schinzel.

1998 J. H. E. Cohn. A conjecture of Erdös on 3-powerful numbers. *Math. of Comp.*, 67:439–440.

1999 A. Kraus. On the equation $X^p + Y^q = Z^r$, a survey. To appear in *Hardy-Ramanujan J.*

1999a P. Ribenboim. *Fermat's Last Theorem for Amateurs.* Springer-Verlag, New York.

1999b P. Ribenboim. Homework.

1999c P. Ribenboim and P. G. Walsh. The *ABC* conjecture and the powerful part of terms in binary recurring sequences. *J. Nb. Th.*, 74:134–147.

2000 P. Ribenboim. *ABC* candies. *J. Nb. Th.*, 81(1):48–60.

10

What Kind of Number Is $\sqrt{2}^{\sqrt{2}}$?*

0   Introduction

Is $\sqrt{2}^{\sqrt{2}}$ a rational number?

A number is rational exactly when its decimal expansion is finite or periodic.

Since a pocket (or even a giant) calculator provides only finitely many decimal digits, it is not useful for deciding whether $\sqrt{2}^{\sqrt{2}}$ is rational or not.

Then what kind of number is $\sqrt{2}^{\sqrt{2}}$, and how to decide?

1   Kinds of numbers

First I recall the various kinds of numbers. There are the integers, which, as KRONECKER said, are "God given" and should serve as basis to build all of Mathematics.

Next, there are the rational numbers, obtained from the integers by divisions.

---

*I am grateful to P. BUNDSCHUH and M. WALDSCHMIDT for advice during the preparation of this text

PYTHAGORAS noted that if the sides of the right-angle of a triangle have measures equal to 1, then the hypotenuse, measured by $\sqrt{2}$, is not a rational number: if $\sqrt{2} = \frac{m}{n}$, then $2 = \frac{m^2}{n^2}$, so $2n^2 = m^2$, the power of 2 in the left-hand side has odd exponent, while in the right-hand side it has even exponent, which is contrary to the uniqueness of factorization of integers into prime factors. This discovery was very perplexing at the time and would demand an important change in the concept of number.

More generally, if p is a prime number, $n \geq 2$, then $\sqrt[n]{p}$ is not a rational number.

So, the extraction of roots may lead to new kinds of numbers. This may be rephrased by stating that the roots of equations $X^n - a = 0$ ($a \geq 1$) need not be rational numbers.

More generally, I examine the roots of polynomial equations with rational coefficients.

Solutions of linear equations are again rational numbers. Solutions of quadratic equations are expressible with square roots. CARDANO showed that solutions of cubic equations, as well as of biquadratic equations, are also expressible with square and cubic roots.

These discoveries led to the following question:

Are solutions of any polynomial equation (with rational coefficients and arbitrary degree) always expressible with radicals?

This problem dominated algebra from about 1750 to 1830 and was the object of important work by LAGRANGE, GAUSS, ABEL, RUFFINI, and GALOIS. This is competently described in NOVÝ's book.

At this stage, all numbers under consideration were real numbers—namely, numbers which correspond to measures of segments. Each such number has a decimal expansion and, as we said above, the rational numbers are those with finite or periodic decimal expansions. Real numbers that are not rational are called *irrational numbers*.

The equation $X^2 + 1 = 0$ cannot have a root which is a real number, since a sum of non-zero squares of real numbers is positive, so not equal to zero.

Thus, it was necessary to invent a new kind of number.

The complex numbers were introduced to insure that all polynomial equations with rational coefficients have solutions.

The complex numbers are those of the form $\alpha = a + bi$ where a, b are real numbers and $i = \sqrt{-1}$ (so $i^2 + 1 = 0$). The complex conjugate of α is $\bar{\alpha} = a - bi$, so α, $\bar{\alpha}$ are solutions of the quadratic equation $X^2 - 2aX + a^2 + b^2 = 0$ which has real coefficients.

D'ALEMBERT and GAUSS proved the fundamental theorem of algebra that says that if $f(X) = 0$ is any polynomial equation with real coefficients (or even with complex coefficients), then it has a root which is a complex number. More precisely, if the polynomial has degree $d \geq 1$, then the equation has d roots which are complex numbers (but need not be all distinct).

For convenience, we recall the usual notations:

$\mathbb{Z}$ = set of all integers
$\mathbb{Q}$ = set of all rational numbers
$\mathbb{R}$ = set of all real numbers
$\mathbb{C}$ = set of all complex numbers

The sets $\mathbb{Q}$, $\mathbb{R}$, and $\mathbb{C}$ are fields, which implies, in particular, that they are closed with respect to division (i.e., solving linear equations), while the fundamental theorem of algebra says that $\mathbb{C}$ is closed with respect to solving polynomial equations. Thus, $\mathbb{C}$ is called an *algebraically closed field*.

The consideration of equations with coefficients in $\mathbb{Z}$ (or in $\mathbb{Q}$) led to the set $\mathbb{Q}^{\mathrm{alg}}$ of all complex numbers which are roots of polynomial equations with coefficients in $\mathbb{Q}$. The set $\mathbb{Q}^{\mathrm{alg}}$ is a field which is algebraically closed and the smallest one containing $\mathbb{Q}$.

Every element of $\mathbb{Q}^{\mathrm{alg}}$ is called an *algebraic number*. Moreover, every algebraic number which is a root of a monic polynomial with coefficients in $\mathbb{Z}$ is called an *algebraic integer*.

If $\alpha \in \mathbb{Q}^{\mathrm{alg}}$ is a root of a polynomial $f(X)$ of degree $d \geq 1$, with coefficients in $\mathbb{Q}$, but of none of smaller degree, then d is called the *degree of α*. The polynomial $f(X)$ is called the *minimal polynomial* of α and it is irreducible over $\mathbb{Q}$. Also, the roots of every irreducible polynomial of degree d are algebraic numbers of degree d.

Thus, α is a rational number exactly when it is an algebraic number of degree 1. Moreover, for every $d \geq 1$ there exist algebraic numbers, and even algebraic integers α of degree d. Equivalently, for every $d \geq 1$ there exist irreducible monic polynomials $f(X) \in Z[X]$ of degree d. For example, if p is any prime number, then $X^d - p$ is irreducible.

In this respect, it should be observed that if $f(X) \in \mathbb{Z}[X]$ and $f(X) = g(X)h(X)$ with $g(X)$, $h(X) \in \mathbb{Q}[X]$, then there exist also

$g_1(X)$, $h_1(X) \in \mathbb{Z}[X]$ of the same degree as $g(X)$, $h(X)$ respectively, such that $f(X) = g_1(X)h_1(X)$. This is a lemma due to GAUSS. Hence, $f(X) \in \mathbb{Z}[X]$ is irreducible over $\mathbb{Q}$ if and only if it is irreducible over $\mathbb{Z}$.

The proof of the irreducibility of $X^d - p$ is essentially the same as the proof of the more general irreducibility criterion due to EISENSTEIN.

If $f(X) = X^d + a_1 X^{d-1} + \cdots + a_{d-1} X + a_d \in \mathbb{Z}[X]$, and if there exists a prime p such that p divides each coefficient a_i, but p^2 does not divide a_d, then $f(X)$ is irreducible.

Every complex number which is not an algebraic number is called a *transcendental number*. Explicitly, α is a transcendental number if there does not exist any polynomial $f(X)$ with coefficients in $\mathbb{Q}$, different from the zero polynomial, such that $f(\alpha) = 0$.

More generally, the numbers α_1, ..., α_n are said to be algebraically independent (over $\mathbb{Q}$) if there does not exist any polynomial $f(X_1, \ldots, X_n)$ with coefficients in $\mathbb{Q}$, different from the zero polynomial, such that $f(\alpha_1, \ldots, \alpha_n) = 0$.

I summarize the above discussion of the various kinds of numbers in Figure 10.1.

It is the moment to evoke some important classical discoveries.

A real number is expressible with quadratic radicals if and only if it is the measure of a segment which is constructible with ruler and compass (beginning with a segment of measure 1). GAUSS showed that the side of the regular polygon with n sides is constructible with ruler and compass (that is, the roots of $X^n - 1 = 0$ are expressible with quadratic radicals) if and only if n is a product of powers of 2

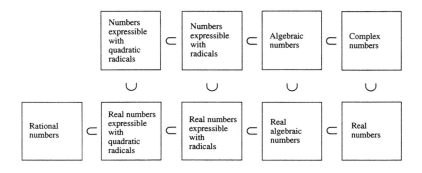

FIGURE 10.1.

and distinct prime Fermat numbers:

$$p = F_m = 2^{2^m} + 1 \quad \text{(with } m \geq 0\text{)}.$$

So, $n = 3$, 5, 17, 257, 65537, ..., or their products with powers of 2.

As a curiosity, I mention that RICHELOT gave in 1832 the explicit formula, with quadratic radicals, for the side of the regular polygon with 257 sides—it filled 83 pages of an article in Crelle's Journal, volume 9, 1832.

ABEL, RUFFINI, and GALOIS showed that if $d \geq 5$ there exist algebraic numbers of degree d which are not expressible by radicals.

More precisely, GALOIS' theorem stated: if $\alpha \in \mathbb{C}$ is the root of an irreducible polynomial $f(X) \in \mathbb{Z}[X]$, then α is expressible by radicals if and only if the Galois group of the polynomial $f(X)$ (that is, the group of automorphisms of the field generated by the roots of $f(X)$) is a solvable group.

ABEL and RUFFINI established that the symmetric group and also the alternating group on $d \geq 5$ letters are not solvable groups. So, any root of an irreducible polynomial of degree $d \geq 5$ with symmetric or alternating group, for example, is not expressible by radicals.

Incidentally, in 1933, VAN DER WAERDEN showed that "almost all" irreducible polynomials have Galois group equal to the symmetric group. Namely, if $f(X) \in \mathbb{Z}[X]$, let $\sqcap f$ denote the maximum of the absolute values of its coefficients. If $d \geq 2$, for every $N \geq 1$ let

$$I_N = \{f(X) \in \mathbb{Z}[X] \mid \deg(f) = d, f(X) \text{ is irreducible}, \sqcap f \leq N\},$$
$$S_N = \{f(X) \in I_N \mid \text{the Galois group if } f(X) \text{ is the symmetric}$$
$$\text{group on } d \text{ letters}\}.$$

Then

$$\lim_{N \to \infty} \frac{\#(S_N)}{\#(I_N)} = 1.$$

Any set which may be put into one-to-one correspondence with the set of natural numbers is said to be *countable*. So, $\mathbb{Z}$ and $\mathbb{Q}$ are countable.

Since each polynomial has only finitely many roots, it follows that $\mathbb{Q}^{\text{alg}}$ is also countable. A famous result of CANTOR is that $\mathbb{R}$ (and therefore $\mathbb{C}$) is uncountable.

So, the sets of irrational numbers and transcendental numbers are uncountable.

2 How numbers are given

The methods to decide whether a given number is transcendental depend on "how well" the number may be approached by rational numbers or algebraic numbers. This, in turn, may become apparent from the form in which the number is given. So it appears useful, as a preliminary step, to discuss how numbers may be presented.

Basically, numbers are defined from known numbers by means of "procedures".

For example, rational numbers are obtained from integers by divisions, algebraic numbers are obtained from rational numbers by solving polynomial equations.

But there are also infinitary procedures, like the following ones:

- writing infinite decimal representations according to some rule, or "randomly"
- limits of sequences
- sums of series
- infinite products
- values of definite integrals
- continued fractions
- values of functions at special points
- mathematical constants
- etc....

Now I proceed to discuss several examples.

Examples

(1) The function $x \mapsto \log x = \int_1^x \frac{dt}{t}$, defined for $0 < x < \infty$, is one-to-one and onto $\mathbb{R}$. The number e is the only real number such that $\log e = 1$.

But e is also given by

$$e = \lim_{n \to \infty} \left(1 + \frac{1}{n}\right)^n,$$

or by

$$e = \sum_{n=0}^{\infty} \frac{1}{n!}.$$

Moreover, EULER gave a simple continued fraction expression for e:

$$e = [2, 1, 2, 1, 1, 4, 1, 1, 6, 1, 1, 8, 1, \ldots]$$

that is,

$$e = 2 + \cfrac{1}{1 + \cfrac{1}{2 + \cfrac{1}{1 + \cfrac{1}{1 + \cfrac{1}{4 + \cfrac{1}{1 + \cfrac{1}{1 + \frac{1}{6} + \cdots}}}}}}}$$

(Continued fractions will be discussed in §4.)

More generally, EULER showed that if $a = 1, 2, 3, \ldots$, then

$$\frac{e^{2/a} + 1}{e^{2/a} - 1} = [a, 3a, 5a, 7a, \ldots].$$

In particular,

$$\frac{e^2 + 1}{e^2 - 1} = [1, 3, 5, 7, \ldots] \quad \text{and} \quad \frac{e + 1}{e - 1} = [2, 6, 10, 14, \ldots].$$

(2) The number π, defined as the ratio

$$\pi = \frac{\text{length of circle}}{\text{diameter of circle}} \quad \text{(for any circle)}$$

is a natural constant. But π is also given in several different ways.

GREGORY's series:

$$\frac{\pi}{4} = 1 - \frac{1}{3} + \frac{1}{5} - \frac{1}{7} + \cdots.$$

VIÈTE's infinite product:

$$\pi = \cfrac{2}{\sqrt{\frac{1}{2}}\sqrt{\frac{1}{2} + \frac{1}{2}\sqrt{\frac{1}{2}}}\sqrt{\frac{1}{2} + \frac{1}{2}\sqrt{\frac{1}{2} + \frac{1}{2}\sqrt{\frac{1}{2}}}} \cdots}.$$

WALLIS' infinite product (1685):

$$\frac{\pi}{2} = \prod_{n=1}^{\infty} \frac{2n}{2n - 1} \times \frac{2n}{2n + 1}.$$

BROUNCKER's continued fraction (published by WALLIS in 1655):

$$\frac{4}{\pi} = 1 + \cfrac{1^2}{2 + \cfrac{3^2}{2 + \cfrac{5^2}{2 + \cfrac{7^2}{2 + \cfrac{9^2}{2 + \cdots}}}}}$$

This is not a simple continued fraction, i.e., the numerators are not all equal to 1.

Using the decimal expansion of π with 35 digits, WALLIS calculated in 1685 the first 34 partial quotients of the simple continued fraction expansion of π:

$$\pi = [3, 7, 15, 1, 292, 1, 1, 1, 2, 1, 3, 1, 14, 2, 1, 1,$$
$$2, 2, 2, 2, 1, 84, 2, 1, 1, 15, 3, 13, 1, 4, 2, 6, 6, 1, \ldots].$$

The first 26 partial quotients were calculated again by LAMBERT in 1770.

This simple continued fraction for π does not reveal any regular pattern. Up to now, no one has found any regular simple continued fraction for any numbers easily related to π.

In a paper of 1878, GLAISHER assembled a collection of series and infinite products for π and its powers. The proofs constitute amusing exercises and, who knows, these formulas may even be useful. Let us quote, for example:

$$\frac{2\pi\sqrt{3}}{27} + \frac{1}{3} = \sum_{j=1}^{\infty} \frac{1}{\binom{2j}{j}},$$

$$\frac{\pi\sqrt{3}}{9} = \sum_{j=1}^{\infty} \frac{1}{j\binom{2j}{j}}.$$

The ubiquity of π is convincingly displayed by CASTELLANOS (1988).

(3) The number $\sqrt{2}^{\sqrt{2}}$ (with which I began the discussion), and, more generally, complex numbers $\alpha^\beta = e^{\beta \log \alpha}$ with $\alpha, \beta \in \mathbb{C}$, $\alpha \neq 0$, are values of an exponential function. The more interesting examples are when α, β are algebraic numbers and β is not rational. I shall return later to this topic.

(4) If $s > 1$, the *Riemann zeta function* is defined by

$$\zeta(s) = \sum_{n=1}^{\infty} \frac{1}{n^s}.$$

It is interesting to consider the values of $\zeta(s)$ when $s = 2, 3, 4, \ldots$. The famous Euler's formula gives

$$\zeta(2k) = (-1)^{k-1} \frac{(2\pi)^{2k}}{2(2k)!} B_{2k}$$

where the numbers B_n $(n \geq 0)$ are the Bernoulli numbers, defined by the formal power series

$$\frac{x}{e^x - 1} = \sum_{n=0}^{\infty} B_n \frac{x^n}{n!}.$$

Thus $B_0 = 1$, $B_1 = -\frac{1}{2}$, $B_2 = \frac{1}{6}$, $B_4 = -\frac{1}{30}$, each B_n is a rational number, and $B_{2n+1} = 0$ (for $n \geq 1$). So, for example, $\zeta(2) = \frac{\pi^2}{6}$, $\zeta(4) = \frac{\pi^4}{90}$.

It follows that $\zeta(2k)/\pi^{2k}$ is a rational number (for $k \geq 1$).

In contrast, much less is known about the values $\zeta(2k + 1)$. RAMANUJAN gave without proof the following formula (see his *Notebooks, Vol. I*, page 259, number 15 and Vol. II, page 177, number 21, published in 1957).

RAMANUJAN's discoveries, incredible formulas mostly left without proofs, have tantalized mathematicians. The work of RAMANUJAN has been the subject of authoritative books by BERNDT containing proofs and insights about likely methods behind the proofs and intuitions governing RAMANUJAN's mind. For these questions, see the books and articles of BERNDT (1974, 1977, 1985, 1989), *Ramanujan's Notebooks*, Part II, page 276 (1989). They contain the relevant references.

Let $\alpha, \beta > 0$, $\alpha\beta = \pi^2$, $k \neq 0$, then

$$\frac{1}{\alpha^k} \left\{ \frac{1}{2} \zeta(2k+1) + \sum_{j=1}^{\infty} \frac{j^{-(2k+1)}}{e^{2\alpha j} - 1} \right\} - \frac{(-1)^k}{\beta^k} \left\{ \frac{1}{2} \zeta(2k+1) + \sum_{j=1}^{\infty} \frac{j^{-(2k+1)}}{e^{2\beta j} - 1} \right\}$$

$$= 2^{2k} \sum_{j=0}^{k+1} (-1)^{j+1} \frac{B_{2j}}{(2j)!} \times \frac{B_{2k+2-2j}}{(2k+2-2j)!} \alpha^{k+1-j} \beta^j.$$

If k is even and $\alpha = \beta = \pi$, the left-hand side is 0, so this formula does not involve the values of Riemann's zeta function.

If k is odd and $\alpha = \beta = \pi$, then

$$\zeta(2k+1) = (2\pi)^{2k}\pi\sum_{j=0}^{k+1}(-1)^{j+1}\frac{B_{2j}}{(2j)!}\times\frac{B_{2k+2-2j}}{(2k+2-2j)!} = 2\sum_{j=1}^{\infty}\frac{j^{-(2k+1)}}{e^{2\pi j}-1}$$

(the last summation is actually a double infinite series involving Bernoulli numbers).

The above special case had been proved by LERCH in 1901.

The general Ramanujan formula was first proved by MALURKAR (1925). Many other mathematicians rediscovered and/or proved these formulas, such as GROSSWALD (1970, 1972) and SMART KATAYAMA, RIESEL, RAO, ZHANG, BERNDT, and SITARAMACHANDARA. This is discussed in BERNDT's book and also by SMART and KATAYAMA in 1973.

A special case is:

$$\zeta(3) = \frac{7\pi^3}{180} - \frac{1}{\pi}\sum_{j=1}^{\infty}\frac{1}{j^4}\times\frac{2\pi j}{e^{2\pi j}-1}.$$

In 1954, MARGRETHE MUNTHE HJORNAES formed the following series expansions for $\zeta(2)$, $\zeta(3)$:

$$\zeta(2) \doteq 3\sum_{j=1}^{\infty}\frac{1}{j^2\binom{2j}{j}},$$

$$\zeta(3) = \frac{5}{2}\sum_{j=1}^{\infty}\frac{(-1)^{j-1}}{j^3\binom{2j}{j}}.$$

MELZAK gave (see page 85 of volume I of his book of 1973) the above formula for $\zeta(2)$ and also the following one for $\zeta(3)$, obtained with telescoping cancellation:

$$\zeta(3) = \sum_{j=1}^{\infty}(-1)^{j+1}\frac{[(j-1)!]^2}{(3j-2)!}\left[\frac{1}{(2j-1)^2}+\frac{5}{12j(3j-1)}\right].$$

The series for $\zeta(2)$ was also given by COMTET (1974), page 89.

These formulas were obtained again (independently) by APÉRY (1979). He used this expansion of $\zeta(3)$ to prove that this number is irrational. This discovery caused a great sensation. In the words of VAN DER POORTEN (1978/9), "a proof that Euler missed...." But one should not miss reading VAN DER POORTEN's paper, written while visiting Queen's University. Another formula of the same kind is:

$$\zeta(4) = \frac{\pi^4}{90} = \frac{36}{17} \sum_{j=1}^{\infty} \frac{1}{j^4 \binom{2j}{j}}.$$

(5) The number

$$\gamma = \lim_{n \to \infty} [(1 + \frac{1}{2} + \cdots + \frac{1}{n}) - \log n]$$

is a mathematical constant, called *Mascheroni's constant* or *Euler's constant*:

$$\gamma = 0.577215665\ldots$$

It is not known if γ is an irrational number. HARDY stated that he would resign his chair in Cambridge if anyone would prove that γ is irrational—another way of saying that this was a very difficult problem (and he felt comfortably seated in Cambridge).

There are many expressions involving γ that may be derived using the gamma-function.

In a letter to GOLDBACH in 1979, EULER defined the gamma-function $\Gamma(z)$ by

$$\Gamma(z) = \frac{1}{z} \prod_{n=1}^{\infty} \left[\left(1 + \frac{1}{n}\right)^z \left(1 + \frac{z}{n}\right)^{-1} \right]$$

(valid for every complex number z, except $0, -1, -2, \ldots$). The gamma-function is analytic everywhere except at the above points, where it has simple poles. The function $\Gamma(x)$ is also given by the integral

$$\Gamma(x) = \int_0^{\infty} e^{-t} t^{x-1} dt,$$

for x real and positive, indicated by EULER. Euler's constant γ is equal to $\gamma = -\Gamma'(1)$.

DIRICHLET gave in 1836 the following integral expression:

$$\gamma = \int_0^{\infty} \left(\frac{-1}{1+t} - \frac{1}{e^t} \right) \frac{1}{t} dt.$$

Euler's constant is also related to Riemann's zeta function:

$$\gamma = \lim_{s \to 1} \frac{\zeta(s) - 1}{s - 1}.$$

On the other hand, MERTENS (1874) related it to the distribution of primes, showing that

$$\gamma = \lim_{x \to \infty} \left[\sum_{p \le x} \frac{1}{\log(1 - \frac{1}{p})} - \log \log x \right]$$

(where the above sum is for all primes $p \le x$). This may be more advantageously written as

$$\frac{e^{-\gamma}}{\log x} \sim \prod_{p \le x} \left(1 - \frac{1}{p} \right) \qquad \text{(asymptotically, as } x \to \infty).$$

For an accessible proof of this formula, see the book of HARDY and WRIGHT (1938). It is perhaps also worthwhile to mention, in connection with the gamma-function, that

$$\pi = \left[\Gamma \left(\frac{1}{2} \right) \right]^2,$$

which is nothing but a special case of the functional equation discovered by EULER:

$$\Gamma(x)\Gamma(1 - x) = \frac{\pi}{\sin \pi x}.$$

(6) GLAISHER indicated curious instances of numbers given in different ways:

$$\sqrt{\frac{1.01000100000100000001\ldots}{1.2002000020000002\ldots}} = \frac{(1.01)(1.0001)(1.000001)\ldots}{(1.1)(1.001)(1.00001)\ldots},$$

$$\frac{1}{11} + \frac{1}{111} + \frac{1}{1111} + \frac{1}{11111} + \cdots$$
$$= \frac{1}{10} + \frac{1}{1100} + \frac{1}{110000} + \frac{1}{111000000} + \frac{1}{111000000000} + \cdots,$$

and

$$\log 2 = 1 - \frac{1}{2} \sum_{j=2}^{\infty} \frac{(-1)^{j-1}}{j} S_j$$

with

$$S_j = \frac{1}{3^j} + \frac{1}{2} \left(\frac{1}{5^j} + \frac{1}{7^j} \right) + \frac{1}{4} \left(\frac{1}{9^j} + \frac{1}{11^j} + \frac{1}{13^j} + \frac{1}{15^j} \right)$$
$$+ \frac{1}{8} \left(\frac{1}{17^j} + \cdots + \frac{1}{31^j} \right) + \cdots.$$

(7) In 1974, SHANKS considered the two numbers

$$\alpha = \sqrt{5} + \sqrt{22 + 2\sqrt{5}}$$

$$\beta = \sqrt{11 + 2\sqrt{29}} + \sqrt{16 - 2\sqrt{29} + 2\sqrt{55 - 10\sqrt{29}}}.$$

To 25 decimal places, these numbers are equal to

$$7.3811759408956579709872266.$$

But are they actually equal? Even though it may seem incredible, $\alpha = \beta$. Namely, $\alpha = \beta = 4x - 1$, where x is the largest root of the polynomial

$$f(X) = X^4 - X^3 - 3X^2 + X + 1.$$

SHANKS advanced the following explanation. The Galois group of $f(X)$ is the octic group of symmetries of the square, which is generated by two elements σ, τ, with relations $\sigma^2 = 1$, $\tau^4 = 1$, $\sigma\tau\sigma = \tau^3$ (here, 1 indicates the identity automorphism).

The resolvent of $f(X)$ is the polynomial

$$g(X) = X^3 - 8X - 7 = (X + 1)(X^2 - X - 7).$$

The polynomials $f(X)$, $g(X)$ have the same discriminant, equal to $5^2 \cdot 29$.

The field $\mathbb{Q}(x)$ contains $\mathbb{Q}(\sqrt{5})$, however it does not contain $\mathbb{Q}(\sqrt{29})$.

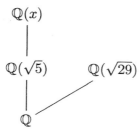

The number x is expressible with any root z of the resolvent $g(X)$. If $z = -1$, then $x = \frac{\alpha + 1}{4}$. If $z = \frac{1 + \sqrt{29}}{2\sqrt{2}}$, then $x = \frac{\beta + 1}{4}$, so $\alpha = \beta$.

In a letter (dated August 23, 1984) AGOH proposed a simpler method to obtain such identities.

Let a, b be integers, $a \geq 0$, $b \geq 0$, such that $a^2 \geq 4b$. Let $y = a - 2\sqrt{b} \geq 0$ and $k = 2ay - y^2$. Then,

$$k = 2ay - y^2 = y(2a - y) = (a - 2\sqrt{b})(a + 2\sqrt{b}) = a^2 - 4b \geq 0.$$

Hence,

$$2a + 2\sqrt{k} = 2a - y + y + 2\sqrt{2ay - y^2} = (\sqrt{2a - y} + \sqrt{y})^2.$$

This gives

$$\sqrt{2a + 2\sqrt{k}} = \sqrt{2a - y} + \sqrt{y}$$

(the minus sign may be disregarded).

Therefore,

$$\sqrt{k} + \sqrt{2a + 2\sqrt{k}} = \sqrt{2a - y} + \sqrt{k} + \sqrt{y} = \sqrt{2a - y} + \sqrt{k + y + 2\sqrt{ky}}.$$

The result now follows from

$$ky = (a^2 - 4b)a - 2(a^2 - 4b)\sqrt{b}\sqrt{a^2 - 4b} + \sqrt{2a + 2\sqrt{a^2 - 4b}}$$

$$= \sqrt{a + 2\sqrt{b}}$$

$$+ \sqrt{a^2 - 4b + a - 2\sqrt{b} + 2\sqrt{(a^2 - 4b)a - 2(a^2 - 4b)\sqrt{b}}}.$$

For example, taking $a = 11$, $b = 29$, one obtains Shanks' identity. Taking $a = 5$, $b = 3$, one gets similarly

$$\sqrt{13} + \sqrt{10 + 2\sqrt{13}} = \sqrt{5 + 2\sqrt{3}} + \sqrt{18 - 2\sqrt{3} + 2\sqrt{65 - 26\sqrt{3}}}.$$

Nested radicals—the ones that are a nightmare to typesetters—are the subject of an article of LANDAU (1994).

3   Brief historical survey

An excellent description of the historical development of the theory of transcendental numbers may be found in WALDSCHMIDT's lecture at the Séminaire d'Historie des Mathématiques, in Paris, 1983.

This brief account reproduces some of the contents of that lecture, in a more succinct presentation.

First Phase. The origin of these studies may be traced to the problem of "squaring the circle". Namely, to construct with ruler and compass the side a of the square with the same area as the circle with radius 1: $a^2 = \pi$, so $a = \sqrt{\pi}$.

A nice and informative report on this problem and properties of π may be found in the special issue of the periodical "Petit Archimède", which is dedicated to the number π (1980).

Another motivation was the discovery by PYTHAGORAS that $\sqrt{2}$ is not a rational number.

LEIBNIZ seems to be the first mathematician who employed the expression "transcendental number" (1704).

In 1737, using continued fractions, EULER proved that e^2, hence also e, is irrational.

If α, β are non-zero algebraic numbers which are multiplicatively independent (that is, if $\alpha^r \beta^s = 1$ with r, s integers, then $r = s = 0$), then $\frac{\log \alpha}{\log \beta}$ is clearly an irrational number; in 1748, EULER stated without proof that $\frac{\log \alpha}{\log \beta}$ is a transcendental number. This was again considered, much later, by HILBERT.

In 1755, EULER conjectured that π is transcendental.

LAMBERT showed in 1761 that π is irrational. He actually proved also that if r is a non-zero rational number, then $\tan r$ and e^r are irrational.

Next, LEGENDRE proved that π^2 is irrational (1794), and FOURIER gave in 1815 an easy proof that e is irrational, using the series expansion of e (see STAINVILLE). In 1840, LIOUVILLE extended this method to show that e and e^2 are irrational and not algebraic of degree 2.

Second Phase.   This phase comprises the first papers using diophantine approximation.

In 1842, DIRICHLET used the pigeon-hole principle to give results on the approximation of irrational numbers by rationals.

In his famous papers of 1844 and 1851, LIOUVILLE constructed a class of transcendental numbers, now called Liouville numbers, including, for instance, the numbers

$$\sum_{n=0}^{\infty} \frac{k_n}{a^{n!}},$$

where $a_n \geq 2$, $0 \leq k_n \leq a - 1$ for every $n \geq 0$, and $k_n \neq 0$ for infinitely many indices n. In particular,

$$\sum_{n=0}^{\infty} \frac{1}{10^{n!}}$$

is a transcendental number.

These results were based on Liouville's inequality, concerning the approximation of algebraic numbers by rationals.

Included also in this phase are the set-theoretical results of CANTOR. He showed in 1874 that $\mathbb{R}$ and $\mathbb{C}$ are not countable sets, while the set of all algebraic numbers is countable. Hence, the set of all transcendental numbers is not countable.

Third Phase. The methods of diophantine approximation were refined to allow the proofs of important results.

In 1873, HERMITE showed that e is transcendental. This was the first number (not constructed ad-hoc) shown to be transcendental.

In 1882, LINDEMANN proved that π is transcendental. This implied, of course, that π is not expressible by radicals, and therefore π and $\sqrt{\pi}$ are not constructible by ruler and compass. So, the long-standing problem of squaring the circle had a negative solution.

In his paper, LINDEMANN stated other results, without proof. They were soon after established by HERMITE and WEIERSTRASS. Lindemann and Hermite's theorem is the following: *If α is a non-zero algebraic number, then e^α is transcendental.*

An equivalent statement is the following: *If α is an algebraic number, $\alpha \neq 0, 1$, then $\log \alpha$ is transcendental.*

The more general Lindemann and Weierstrass's theorem states: *If $\alpha_1, \ldots, \alpha_n$ are algebraic numbers linearly independent over $\mathbb{Q}$, then $e^{\alpha_1}, \ldots, e^{\alpha_n}$ are algebraically independent.*

These theorems were only stated by LINDEMANN, who did not prove them. Instead, he turned his attention to Fermat's Last Theorem and published a book with a general proof of the theorem. Unfortunately, his proof was wrong.

In 1886, WEIERSTRASS considered the question whether a transcendental function (like the exponential or circular functions) takes transcendental values at algebraic points (excluding a few exceptional points). He showed that this is true for special functions, but not true for arbitrary entire transcendental functions.

Fourth Phase. In 1900, in his 7th problem, HILBERT asked whether the following is true: if α is any algebraic number ($\alpha \neq 0, 1$), and if β is any irrational algebraic number, then α^β is transcendental.

This was considered by HILBERT to be a very difficult problem. Indeed, in a seminar in Göttingen around 1920, HILBERT went so far as to say that no one present would live long enough to see this question settled.

Yet, in 1934, GEL'FOND and SCHNEIDER independently, and with different methods, solved Hilbert's 7th problem in the affirmative.

In particular, $e^{\pi} = (-1)^{-i}$ and the number $\sqrt{2}^{\sqrt{2}}$ are transcendental numbers.

At the same time, important progress was made in the theory of diophantine approximation, by THUE (1909), SIEGEL (1921), and ROTH (1955), with applications to diophantine equations and transcendental numbers.

Quite recently, beginning in 1968, BAKER published a series of fundamental papers on algebraic independence of logarithms, which have, as corollaries, most of the classical results (see his book, 1975).

The classification of transcendental numbers was initiated by MAHLER in 1932, but much has still to be done.

For the convenience of the reader, I summarize some of the results concerning the numbers considered above.

Number	Irrational	Transcendental
e	Yes, proved by EULER (1737)	Yes, proved by HERMITE (1873)
π	Yes, proved by LAMBERT (1761)	Yes, proved by LINDEMANN (1882)
γ	?	?
$\sqrt{2}^{\sqrt{2}}$	Yes	Yes, special case of GEL'FOND and SCHNEIDER's theorem (1934)
$\zeta(3)$	Yes, proved by APÉRY (1979)	?

4 Continued fractions

Continued fractions were been first introduced by BOMBELLI in 1572, in connection with the approximate calculation of square roots of numbers which are not perfect squares. They play a rather fundamental role in the approximation of numbers by rational numbers, so it is useful to summarize the relevant definitions and properties. The proofs of all statements are easily found in textbooks such as PERRON (1910, 1913), KHINTCHINE (1935), NIVEN (1957), and OLDS (1963).

A.   Generalities

Let α be a positive real number. I shall define the simple continued fraction of α. Let $a_0 \geq 0$ be the unique integer such that $a_0 \leq \alpha < a_0 + 1$, that is, $a_0 = [\alpha]$ is the integral part of α. If α is not an integer, then $0 < \alpha - a_0 < 1$. Let $\alpha_1 = \frac{1}{\alpha - a_0}$. The process is repeated with α_1, leading successively to numbers α_1, α_2, ... It terminates in a finite number of steps if and only if α is a rational number. The notation $\alpha = [a_0, a_1, a_2, \ldots]$ means that

$$\alpha = a_0 + \cfrac{1}{a_1 + \cfrac{1}{a_2 + \cdots}}.$$

This is the *simple continued fraction expansion of* α. [It is called "simple" because the "numerators" are all equal to 1; I shall not considered continued fractions which are not simple, except for some examples given in §2.]

Conversely, if $a_0 \geq 0$, a_1, a_2, ..., are positive integers, let $r_n = \frac{h_n}{k_n} = [a_0, a_1, \ldots, a_n]$ where $1 \leq h_n, k_n$, $\gcd(h_n, k_n) = 1$. Then

$$r_0 < r_2 < r_4 < \cdots \qquad \cdots < r_5 < r_3 < r_1,$$

and the following limits exist and are equal to some irrational number $\alpha = \lim r_{2n} = \lim r_{2n-1}$. The simple continued fraction expansion of α turns out to be

$$\alpha = [a_0, a_1, a_2, \ldots].$$

The convergents $r_n = \frac{h_n}{k_n}$ of α have important properties of a good approximation to α. More precisely:

(4.1) For every $n \geq 1$: $|\alpha - \frac{h_n}{k_n}| < \frac{1}{k_n k_{n+1}} < \frac{1}{k_n^2}$.

(4.2) For every $n \geq 1$: $|\alpha k_n - h_n| < |\alpha k_{n-1} - h_{n-1}|$, hence

$$\left| \alpha - \frac{h_n}{k_n} \right| < \left| \alpha - \frac{h_{n-1}}{k_{n-1}} \right|.$$

(4.3) For every $n \geq 1$, the convergent $\frac{h_n}{k_n}$ is the "best approximation" with denominator at most k_n, that is, if $|b\alpha - a| < |k_n \alpha - h_n|$, then $b > k_n$ and so if $|\alpha - \frac{a}{b}| < |\alpha - \frac{h_n}{k_n}|$, then $b > k_n$.

Conversely,

(4.4) If $\frac{a}{b}$ is a best approximation of α, then $\frac{a}{b}$ is equal to a convergent $\frac{h_n}{k_n}$, with $n \geq 0$.

B. Periodic continued fractions

An infinite simple continued fraction

$$\alpha = [a_0, a_1, a_2, \ldots]$$

is *periodic* if there exists $n_0 \geq 0$, $t > 0$ such that $a_{n+t} = a_n$ for every $n \geq n_0$. Choosing the smallest such t and n_0, the following notation is used:

$$\alpha = [a_0, \ldots, a_{n_0-1}, \overline{a_{n_0}, a_{n_0+1}, \ldots, a_{n_0+t-1}}].$$

$(a_0, \ldots, a_{n_0-1})$ is the *pre-period*, n_0 is the length of the pre-period, $(a_{n_0}, a_{n_0+1}, \ldots, a_{n_0+t-1})$ is the *period*, and t is the length of the period. If $n_0 = 0$, then the continued fraction is *purely periodic*. By the minimal choice of n_0, $a_{n_0-1} \neq a_{n_0+t-1}$.

Now I study the continued fraction expansion of real quadratic irrational numbers α. Each such number may be written in the form $\alpha = \frac{p \pm \sqrt{D}}{q}$, where p, $q \neq 0$, $D > 1$ are integers, and D is not a square. From $\frac{p-\sqrt{D}}{q} = \frac{-p+\sqrt{D}}{-q}$, it may be always assumed that α is in the form $\frac{p+\sqrt{D}}{q}$.

Moreover, if $\frac{D-p^2}{q}$ is not an integer, say $\frac{D-p^2}{q} = \frac{c}{d}$, then $\alpha = \frac{dp+\sqrt{Dd^2}}{dq}$ and now $\frac{Dd^2-d^2p^2}{dq} = d\frac{D-p^2}{q} = c$ is an integer.

So, there is no loss of generality to assume that $\alpha = \frac{p+\sqrt{D}}{q}$ and q divides $D - p^2$.

EULER proved (1737):

(4.5) If α has an infinite periodic simple continued fraction expansion, then α is a real quadratic irrational number.

The most important result about periodic continued fractions is the converse. It was proved by LAGRANGE in 1770:

(4.6) The continued fraction expansion of every real quadratic irrational number α is periodic.

For example, the golden number $\frac{\sqrt{5}+1}{2}$ and $\sqrt{2}$ have the following continued fraction expansions:

$$\frac{\sqrt{5}+1}{2} = [1, 1, 1, \ldots],$$
$$\sqrt{2} = [1, 2, 2, 2, \ldots].$$

It is important to note that the simple continued fraction expansions of any real algebraic number of degree higher than two seem to have random quotients which do not remain bounded. For extensive numerical calculations and statistical analysis, see the paper of BRENT, VAN DER POORTEN, and TE RIELE (1996) .

The next result is about purely periodic continued fractions.

The conjugate of $\alpha = \frac{p+\sqrt{D}}{q}$ is denoted by $\alpha' = \frac{p-\sqrt{D}}{q} = \frac{-p+\sqrt{D}}{-q}$.

(4.7) The simple continued fraction expansion of the real quadratic irrational number α is purely periodic if and only if $1 < \alpha$ and $-1 < \alpha' < 0$. Moreover, if $1 < \alpha$ and $\alpha' < -1$, then the pre-period has only one element.

In 1828, GALOIS proved:

(4.8) Let $\alpha = \frac{p+\sqrt{D}}{q}$ with $p, q \neq 0$, $D > 0$ integers, D not a square, and q dividing $D^2 - p$. Let $\alpha' = \frac{-p+\sqrt{D}}{-q} = \frac{p-\sqrt{D}}{q}$ be its conjugate. If $\alpha = [\overline{a_0, a_1, \cdots, a_{t-1}}]$, then $\frac{1}{\alpha'} = [\overline{a_{t-1}, a_{t-2}, \ldots, a_1, a_0}]$.

In 1828, LEGENDRE gave the following result for the simple continued fraction expansion of $\sqrt{D}$, where D is a positive integer that is not a square:

(4.9) $\sqrt{D} = [a_0, \overline{a_1, a_2, \ldots, a_2, a_1, 2a_0}]$, that is, the pre-period has length 1, the period consists of a symmetric part followed by the double of the term in the pre-period (note that the number of terms in the period may be even or odd).

The following result is interesting:

(4.10) If the continued fraction expansion of $\sqrt{D}$ has period with an odd number of terms, then D is the sum of two squares.

FERMAT considered the equation $X^2 - DY^2 = 1$ (where $D > 0$ is a square-free integer) and he stated that it has infinitely many solutions in natural numbers. This was first proved by LAGRANGE in 1770 using the theory of continued fractions.

(4.11) Let $D > 0$ be a square-free integer. Let $\frac{h_n}{k_n}$ be the convergents of the continued fraction of $\sqrt{D}$, and t the length of the period.

(1) The solutions in natural numbers of $X^2 - DY^2 = 1$ are $(1,0)$ and (h_{nt-1}, k_{nt-1}) when t is even, and (h_{2nt-1}, k_{2nt-1}) when t is odd, for all $n \geq 1$. Thus, the equation has infinitely many solutions.

(2) If t is even, the equation $X^2 - DY^2 = -1$ has no solution in natural numbers, while if t is odd then its solutions in natural numbers are (h_{nt-1}, k_{nt-1}) for all odd $n \geq 1$.

(3) For all $n \geq 1$: $h_{nt-1} + k_{nt-1}\sqrt{D} = (h_{t-1} + +k_{t-1}\sqrt{D})^n$.

C. Simple continued fractions of π and e

Now I turn my attention to the numbers π and e.

As already indicated in §2, the simple continued fraction expansion of π is

$$\pi = [3, 7, 15, 1, 292, 1, 1, 1, 2, 1, 3, 1, \ldots].$$

The convergents are

$$\frac{3}{1}, \frac{22}{7}, \frac{133}{106}, \frac{355}{113}, \frac{103993}{33102}, \frac{104348}{33215}, \frac{208341}{66317}, \frac{3123689}{99532}, \ldots$$

By **(4.3)**, the convergents are the best approximating for π. For some convergents the actual approximation is much better than expected. Thus

$$\left| \pi - \frac{22}{7} \right| \approx \frac{1}{10^3},$$

$$\left| \pi - \frac{333}{106} \right| \approx \frac{8}{10^5},$$

$$\left| \pi - \frac{355}{113} \right| \approx \frac{26}{10^8}.$$

The value $\frac{22}{7}$ was already known by ARCHIMEDES, while ADRIANUS METIUS (1571–1635) knew the values $\frac{133}{106}, \frac{355}{113}$. Already in 1685 WALLIS had computed the 34th convergent. It should also be noted that the convergents

$$\frac{h_{12}}{k_{12}} = \frac{5419351}{1725033}, \qquad \frac{h_{27}}{k_{27}} = \frac{428224593349304}{136308121570117}$$

were given by R. ARIMA, Lord of Kurume in Japan, by 1769, and these provide an approximation to π with an error of about 10^{-29}.

As already stated in §2, EULER gave simple infinite continued fractions for

$$\frac{e^{2/a}+1}{e^{2/a}-1} \quad (\text{for } a \geq 1)$$

and also for e. I shall give the proof which is pretty.

(4.12) If $a \geq 1$ is any integer, then

$$\frac{e^{2/a}+1}{e^{2/a}-1} = [a, 3a, 5a, 7a, \ldots].$$

In particular,

$$\frac{e^2+1}{e^2-1} = [1, 3, 5, 7, \ldots],$$

$$\frac{e+1}{e-1} = [2, 6, 10, 14, \ldots].$$

PROOF.   To establish this expansion I consider, for every $m \geq 0$, the series

$$S_m = \sum_{i=0}^{\infty} \frac{2^m(m+i)!}{i!(2m+2i)!}\left(\frac{1}{a}\right)^{2i+m}.$$

It converges, as seen by comparing it with the series

$$\sum_{i=0}^{\infty} \frac{2^m}{i!}\left(\frac{1}{a}\right)^{2i+m} = \left(\frac{2}{a}\right)^m e^{1/a^2}.$$

Note that

$$S_0 = \sum_{i=0}^{\infty} \frac{1}{(2i)!}\left(\frac{1}{a}\right)^{2i} = \frac{e^{1/a}+e^{-1/a}}{2},$$

$$S_1 = \sum_{i=0}^{\infty} \frac{1}{(2i+1)!}\left(\frac{1}{a}\right)^{2i+1} = \frac{e^{1/a}-e^{-1/a}}{2}.$$

By a simple calculation, one sees that

$$S_m - (2m+1)aS_{m+1} = S_{m+2}$$

for every $m \geq 0$.

Let $R_m = \frac{S_m}{S_{m+1}}$, so

$$R_0 = \frac{e^{1/a}+e^{-1/a}}{e^{1/a}-e^{-1/a}} = \frac{e^{2/a}+1}{e^{2/a}-1}.$$

Also, $R_m = (2m+1)a + \frac{1}{R_{m+1}}$, thus, in particular,

$$R_0 = a + \frac{1}{R_1}, \quad R_1 = 3a + \frac{1}{R_2}, \quad R_2 = 5a + \frac{1}{R_3}, \quad \ldots$$

This shows, as required, that

$$\frac{e^{2/a}+1}{e^{2/a}-1} = [a, 3a, 5a, 7a, \ldots]. \qquad \square$$

If α is any positive real number, and if a_0, a_2, ... are positive integers, one defines $[\alpha] = \alpha$, and by induction,

$$[a_0, a_1, \ldots, a_n, \alpha] = [a_0, \ldots, a_{n-1}, a_n + \frac{1}{\alpha}].$$

EULER first discovered by explicit calculation, and then gave a proof of, the continued fraction expansion of the number e; the simple proof below is due to HURWITZ (1891b):

(4.13) $e = [2, 1, 2, 1, 1, 4, 1, 1, 6, \ldots]$

PROOF.   From $\frac{e+1}{e-1} = [2, 6, 10, 14, \ldots]$ it follows that

$$\frac{2}{e-1} = \frac{e+1}{e-1} - 1 = [1, 6, 10, 14, \ldots],$$

hence $\frac{e-1}{2} = [0, 1, 6, 10, 14, \ldots]$. One now needs to express $2 \times [0, 1, 6, 10, 14, \ldots]$ as a continued fraction.

If α is any real number, then $2 \times [0, 2a+1, \alpha] = [0, a, 1, 1, \frac{\alpha-1}{2}]$. Indeed,

$$2 \times \frac{1}{(2a+1) + \frac{1}{\alpha}} = \frac{1}{a + (\frac{1}{2} + \frac{1}{2\alpha})} = \frac{1}{a + \frac{1}{\frac{2\alpha}{\alpha+1}}}$$

$$= \frac{1}{a + \frac{1}{1+\frac{\alpha-1}{\alpha+1}}} = \frac{1}{a + \frac{1}{1+\frac{1}{1+\frac{\alpha-1}{2}}}} = [0, a, 1, 1, \frac{\alpha-1}{2}].$$

I shall repeatedly use this formula.
Let $\alpha = [6, 10, 14, \ldots]$, then

$$2 \times [0, 1, \alpha] = [0, 0, 1, 1, \frac{\alpha-1}{2}] = [1, 1, \frac{\alpha-1}{2}].$$

But

$$\frac{\alpha - 1}{2} = \frac{1}{2} \times [5, 10, 14, \ldots] = \frac{1}{2} \times [0, 0, 5, 10, 14, \ldots]$$
$$= [0, 2 \times [0, 5, 10, 14, \ldots]].$$

Now let $\beta = [10, 14, 18, \ldots]$ and compute

$$2 \times [0, 5, \beta] = [0, 2, 1, 1, \frac{\beta - 1}{2}].$$

Again,

$$\frac{\beta - 1}{2} = \frac{1}{2}[9, 14, 18, \ldots] = [0, 2 \times [0, 9, 14, 18, \ldots]].$$

So one has already $e = 1 + [1, 1, 0, 0, 2, 1, 1, \frac{\beta-1}{2}] = [2, 1, 2, 1, 1, \frac{\beta-1}{2}]$. More generally, if $\gamma = [4m + 2, 4(m + 1) + 2, \ldots]$, then

$$2 \times [0, 4(m - 1) + 1, \gamma] = [1, 1, \frac{\gamma - 1}{2}]$$

and by induction,

$$e = [2, 1, 2, 1, 1, 4, 1, 1, \ldots, 2m, 1, 1, \ldots],$$

concluding the proof. $\qquad\qquad\qquad\qquad\qquad\qquad\qquad\qquad\qquad\qquad\square$

With the same method, HURWITZ also proved that

$$e^2 = [7, 2, 1, 1, 3, 18, 5, 1, 1, 6, 30, 8, 1, 1, 9, 42, \ldots],$$

and the pattern of quotients is easily recognized as being $a_{5(m-1)+1} = 3m - 1$, $a_{5(m-1)+2} = 1$, $a_{5(m-1)+3} = 1$, $a_{5(m-1)+4} = 3m$, $a_{5m} = 12m + 6$, for $m = 1, 2, 3, \ldots$.

With self-explanatory notation I write

$$e = [2, \overline{1, 2m, 1}]_{m > 1} \quad \text{and} \quad e^2 = [7, \overline{3m - 1, 1, 1, 3m, 12m + 6}]_{m \geq 1}.$$

From (4.7) it follows at once that e and e^2 are not roots of quadratic equations with integer coefficients.

5 Approximation by rational numbers

The kind of an irrational number, whether it is algebraic or transcendental, depends on how well it may be approximated by rational numbers. Thus, the concepts of approximation are central in the study of irrational numbers.

The leading idea in this section goes back to LIOUVILLE and DIRICHLET. In 1909, THUE considered the order of approximation of real algebraic numbers by rational numbers, while studying the solution of certain types of diophantine equations. I shall describe later this relationship.

A.   The order of approximation

In the next considerations, the rational numbers are written $\frac{a}{b}$ where $b \geq 1$ and $\gcd(a, b) = 1$.

Let $\alpha \in \mathbb{R}$, $\nu \in \mathbb{R}$, $\nu \geq 1$. The number α is said to be *approximable by rational numbers to the order* $\nu \geq 1$ when there exists $C > 0$ (depending on α, ν) and infinitely many rational numbers $\frac{a}{b}$ such that

$$\left| \alpha - \frac{a}{b} \right| < \frac{C}{b^\nu}.$$

Clearly, if α is approximable by rational numbers to the order ν and $\nu \geq \nu' \geq 1$, then α is also approximable to the other ν'.

Let $\nu(\alpha) = \sup\{\nu \in \mathbb{R} \mid \alpha$ is approximable by rational numbers to the order $\nu\}$. Thus, $1 \leq \nu(\alpha) \leq \infty$.

One deduces at once:

(5.1) Let $\alpha \in \mathbb{R}$.

(1) For every $\epsilon > 0$ there exists an integer $b_0 \geq 1$ such that if $\frac{a}{b}$ is a rational number with denomonator $b \geq b_0$, then $|\alpha - \frac{a}{b}| > \frac{1}{b^{\nu(\alpha)+\epsilon}}$.

(2) For every $\epsilon > 0$ there exists $C(\alpha, \epsilon) = C > 0$ such that $0 < C < 1$ and $|\alpha - \frac{a}{b}| > \frac{C}{b^{\nu(\alpha)+\epsilon}}$, for all $\frac{a}{b} \neq \alpha$.

A first easy remark is the following:

(5.2) Every rational number is approximable by rational numbers to the order 1 (using any constant $C > 1$), but not to any order $1 + \epsilon$ ($\epsilon > 0$); thus, $\nu(\alpha) = 1$ for every $\alpha \in \mathbb{Q}$.

I shall indicate later the theorem of LIOUVILLE on the approximation of irrational algebraic numbers by rational numbers (see **(5.9)**).

The pigeon-hole principle is a very simple idea: if there are more pigeons than holes, then some hole will have at least two pigeons. DIRICHLET put pigeons to good use in (1842). His result actually follows from **(4.2)**:

(5.3) If α is a real irrational number, then α is approximable by rational numbers to the order 2 (using $C = C(\alpha, 2) = 1$); explicitly, there exist infinitely many rational numbers $\frac{a}{b}$ such that $|\alpha - \frac{a}{b}| < \frac{1}{b^2}$.

Thus, with the present notation, if $\alpha \in \mathbb{R} \setminus \mathbb{Q}$, then $\nu(\alpha) \geq 2$.

In this respect, HURWITZ determined in 1891 that $\frac{1}{\sqrt{5}}$ is the best constant in Dirichlet's theorem. A simple proof is in NIVEN's book, 1963.

(5.4) (1) For every real irrational number α, there exists infinitely many rational numbers $\frac{a}{b}$ such that

$$|\alpha - \frac{a}{b}| < \frac{1}{\sqrt{5}b^2}.$$

(2) However, if $0 < C < \frac{1}{\sqrt{5}}$ and $\alpha = \frac{1+\sqrt{5}}{2}$ (the golden number), then there are only finitely many rational numbers $\frac{a}{b}$ satisfying $|\alpha - \frac{a}{b}| < \frac{C}{b^2}$.

B.   The Markoff numbers

For each irrational number α, PERRON introduced in 1921 an invariant $M(\alpha)$. A closely related concept had been studied by MARKOFF already in 1879.

Let S_α be the set of all positive numbers λ such that there exist infinitely many rational numbers $\frac{a}{b}$ satisfying the inequality $|\alpha - \frac{a}{b}| < \frac{1}{\lambda b^2}$.

Clearly, if $\lambda \in S_\alpha$ and $0 < \lambda' < \lambda$, then $\lambda' \in S_\alpha$.

Let $M(\alpha) = \sup\{\lambda \mid \lambda \in S_\alpha\}$. By **(5.4)**, $\sqrt{5} \in S_\alpha$; thus $\sqrt{5} \leq M(\alpha)$ for every irrational number α. Also, for the golden number,

$$M\left(\frac{\sqrt{5}+1}{2}\right) = \sqrt{5}.$$

The following result is easy to show:

(5.5) Let $\alpha = [a_1, a_1, a_2, \ldots]$. Then $M(\alpha) < \infty$ if and only if the sequence $(a_n)_{n \geq 0}$ is bounded.

The above result says that irrational numbers having continued fraction expansions with unbounded quotients admit arbitrarily close approximation by convergents.

Now I turn my attention to the possible values of $M(\alpha)$, for all irrational numbers α.

The real numbers α and α' are said to be *equivalent* ($\alpha \sim \alpha'$) when there exist integers a, b, c, and d such that $ad - bc = \pm 1$ and $\alpha' = \frac{a\alpha+b}{c\alpha+d}$. It follows that $\alpha = \frac{-d\alpha'+b}{c\alpha'-a}$, so this is indeed an equivalence relation. Moreover, each equivalence class is either finite or countably infinite.

HURWITZ established in 1891 the following proposition:

(5.6) If $\alpha \sim \alpha'$, then $M(\alpha) = M(\alpha')$.

In general, the converse of **(5.6)** is not true. Yet, it is true for the golden number:

(5.7) If α is an irrational number which is not equivalent to the golden number, then $M(\alpha) \geq \sqrt{8}$.

A full description of the values assumed by $M(\alpha)$ is still incomplete. The study of the values $M(\alpha) < 3$ depends on Markoff's equation

$$X^2 + Y^2 + Z^2 = 3XYZ.$$

MARKOFF showed in 1879 that there exist infinitely many natural numbers x with the property that there exist natural numbers y, z such that (x, y, z) is a solution of the above equation.

These numbers are

$$x = 1, 2, 5, 13, 29, 34, 89, 169, 194, 233, 433, \ldots$$

(the *Markoff numbers*).

PERRON showed:

The values $M(\alpha)$ which are less than 3 are precisely the numbers $\frac{\sqrt{9x^2-4}}{x}$ for all Markoff numbers $x = 1, 2, 5, \ldots$. Thus they are

$$\sqrt{5} < \sqrt{8} < \frac{\sqrt{221}}{5} < \frac{\sqrt{1521}}{13} < \frac{\sqrt{7569}}{29} < \cdots$$

and

$$\lim_{x \to \infty} \frac{\sqrt{9x^2-4}}{x} = 3.$$

Furthermore, $M(\alpha) = \frac{1}{x}\sqrt{9x^2 - 4}$ exactly if α is equivalent to $\frac{1}{2x}(\sqrt{9x^2 - 4} + x + \frac{2y}{z})$ where (x, y, z) is a solution of Markoff's equation. It follows that if $M(\alpha) < 3$, then $\alpha \sim \alpha'$ if and only if $M(\alpha) = M(\alpha')$. Also, if α is not a quadratic irrational number, then $M(\alpha) \geq 3$.

The numbers with $M(\alpha) = 3$ are those equivalent to

$$[2, 2, 1, 1, \ldots, 1, 2, 2, 1, 1, \ldots, 1, 2, 2, 1, 1, \ldots, 1, \ldots]$$

with blocks of quotients containing m_1, m_2, m_3, $\ldots$ quotients equal to 1, and $m_1 < m_2 < m_3 < \cdots$. Since this set is uncountable, there exists uncountably many pairwise non-equivalent transcendental numbers α with $M(\alpha) = 3$.

The study of the values $M(\alpha) > 3$ is much more elaborate. For example, $M(\alpha)$ cannot be in the open interval between $\sqrt{12}$ and $\sqrt{13}$. But there are uncountably many α such that $M(\alpha) = \sqrt{12}$; on the other hand, $M(\alpha) = \sqrt{13}$ exactly if $\alpha \sim \frac{3+\sqrt{13}}{2}$. Next, $M(\alpha)$ cannot be in the open interval between $\sqrt{13}$ and $\frac{9\sqrt{13}+65}{22} = 3.6631\ldots$; the set of all α such that $M(\alpha) = 3.6631\ldots$ is uncountable.

These classical results are explained well in the book of KOKSMA (1936) who refers in particular to the work of SHIBATA (1929).

More recently (1982), ZAGIER studied the distribution of Markoff numbers. Let $z > 0$ and $Z(z) = \{x \mid x \leq z, x \text{ is a Markoff number}\}$. ZAGIER proved that $\#Z(z) = C\log^2 3x + O(\log x \log\log^2 x)$ with $C = 0.1807\ldots$ Numerical calculations indicate that the error should be even smaller.

C. Measures of irrationality

Let α be an irrational number; the number $\nu \geq 1$ is a *measure of irrationality* of α when for every $\epsilon > 0$ there exist only finitely many rational numbers $\frac{a}{b}$ such that $|\alpha - \frac{a}{b}| < \frac{1}{b^{\nu+\epsilon}}$.

So $\nu(\alpha) \leq \nu$ for every measure of irrationality ν of α. In order to determine or estimate $\nu(\alpha)$ one aims to find as small a measure of irrationality of α as possible.

Sometimes it may be simpler to determine values of ν which are *not* orders of approximation for α than values which *are* orders of approximation.

Here is a criterion to show that a number ν is an irrationality measure for α:

(5.8) Let α be an irrational number. Suppose that $\frac{p_n}{q_n}$ is a sequence of rational numbers such that for every $n \geq 1$, $q_{n+1} = q_n^{1+s_n}$ where $s_n > 0$ and $\lim_{n \to \infty} s_n = 0$. If there exist λ, $0 < \lambda < 1$, and $C > 0$ such that $|\alpha - \frac{p_n}{q_n}| < \frac{C}{q_n^{1+\lambda}}$ for every $n \geq 1$, then $\nu = 1 + \frac{1}{\lambda}$ is a measure of irrationality for α.

The proof of this criterion is simple; see for example ALLADI (1979), who gives another similar criterion.

In the next section I shall indicate measures of irrationality for some special numbers.

D. Order of approximation of irrational algebraic numbers

Let α be an algebraic number of degree $d \geq 1$, and let

$$f(X) = a_0 X^d + a_1 X^{d-1} + \cdots + a_d \in \mathbb{Z}[X]$$

be the minimal polynomial of α over $\mathbb{Q}$, with $\gcd(a_0, a_1, \ldots, a_d) = 1$, $a_0 \neq 0$.

Let the *height* of α be defined as $H(\alpha) = \max_{0 \leq i \leq d}\{|a_i|\}$, so $H(\alpha) = \sqcap f$ as in Chapter 1.

Let

$$C(\alpha) = \begin{cases} \frac{1}{d(d+1)H(\alpha)(|\alpha|+1)^{d-1}} & \text{if } \alpha \in \mathbb{R}, \\ \frac{|\gamma|}{2} & \text{if } \alpha = \beta + \gamma i \text{ with } \beta, \gamma \in \mathbb{R}, \gamma \neq 0. \end{cases}$$

LIOUVILLE proved in 1844:

(5.9) If α is an algebraic number of degree $d \geq 1$, then $|\alpha - \frac{a}{b}| > \frac{C(\alpha)}{b^d}$ for every $\frac{a}{b} \in \mathbb{Q}$, $\frac{a}{b} \neq \alpha$.

Hence α is not approximable by rationals to any order $d + \epsilon$ ($\epsilon > 0$). So, $\nu(\alpha) \leq d$.

It will be seen in §7 that this result has been sharpened by ROTH to the best possible result.

PROOF. If $\alpha = \beta + \gamma i$ with $\beta, \gamma \in \mathbb{R}$, $\gamma \neq 0$, then

$$|\alpha - \frac{a}{b}| = |(\beta - \frac{a}{b}) + \gamma i| \geq |\gamma| > \frac{|\gamma|/2}{b^d}$$

for every $\frac{a}{b} \in \mathbb{Q}$.

Now assume that α is real of degree $d \geq 1$ and has minimal polynomial $f(X) = \sum_{i=0}^{d} a_i X^{d-i}$.

If $\frac{a}{b} \in \mathbb{Q}$ and $\frac{a}{b} \neq \alpha$, then $f(\frac{a}{b}) \neq 0$ because if $d \geq 2$, then $f(X)$ is irreducible, so it has no rational roots. Then $b^d f(\frac{a}{b})$ is an integer different from 0 and therefore $|b^d f(\frac{a}{b})| \geq 1$, so $|f(\frac{a}{b})| \geq \frac{a}{b^d}$.

From $f(\alpha) = 0$ it follows that

$$\frac{1}{b^d} \leq \left| f\left(\frac{a}{b}\right) \right| = |f\left(\frac{a}{b}\right) - f(\alpha)| = |\alpha - \frac{a}{b}||f'(\xi)|$$

for some real number ξ such that $|\xi - \alpha| < |\alpha - \frac{a}{b}|$.

First case: $|\alpha - \frac{a}{b}| \geq 1 \geq \frac{1}{b^d} > \frac{C(\alpha)}{b^d}$, because $C(\alpha) < 1$.

Second case: $|\alpha - \frac{a}{b}| < 1$, so $|\xi - \alpha| < 1$ and $|\xi| < |\alpha| + 1$. Then

$$\frac{1}{b^d} \leq |\alpha - \frac{a}{b}| \sum_{|\xi - \alpha| < 1} |f'(\xi)|.$$

But $f'(\xi) = \sum_{i=0}^{d-1} (d-i) a_i \xi^{d-i-1}$, so

$$|f'(\xi)| \leq \sum_{i=0}^{d-1} (d-i)|a_i||\xi|^{d-i-1}.$$

If $|\xi| \geq 1$, then

$$|f'(\xi)| \leq [\sum_{i=0}^{d-1} (d-i)] H(\alpha)|\xi|^{d-1}$$

$$< \frac{d(d+1)}{2} H(\alpha)(|\alpha| + 1)^{d-1} < \frac{1}{C(\alpha)}.$$

If $|\xi| < 1$, then

$$|f'(\xi)| \leq [\sum_{i=0}^{d-1} (d-i)] H(\alpha)$$

$$= \frac{d(d+1)}{2} H(\alpha) = d(d+1)H(\alpha)(|\alpha| + 1)^{d-1} = \frac{1}{C(\alpha)}.$$

Thus, $|\alpha - \frac{a}{b}| > \frac{C(\alpha)}{b^d}$, concluding the proof. $\square$

As was mentioned in the historical survey, Liouville's theorem was one of the keystones in the building of the theory of approximation by rational numbers. Even though the inequality provided by the theorem is rather loose, it may still be used to prove various results about the irrationality of specific numbers. This will be illustrated in the next section.

6 Irrationality of special numbers

In this section I prove that several interesting numbers are irrational, beginning with the number e.

(6.1) e and e^2 are irrational numbers, which, moreover, are not algebraic of degree 2. (And, in fact, these numbers are transcendental.)

PROOF. As was shown in subsection C., EULER gave the continued fraction expansions

$$\alpha = \frac{e+1}{e-1} = [2, 6, 10, 14, \ldots],$$

$$\beta = \frac{e^2+1}{e^2-1} = [1, 3, 5, 7, \ldots].$$

So, α and β are irrational numbers, hence e and e^2 are irrational numbers which are not algebraic of degree 2, as follows from LAGRANGE's proposition **(4.7)**. □

Another proof of the irrationality of e, independent of the continued fraction expansion, was given by FOURIER, and appeared in STAINVILLE's book in 1815. It is based on a simple criterion:

(6.2) Let α be a real number, let $(f(n))_{n \geq 0}$ be a sequence of positive real numbers such that $\liminf_{n \to \infty} f(n) = 0$, and assume that there exists n_0 such that for every $n \geq n_0$ there exist integers a_n and b_n such that $0 < |b_n \alpha - a_n| \leq f(n)$. Then α is irrational.

This was FOURIER's proof:

PROOF. It is equivalent to show that $\alpha = e - 2 = \sum_{n=2}^{\infty} \frac{1}{n!}$ is irrational. For every $n \geq 2$, $n!\alpha = k_n + s_n$ with $k_n = n!(\sum_{i=2}^{n} \frac{1}{i!})$ is an integer, and

$$0 < s_n = \frac{1}{n+1} + \frac{1}{(n+1)(n+2)} + \cdots$$
$$< \frac{1}{(n+1)} + \frac{1}{(n+1)^2} + \frac{1}{(n+1)^3} + \cdots = \frac{1}{n}.$$

By **(6.2)**, α is irrational.   □

The criterion **(6.2)** also implies that if $p_1 < p_2 < p_3 < \cdots$ is a sequence of prime numbers, then

$$\alpha = \sum_{i=1}^{\infty} \frac{1}{p_1 p_2 \cdots p_i}$$

is an irrational number.

Other irrationality criteria may be obtained using representations of positive real numbers which generalize the decimal representation. There are several noteworthy representations, as explained for example, in PERRON's book *Irrationalzahlen*.

In 1869, CANTOR proved:

(6.3) Let $a_1, a_1, a_3, \ldots \geq 2$ be integers. Then every real number α has a unique representation.

$$\alpha = c_0 + \sum_{i=1}^{\infty} \frac{c_i}{a_1 a_2 \cdots a_i}$$

where each c_i is an integer, and for $i \geq 1$, $0 \leq c_i \leq a_i - 1$ and $c_i < a_i - 1$ for infinitely many indices $i \geq 1$.

Conversely, every series of the above kind is convergent. Taking $a_i = 10$ for each i, one obtains the ordinary decimal representation.

This gives the following irrationality criterion:

(6.4) If each prime p divides infinitely many a_i, then α is irrational if and only if $c_i \geq 1$ for infinitely many indices i.

Note the following special cases:
  (a) If each $a_i = i + 1$ and each $c_i = 1$, then $e = 2 + \sum_{n=2}^{\infty} \frac{1}{n!}$ is irrational.

(b) If F_1, F_2, ..., F_i, ... is the sequence of Fibonacci numbers, then

$$\alpha = \sum_{i=1}^{\infty} \frac{1}{F_1 F_2 \cdots F_i} = 2 + \sum_{i=3}^{\infty} \frac{1}{F_3 \cdots F_i}$$

is an irrational number.

In the same spirit, I describe now the representations given by SYLVESTER, LÜROTH, and ENGEL.

In 1880, SYLVESTER showed:

(6.5) Every real number α has a unique representation

$$\alpha = c + \sum_{i=1}^{\infty} \frac{1}{a_i}$$

where c is an integer and a_1, a_2, a_3, ... ≥ 2 are integers, such that $a_{i+1} > (a_i - 1)a_i$.

Conversely, every series of this kind is convergent and its sum α is irrational if and only if $a_{i+1} > (a_i - 1)a_i + 1$ for infinitely many indices i.

For example, $\alpha = \sum_{i=0}^{\infty} \frac{1}{2^{2^i}}$ is an irrational number.

In 1883, LÜROTH proved:

(6.6) Every real number α has a unique representation

$$\alpha = c + \frac{1}{a_1} + \sum_{i=1}^{\infty} \frac{1}{(a_1 - 1)a_1(a_2 - 1)a_2 \cdots (a_i - 1)a_i} \times \frac{1}{a_{i+1}}$$

where c is an integer and a_1, a_2, a_3, ... ≥ 2 are integers.

Conversely, every such series is convergent and its sum α is irrational if and only if the sequence a_1, a_2, a_3, ... is not periodic.

In 1913, ENGEL established the following result:

(6.7) Every real number α has a unique representation

$$\alpha = c + \sum_{i=1}^{\infty} \frac{1}{a_1 a_2 \cdots a_i}$$

where c is an integer and $2 \leq a_1 \leq a_2 \leq a_3 \leq \cdots$ is a sequence of integers.

Conversely, every such sequence is convergent and its sum is irrational if and only if $\lim_{i \to \infty} a_i = \infty$.

I note the following special cases:

(a) If $p_1 < p_2 < p_3 < \cdots$ is the sequence of prime numbers, then $\alpha = \sum_{i=1}^{\infty} \frac{1}{p_1 p_2 \cdots p_i}$ is an irrational number (this example was already mentioned).

(b) Let E_2, E_4, E_6, ... be the sequence of Euler numbers, which are also called the secant coefficients because they are defined by

$$\sec x = 1 - \frac{E_2}{2!} x^2 + \frac{E_4}{4!} x^4 - \frac{E_6}{6!} x^6 + \cdots \qquad (\text{for } |x| < \tfrac{\pi}{2}).$$

These numbers are integers satisfying the recurrence relation

$$E_{2n} + \binom{2n}{2n-2} E_{2n-2} + \binom{2n}{2n-4} E_{2n-4} + \cdots + \binom{2}{2} E_{2n} + 1 = 0.$$

Moreover, $(-1)^n E_{2n} > 0$.

Thus, it follows that

$$\alpha = \sum_{n=1}^{\infty} \frac{1}{|E_2 E_4 \cdots E_{2n}|}$$

is an irrational number.

I shall now consider examples of other numbers defined as sums of series.

Let $(f(n))_{n \geq 0}$ be a strictly increasing sequence of positive integers, let $d \geq 2$ be an integer, and let

$$\alpha = \sum_{n=0}^{\infty} \frac{1}{d^{f(n)}}.$$

I now investigate what kind of number α is.

(1) If $f(n) = n$, then

$$\alpha = \sum_{n=0}^{\infty} \frac{1}{d^n} = 1 + \frac{1}{d} + \frac{1}{d^2} + \frac{1}{d^3} + \cdots = \frac{1}{1 - \frac{1}{d}} = \frac{d}{d-1},$$

so α is rational.

(2) Let $s \geq 2$ be an integer and $f(n) = n^s$ (for $n \geq 0$).

Then $\alpha = \sum_{n=0}^{\infty} \frac{1}{d^{n^s}}$ is an irrational number. This follows from **(6.7)**.

By **(5.3)**, α is therefore approximable by rational numbers at least to the order 2.

(3) Let $s \geq 2$ be an integer and $f(n) = s^n$ (for $n \geq 0$). Then α is approximable by rational numbers to the order s. Thus, α is irrational.

Note that the irrationality of α follows from **(6.7)**.

(4) Let $\limsup_{n \to \infty} \frac{f(n+1)}{f(n)} = \mu > 2$ and $\alpha = \sum_{n=0}^{\infty} \frac{1}{d^{f(n)}}$. Then for every $0 < \epsilon < \mu - 2$, the number α is approximable by rational numbers to the order $\mu - \epsilon$, so α is irrational. Note that the irrationality of α follows once again from **(6.7)**.

Now I turn my attention to the number π.

(6.8) π^2, hence also π, is an irrational number.

PROOF.   The first proof that π is irrational was given by LAMBERT, while LEGENDRE proved that π^2 is irrational, as already stated in the historical survey. The proof given below may be found in NIVEN's book. It is a modification of his own proof that π is irrational.

I need the following lemma:

Lemma. *Let* $g \in \mathbb{Z}[X]$ *and* $h(X) = \frac{X^n g(X)}{n!}$ *where* $n \geq 1$. *Then* $h^{(j)}(0) \in \mathbb{Z}$ *for every* $j \geq 0$.

PROOF.   Indeed, let $X^n g(X) = \sum_{j \geq 0} c_j X^j$, so $c_j = 0$ for $j < n$. From $h^{(j)}(0) = c_j \frac{j!}{n!}$ it follows that $h^{(j)}(0) \in \mathbb{Z}$ for every $j \geq 0$.   □

Now I show that π^2 is irrational.

Let $h(X) = \frac{X^n (1-X)^n}{n!}$ (where n is a positive integer, to be chosen later); then if $0 < x < 1$, then $0 < h(x) < \frac{1}{n!}$.

Noting that $h^{(j)}(0)$ is an integer for $j \geq 0$, from $h(1 - X) = h(X)$, it follows that $h^{(j)}(1)$ is also an integer for $j \geq 0$.

If $\pi^2 = \frac{a}{b}$ with $a, b > 0$ relatively prime integers, let

$$f(X) = b^n [\pi^{2n} h(X) - \pi^{2n-2} h^{(2)}(X) + \pi^{2n-4} h^{(4)}(X) - \cdots + (-1)^n h^{(2n)}(X)],$$

so $f(0)$, $f(1)$ are integers. Moreover,

$$\frac{d}{dx}[f'(x) \sin \pi x - \pi f(x) \cos \pi x] = [f''(x) + \pi^2 f(x)] \sin \pi x$$
$$= b^n \pi^{2n+2} h(x) \sin \pi x$$
$$= \pi^2 a^n h(x) \sin \pi x,$$

hence

$$\pi a^n \int_0^1 h(x) \sin \pi x \, dx = \left[\frac{f'(x) \sin \pi x}{\pi} - f(x) \cos \pi x \right]_0^1$$
$$= f(1) + f(0) \in \mathbb{Z}.$$

So,

$$0 < \pi a^n \int_0^1 h(x) \sin \pi x \, dx < \frac{\pi a^n}{n!} < 1$$

when n is sufficiently large. This is a contradiction.   □

The same method was successfully used by NIVEN to prove older results about the values of trigonometric functions and hyperbolic functions.

(6.9) If r is a non-zero rational number, then $\cos r$ is irrational.

As a corollary, π is irrational because $\cos \pi = -1$ is rational.

It follows from $1 - 2 \sin^2 r = \cos 2r$ and $\cos 2r = \frac{1-\tan^2 r}{1+\tan^2 r}$ that if r is rational, $r \neq 0$, then $\sin r$, $\tan r$ are irrational, hence also $\sec r$, $\csc r$, and $\cot r$ are all irrational.

Similarly, if r is rational, then $\arccos r$ (when $r \neq 1$), and $\arcsin r$, $\arctan r$ (when $r \neq 0$), are irrational numbers.

(6.10) If r is a non-zero rational number, then $\cosh r$ is irrational.

Since $\cosh 2r = 2 \sinh^2 r + 1 = \frac{1+\tanh^2 r}{1-\tanh^2 r}$, then $\sinh r$ and $\tanh r$ are irrational. It follows also that the values of the inverse hyperbolic functions calculated at rational numbers are either 0 or irrational numbers.

In this way, one obtains immediately a new proof of LAMBERT's result:

(6.11) (1) If r is a non-zero rational number, then e^r is irrational.
(2) If r is a positive rational number, $r \neq 1$, then $\log r$ is irrational.

PROOF.   (1) If e^r is rational so is e^{-r}, hence also $\cosh r = \frac{e^r + e^{-r}}{2}$, contrary to **(6.10)**.

(2) If $r \neq 1$ and $\log r$ is rational, it is not zero, so by (1), $r = e^{\log r}$ would be irrational.   □

The following remark is obvious: If r is a positive rational number and $a > 1$ is an integer, then $\log_a r = \frac{\log r}{\log a}$ is an irrational number if and only if $r \neq a^s$ (with s rational).

Thus, for example, $\frac{\log 3}{\log 2}$ is irrational. Also $\log_{10} r$ is irrational, unless $r^m = 10^n$ for some integers m and n.

On the other hand, it is easy to show:

(6.12) If r is a rational number, then the values of the trigonometric functions at $r\pi$ are algebraic numbers.

Now I shall consider the question of irrationality of the values $\zeta(k) = \sum_{j=1}^{\infty} \frac{1}{j^k}$, the Riemann zeta function (for every integer $k \geq 2$).

Since $\zeta(2) = \frac{\pi^2}{6}$, and π^2 is irrational, then so is $\zeta(2)$. Euler's formula for $\zeta(2k)$, indicated in §2, states that $\zeta(2k) = \pi^{2k} r_{2k}$, where r_{2k} is a rational number. It will be indicated in §2 that π, hence also π^{2k}, is transcendental, therefore $\zeta(2k)$ is not only irrational but also transcendental.

The situation is very different for $\zeta(2k+1)$. It was an open problem for a long time to decide whether $\zeta(3)$ was irrational. This problem was solved in the affirmative by APÉRY as mentioned in the historical survey. His ingenious method was applicable also for $\zeta(2)$.

(6.13) $\zeta(2)$ and $\zeta(3)$ are irrational numbers.

A very different proof that $\zeta(2)$ and $\zeta(3)$ are irrational numbers was given by BEUKERS in 1979.

Other numbers which have been considered are sums of series built in terms of binary recurrences. Thus ANDRÉ-JEANNIN proved in 1991:

(6.14) If $(F_n)_{n \geq 0}$ denotes the sequence of Fibonacci numbers, then $\sum_{n=1}^{\infty} \frac{1}{F_n}$ is an irrational number.

Another classical result concerning Fibonacci numbers was obtained by GOOD (1974) and HOGGATT (1976):

(6.15) The sum

$$\sum_{n=0}^{\infty} \frac{1}{F_{2^n}} = \frac{7 - \sqrt{5}}{2}.$$

A corollary of the results of BECKER and TÖPFER (1994) is applicable, for example, to Lucas numbers $L_0 = 2$, $L_1 = 1$, $L_n = L_{n-1} + L_{n-2}$ (for $n \geq 2$).

(6.16) The number

$$\sum_{n=0}^{\infty} \frac{1}{L_{2^n}}$$

is irrational.

The following is indeed trivial. Let $n \geq 3$. The following statements are equivalent:

(1) For every rational number $x > 0$, the number $(1 + x^n)^{1/n}$ is irrational.

(2) Fermat's Last Theorem is true for the exponent n.

Now, everyone knows (even laymen!) that WILES proved Fermat's Last Theorem. Therefore, (1) holds for every $n \geq 3$. But suppose that someone—very intelligent—would find a direct proof of (1) with methods of diophantine approximations. This would constitute a new proof of Fermat's Last Theorem. I have no ideas in this direction.

The irrationality measures of many irrational numbers have been estimated and, in some cases, explicitly computed.

Thus $\nu(e) = \nu(e^2) = 2$. ALLADI showed in 1979 that $\nu(e^r) = 2$ for every non-zero rational r.

APÉRY made calculations in 1979 that gave the following bounds for the measure of irrationality of $\zeta(2)$ and $\zeta(3)$:

$$\nu(\zeta(2)) < 11.85,$$
$$\nu(\zeta(3)) < 13.42.$$

It is very important to give effective lower bounds for the distance between a given irrational number and rational numbers. Some examples to illustrate are the following:

(a) ALLADI (1979), improving a previous method of BAKER,

$$\left| \log 2 - \frac{a}{b} \right| > \frac{1}{10^{10} b^{5.8}}$$

for every rational number $\frac{a}{b}$.

(b) BAKER (1964b):

$$\left| \sqrt[3]{2} - \frac{a}{b} \right| > \frac{C}{b^{296}},$$

where C is an explicit constant, for every rational number $\frac{a}{b}$.

(c) MAHLER (1953) proved the striking results

$$\left|\pi - \frac{a}{b}\right| > \frac{1}{b^{42}},$$

for every rational number $\frac{a}{b}$, and

$$\left|\pi - \frac{a}{b}\right| > \frac{1}{b^{30}},$$

for every rational number $\frac{a}{b}$ with sufficiently large denominator. MIGNOTTE (1974) showed that

$$\left|\pi - \frac{a}{b}\right| > \frac{1}{b^{20.6}}$$

for every rational number $\frac{a}{b}$, and, if $b > q > 96$, then

$$\left|\pi - \frac{a}{b}\right| > \frac{1}{b^{20}}.$$

So, $\nu(\pi) \leq 20$. Also, $\nu(\pi^2) \leq 17.8$.

This means that $\nu = 30$ is a measure of irrationality of π. Many authors worked on this question and obtained improved irrationality measures for π. I just note (without giving any hint of the method): MIGNOTTE, $\nu = 20$ (in 1974), followed by the work of G. V. CHUDNOVSKY and D. CHUDNOVSKY, F. BEUKERS, C. VIOLA, G. RHIN, R. DVORNICICH, E. A. BUKHADZE, and M. HATA, who has the best result up to 1993:

$$\left|\pi - \frac{a}{b}\right| > \frac{1}{b^{8.0161}}$$

for all sufficiently large b.

The calculations involved are lengthy and sharp; the verification requires time that only specialists can spare.

7   Transcendental numbers

CANTOR proved that the set $\mathbb{R}$ of real numbers is uncountable. This was, at the time, a very striking discovery. It is easier to show that the set of all algebraic numbers is countable. Therefore, the set of transcendental numbers is also uncountable, yet it is not so easy to produce infinite families of transcendental numbers, nor it is easy in general to show that numbers of a given type are transcendental.

Once again, it will be important to consider how well the number may be approximated by rational numbers.

A.   Liouville numbers

LIOUVILLE considered the following numbers: $\alpha \in \mathbb{R} \setminus \mathbb{Q}$ is called a *Liouville number* if for every integer $n \geq 2$ there exist $\frac{a_n}{b_n} \in \mathbb{Q}$, with $b_n \geq 2$, such that $\left| \alpha - \frac{a_n}{b_n} \right| < \frac{1}{b_n^n}$.

(7.1) The real number α is a Liouville number if and only if α is approximable by rational numbers to any order $\nu \geq 1$.

In particular, α is transcendental (by LIOUVILLE's own theorem **(5.9)**). Thus α is a Liouville number if and only if $\nu(\alpha) = \infty$. Let $\mathbb{L}$ denote the set of Liouville numbers.

Here are some examples of Liouville numbers: Let $d \geq 2$, $0 \leq k_n \leq d - 1$ and $k_n \neq 0$, for infinitely many indices n. Then $\alpha = \sum_{n=0}^{\infty} \frac{k_n}{d^{n!}}$ is a Liouville number. It suffices to show that α is approximable by rational numbers to every order $n \geq 1$. Let $n \geq 1$, $m \geq n$ and

$$b_m = d^{m!} a_m = \left(\sum_{i=0}^{m} \frac{k_i}{d^{i!}} d^{m!} \right).$$

Thus,

$$0 < \alpha - \frac{a_m}{b_m} = \sum_{i=m+1}^{\infty} \frac{k_i}{d^{i!}} \leq (d-1) \sum_{i=m+1}^{\infty} \frac{1}{d^{i!}}$$

$$< \frac{d-1}{d^{(m+1)!} - 1} < \frac{1}{(d^{m!})^m} = \frac{1}{b_m^m} \leq \frac{1}{b_m^n},$$

because if $c = d^{m!}$, then

$$\frac{1}{d^{(m+1)!}} + \frac{1}{d^{(m+2)!}} + \frac{1}{d^{(m+3)!}} + \cdots < \frac{1}{c^{m+1}} + \frac{1}{c^{(m+1)^2}} + \frac{1}{c^{(m+1)^3}} + \cdots$$

$$< \frac{1}{c^{m+1}} + \frac{1}{c^{2(m+1)}} + \frac{1}{c^{3(m+1)}} + \cdots$$

$$= \frac{1}{c^{m+1}} \cdot \frac{1}{1 - \frac{1}{c^{m+1}}}$$

$$= \frac{1}{c^{m+1} - 1}$$

$$= \frac{1}{d^{(m+1)!} - 1},$$

and $d^{m!m+1} + 1 < d^{(m+1)!} + d^{m!m}$, so

$$\frac{d-1}{d^{(m+1)!} - 1} < \frac{1}{(d^{m!})^m}.$$

Since there exist infinitely many n such that $k_n \neq 0$, then there exist infinitely many distinct rationals $\frac{a_m}{b_m}$ satisfying $0 < \alpha - \frac{a_m}{b_m} < \frac{1}{b_m^n}$; thus $\alpha \in \mathbb{L}$.

In particular, $\sum_{n=0}^{\infty} \frac{1}{10^{n!}} \in \mathbb{L}$.

These were the first known examples of transcendental numbers.

Liouville numbers have been studied by MAILLET (see his book, 1906) and have also been the subject of a chapter in SCHNEIDER's book (1959).

Some properties:

(7.2) The set of Liouville numbers is uncountable.

(7.3) The set of Liouville numbers is dense in $\mathbb{R}$.

(7.4) The set of Liouville numbers has measure 0.

Thus "almost all" transcendental numbers are not Liouville numbers.

Since the set of Liouville numbers is uncountable, there are uncountably many Liouville numbers which are algebraically independent over $\mathbb{Q}$.

PERRON (1932) and later SCHMIDT (1962) gave the following countable set of algebraically independent Liouville numbers:

(7.5) For every $i \geq 1$, let $\alpha_i = \sum_{n=1}^{\infty} \frac{1}{2^{(in)!}}$. Then α_1, α_2, α_3, ... are algebraically independent over $\mathbb{Q}$ (that is, for every $m \geq 1$ the numbers α_1, α_2, ..., α_m are algebraically independent over $\mathbb{Q}$).

It is known that e, π, and $\log 2$ are not Liouville numbers (see §6), however, it is unknown whether e^{π} is a Liouville number.

Even though it may at first sight seem paradoxical (because the set of Liouville numbers has measure 0), ERDÖS proved in 1962 that every non-zero real number is the sum, and also the product, of two Liouville numbers.

B. Approximation by rational numbers: sharper theorems

The theorem of Liouville **(5.9)** about the approximation of algebraic numbers by rational numbers gives a loose order of approximation which depends on the degree of the algebraic number; on the other hand, the constant in the inequality is explicit.

The modern theory of diophantine approximation was initiated by THUE (1909) who established a new relationship between the solutions of certain diophantine equations and the approximation of algebraic numbers by rational numbers.

I shall describe these ideas very succinctly.

Let $d \geq 3$; let $F(X, Y) = a_0 X^d + a_1 X^{d-1} Y + a_2 X^{d-2} Y^2 + \cdots + a_{d-1} XY^{d-1} + a_d Y^d$ be a homogeneous polynomial of degree d with integer coefficients, and assume that $a_0 \neq 0$. For convenience, assume also that $f(X) = a_0 X^d + a_1 X^{d-1} + \cdots + a_d$ is an irreducible polynomial.

If a is any integer, we consider the equation $F(X, Y) = a$. The aim is to show that this equation has only finitely many solutions in integers.

Let $\alpha_1, \ldots, \alpha_d \in \mathbb{C}$ be the roots of $f(X) = 0$. Since $f(X)$ is irreducible, it follows that $\alpha_1, \ldots, \alpha_d$ are distinct algebraic numbers of degree d.

If x and y are integers such that $F(x, y) = a$, then $a_0 \prod_{k=1}^{d}(x - \alpha_k y) = a$.

There are two cases. If $a = 0$, there exists j with $1 \leq j \leq d$ such that $x - \alpha_j y = 0$; but $\alpha_j \notin \mathbb{Q}$, so $y = 0$ and necessarily $x = 0$. Therefore, if $a = 0$, then the only solution is $(0, 0)$.

On the other hand, if $a \neq 0$, then

$$\prod_{k=1}^{d}(x - \alpha_k y) = \frac{|a|}{|a_0|}.$$

Let $C_1 = (\frac{|a|}{|a_0|})^{1/d}$, so there exists an index, therefore a smallest index, j such that $|x - \alpha_j y| \leq C_1$.

Let $C_2 = \min_{h \neq k} |\alpha_h - \alpha_k|$; so $C_2 > 0$. Note that for each integer y_0 there exists at most d integers x such that (x, y_0) is a solution of the equation.

In order to show that $F(X, Y) = a$ has only finitely many solutions in integers, it suffices to show that the set $S = \{$solutions $(x, y) : |y| > \frac{2C_1}{C_2}\}$ is finite.

If $(x, y) \in S$, then for every $k \neq j$ with j as above,

$$0 < \frac{C_2|y|}{2} < C_2|y| - C_1 \leq |\alpha_k - \alpha_j| \cdot |y| - |x - \alpha_j y|$$
$$\leq |(\alpha_k - \alpha_j)y - (x - \alpha_j y)| = |x - \alpha_k y|.$$

So,

$$|x - \alpha_k y| = \frac{|a|}{|a_0| \prod_{k \neq j} |x - \alpha_k y|} < \frac{2^{d-1}|a|}{|a_0| C_2^{d-1} |y|^{d-1}} = \frac{C_3}{|y|^{d-1}}$$

where $C_3 = \frac{2^{d-1}|a|}{|a_0| C_2^{d-1}}$ (note that C_3, like C_1 and C_2, depends only on the given equation and not on the solution (x, y) being considered). Writing $\text{sgn}(y) = \frac{|y|}{y}$, then

$$\left| \alpha_j - \frac{x\,\text{sgn}(y)}{|y|} \right| = \left| \alpha_j - \frac{x}{y} \right| < \frac{C_3}{|y|^n}.$$

For every $k = 1, \ldots, d$, let

$$T_k = \left\{ \frac{m}{n} \in \mathbb{Q} \mid \gcd(m, n) = 1 \quad \text{and} \quad \left| \alpha_k - \frac{m}{n} \right| < \frac{C_3}{n^d} \right\}.$$

It was seen that if $(x, y) \in S$, then the rational number $\frac{x\,\text{sgn}(y)}{|y|}$ belongs to $\bigcup_{k=1}^{d} T_k$. This defines a mapping Φ from S into the preceding set. Moreover, Φ is one-to-one. Indeed, if (x, y) and (x', y') are in S, and $\frac{x\,\text{sgn}(y)}{|y|} = \frac{x'\,\text{sgn}(y')}{|y'|}$, then $|y'| = r|y|$ and $x'\,\text{sgn}(y') = rx\,\text{sgn}(y)$ (for some non-zero rational number r), hence $a = F(x', y') = \pm r^d F(x, y) = \pm r^d a$, which implies that $r = \pm 1$, so $r = 1$.

Thus to show that S is finite it suffices to show that each set $\Phi(S) \cap T_k$ is finite.

Let $\frac{x\,\text{sgn}(y)}{|y|} \in T_k$, where $(x, y) \in S$. If $\alpha_k \notin \mathbb{R}$, let $\alpha_k = \beta + i\gamma$, with $\beta, \gamma \in \mathbb{R}$, and $\gamma \neq 0$. Then

$$|\gamma| \leq \left| (\beta - \frac{x}{y}) + i\gamma \right| = \left| \alpha_k - \frac{x\,\text{sgn}(y)}{|y|} \right| < \frac{C_3}{|y|^d},$$

hence

$$|y|^d < \frac{C_3}{|\gamma|}.$$

So $|y|$ may have only finitely many values, and by a previous remark, the same is true for x, so $\Phi(S) \cap T_k$ is finite.

If $\alpha_k \in \mathbb{R}$, one is led to study whether the set T_k is finite.

This established the link between the approximation of algebraic numbers by rational numbers and the finiteness of the number of solutions of the equation considered above. For a survey on some fundamental methods on the theory of diophantine equations, see RIBENBOIM (1986).

Sharper and effective inequalities will therefore have important indications on the the number and size of solutions. Thus, important mathematicians considered this problem.

For this purpose, Liouville's theorem was improved successively by THUE (1909), SIEGEL (1921), DYSON (1947), and SCHNEIDER (1949); finally, ROTH proved in 1955 the best possible result:

(7.6) If α is any algebraic number, for every $\epsilon > 0$ there exists $C = C(\alpha, \epsilon) > 0$ such that for every rational number $\frac{a}{b} \neq \alpha$ one has $|\alpha - \frac{a}{b}| > \frac{C}{b^{2+\epsilon}}$.

So, α is not approximable by rational numbers to the order $2 + \epsilon$ for every $\epsilon > 0$. Therefore, $\nu(\alpha) \leq 2$. Thus, if $\alpha \notin \mathbb{Q}$, then $\nu(\alpha) = 2$ (as follows from Dirichlet's theorem).

A corollary is the following transcendence criterion:

(7.7) If α is a real number which is approximable by rational numbers to an order $\nu > 2$, then α is transcendental.

If α has degree at least 3, the number $C(\alpha, \epsilon)$ in **(7.6)** is not effectively computable; the proposition only asserts its existence.

Consider the following statement, which is sharper than Roth's theorem: If α is any algebraic number, there exists $C(\alpha) > 0$ such that for every rational number $\frac{a}{b} \neq \alpha$ one has $|\alpha - \frac{a}{b}| > \frac{C(\alpha)}{b^2}$. It may be shown that the above statement is equivalent to the following assertion, which is as yet unproven:

If α is any real algebraic number and $\alpha = [a_0, a_1, a_2, \ldots]$ is its simple continued fraction expansion, then there exists a number $M = M(\alpha) > 0$ such that $|a_0|, a_1, a_2, \ldots < M$.

However, it is generally believed that the above statement is false. Rather, on the contrary, it is quite possible that if α is a real algebraic number of degree $d \geq 3$ then, using the above notation, it follows that $\sup\{a_i \mid i \geq 1\} = \infty$.

Now I consider some examples.

Example 1. Up to now it is not known whether $\alpha = \sum_{n=0}^{\infty} \frac{1}{d^{n^s}}$ is a transcendental number (where $d \geq 2$, $s \geq 2$). For $s = 2$ this may require a deeper study of theta functions.

The function $f(z) = \sum_{n=0}^{\infty} \frac{z^n}{2^{n(n-1)}}$ satisfies the functional equation $f(z) = 1 + zf(\frac{z}{4})$; this is used to show that $f(\frac{1}{2}) = \sum_{n=0}^{\infty} \frac{1}{2^{n^2}}$ is not a Liouville number. More generally, BUNDSCHUH showed (1970): if $d \geq 2$ then $\sum_{n=0}^{\infty} \frac{1}{d^{n^2}}$ is not a Liouville number.

Example 2. Let $s \geq 3$ be an integer, $d \geq 2$, and $\alpha = \sum_{n=0}^{\infty} \frac{1}{d^{s^n}}$. Then α is transcendental.

Indeed, it was seen in §6 that α is approximable by rational numbers to the order s. Since $s > 2$, by **(7.7)**, α is transcendental.

More generally:

Example 3. Let $(f(n))_{n \geq 1}$ be a sequence of positive integers such that $\lim_{n \to \infty} \frac{f(n+1)}{f(n)} = \mu > 2$. Then for every integer $d \geq 2$, the number $\alpha = \sum_{n=0}^{\infty} \frac{1}{d^{f(n)}}$ is transcendental.

Indeed, as was seen in §6, if $0 < \epsilon < \mu - 2$, α is approximable by rational numbers to the order $\mu - \epsilon > 2$. By **(7.7)**, α is transcendental.

The following was first proved by MAHLER (1929) and may also be proved with a variant of Roth's theorem:

Example 4. $\alpha = \sum_{n=0}^{\infty} \frac{1}{d^{2^n}}$ (with $d \geq 2$) is transcendental.

More generally:

Example 5. Let $r \in \mathbb{Q}$, $r \neq 0$, $C \geq 1$, $d \geq 2$, $s \geq 2$ be integers. For every n let c_n be an integer such that $|c_n| \leq C^n$ and $c_n \neq 0$ for infinitely many $n \geq 1$. Then $\sum_{n=0}^{\infty} \frac{c_n r^n}{d^{s^n}}$ is transcendental.

MAHLER extended this construction in 1976. Let f be any function defined on the integers. Let α be the number between 0 and 1 whose decimal expression is the following: $f(1)$ times the digit 1, followed by $f(1)$ times the digit 2, ..., followed by $f(1)$ times the digit 9, followed by $f(2)$ times the 2-digit number 10, ..., followed by $f(2)$ times the 2-digit number 99, followed by $f(3)$ times each of the numbers 100, 101, ..., 999 in succession, etc. The resulting numbers α are transcendental numbers but not Liouville numbers.

For every $\mu > 2$ let $\mathbb{R}_\mu = \{\alpha \in \mathbb{R} \mid \nu(\alpha) \geq \mu\}$. Thus every $\alpha \in \mathbb{R}_\mu$ is transcendental. Moreover, $\mathbb{L} = \bigcap_{\mu > 2} \mathbb{R}_\mu$.

Since $\mathbb{L}$ is uncountable and dense in $\mathbb{R}$, then each $\mathbb{R}_\mu$ is uncountable and dense, too.

MAHLER (1937) gave a class of transcendental numbers which are not Liouville numbers:

(7.8) Let $f(X) \in \mathbb{Z}[X]$ be a polynomial of degree at least 1, such that $f(k) \geq 1$ for every $k \geq 1$. Let $a = 0.a_1a_2a_3a_4\ldots$ with $a_k = f(k)$ (the notation becomes clear in the example which follows). Then α is transcendental, but not a Liouville number.

For example, if $f(X) = X$ this gives: $0.12345678910111213\ldots$ is transcendental but not a Liouville number.

In 1924, KHINTCHINE proved a general theorem on approximation by rational numbers (see also his book (1935)). A special case is the following:

(7.9) For every $\mu > 2$, the set $\mathbb{R}_\mu$ has measure 0.

Example 6. (KNUTH (1964)) Let $a \geq 2$ be an integer and let ξ_a be the irrational number with simple continued fraction expansion

$$\xi_a = [a^{F_0}, a^{F_1}, a^{F_2}, \ldots]$$

where $(F_n)_{n\geq 0}$ is the sequence of Fibonacci numbers. Then ξ_a is approximable by rational numbers to the order $\alpha+1$, where $\alpha = \frac{1+\sqrt{5}}{2}$ is the golden number. Hence, $\xi_a \in \mathbb{R}_{\alpha+1}$ and is transcendental.

C.   Hermite, Lindemann, and Weierstrass

Now I give the important and classical results of HERMITE and LINDEMANN.

In 1873, HERMITE showed

(7.10) e is a transcendental number.

A rather simple proof of this theorem, given by HURWITZ in 1893, is reproduced in NIVEN's book.

In 1882, LINDEMANN proved the following equivalent results (this improves **(6.11)**):

(7.11) (1) If log is any determination of the complex logarithmic function, and if r is a rational number, $r \neq 0$, then $\log r$ is 0 or transcendental.

(2) If α is an algebraic number, $\alpha \neq 0$, then e^α is irrational.

The equivalence of these assertions is obvious. In particular, LINDEMANN obtained:

(7.12) π is transcendental.

If π is algebraic, so is $i\pi \neq 0$, hence $e^{i\pi} = -1$ would be irrational, a contradiction.

This important result gave a negative solution to the problem of squaring the circle.

For other proofs that e and π are transcendental, see for example, POPKEN (1929a,b), VEBLEN (1904), and SCHENKMAN (1970).

HERMITE gave a proof of the following theorem which extends **(7.11)** and was stated by LINDEMANN:

(7.13) (1) If log is any determination of the complex logarithmic function, and if α is a non-zero algebraic number, then $\log \alpha$ is either 0 or transcendental.

(2) If α is an algebraic number, $\alpha \neq 0$, then e^{α} is transcendental.

The following results which improve previous statements follow as consequences:

(7.14) If α is an algebraic number, $\alpha \neq 0$, then $\cos \alpha$, $\sin \alpha$, $\tan \alpha$, $\cosh \alpha$, $\sinh \alpha$, $\tanh \alpha$ are transcendental numbers.

WEIERSTRASS gave a proof of the following theorem stated by LINDEMANN:

(7.15) If α_1, ..., α_n are distinct algebraic numbers, then e^{α_1}, ..., e^{α_n} are linearly independent over the field of all algebraic numbers.

For example, taking $n = 2$, $\alpha_2 = 0$, one obtains the theorem of LINDEMANN and HERMITE.

This theorem admits the following equivalent formulation, which is a result of algebraic independence.

(7.16) If α_1, ..., α_n are algebraic numbers which are linearly independent over $\mathbb{Q}$, then e^{α_1}, ..., e^{α_n} are algebraically independent over $\mathbb{Q}$.

A very useful tool in the modern proof of the theorems of LINDEMANN and WEIERSTRASS is the following result of linear algebra, known as Siegel's lemma.

Let K be a number field, with degree $n = [K : \mathbb{Q}]$, and let σ_1, ..., σ_n be the isomorphisms from K into $\mathbb{C}$. For each $\alpha \in K$, let $\|\alpha\| = \max_{1 < i < n}\{|\sigma_i(\alpha)|\}$.

Given the real number $A > 0$, the integer $d > 0$ and the integer r, $1 \le r < n$, let $G_K(d, A, r)$ be the set of systems of linear equations

$$\sum_{j=1}^{n} \alpha_{ij} X^j = 0 \quad (i = 1, \dots, r)$$

with the following properties:

(1) each $\alpha_{ij} \in K$;

(2) for every $i = 1, \dots, r$ there exists an integer d_i, $0 < d_i \le d$ such that $d_i \alpha_{ij}$ is an algebraic integer (for every $j = 1, \dots, n$);

(3) $\max_{i,j}\{\|\alpha_{ij}\|\} \le A$.

Siegel's lemma is the following:

(7.17) Let K, d, A, r be as above. Then there exists a real number $c_K > 0$ such that every system of linear equations in the set $S_K(d, A, r)$ has a non-trivial solution $(\zeta_1, \dots, \zeta_n)$ where each ζ_j is an algebraic integer and

$$\max_{1 \le j \le n} \|\zeta_j\| \le c_K + c_K (c_K n d A)^{\frac{r}{n-r}}.$$

D.   A result of Siegel on exponentials

The following interesting result, which is a special case of a theorem of SIEGEL, is evoked in a paper of HALBERSTAM. It was proposed in the 1972 Putnam Competition and not one of the 2000 candidates could solve it. I give its proof below, following HALBERSTAM's article.

(7.18) If α is a real positive number and 2^α, 3^α, 5^α, ..., p^α, ... are integers (for every prime p), then α is an integer.

PROOF.   From the hypothesis, it follows that n^α is an integer for every integer n.

If α is not an integer, let $k = [\alpha]$, so $0 \le k < \alpha < k + 1$. The method of proof involves finite differences.

If $f(x)$ is an indefinitely differentiable function of the real variable x, let

$$\Delta f(x) = f(x + 1) - f(x).$$

So there exists θ_1, $0 < \theta_1 < 1$, such that

$$\Delta f(x) = f'(x + \theta_1).$$

(Note that θ_1 depends on x.)

Similarly, let

$$\Delta^2 f(x) = \Delta(\Delta f(x)) = \Delta f(x+1) - \Delta f(x) = f(x+2) - 2f(x+1) + f(x).$$

It is also equal to

$$\Delta^2 f(x) = f'(x + 1 + \theta_1) - f'(x + \theta_1) = f''(x + \theta_2),$$

where $0 < \theta_1 < \theta_2 < 2$ (θ_2 depends on x, θ_1).

More generally, if $r \geq 1$, let $\Delta^r f(x) = \Delta(\Delta^{r-1} f(x))$, so

$$\Delta^r f(x) = \sum_{i=0}^{r} (-1)^{r-i} \binom{r}{i} f(x + i) = f^{(r)}(x + \theta_r),$$

with $0 < \theta_r < r$.

Now take $f(x) = x^\alpha$, so

$$\Delta^r x^\alpha = \sum_{i=0}^{r} \binom{r}{i} (-1)^{r-i} (x + i)^\alpha,$$

hence $\Delta^r x^\alpha$ is an integer for every integer $x > 0$. But it is also equal to

$$\Delta^r x^\alpha = \alpha(\alpha - 1) \cdots (\alpha - r + 1)(x + \theta)^{\alpha - r},$$

where $0 < \theta < r$. Taking $r = k + 1$, $x = n$, and writing $(\Delta^r x^\alpha)(n) = \Delta^r n^\alpha$, one has

$$\Delta^{k+1} n^\alpha = \alpha(\alpha - 1) \cdots (\alpha - k + 1)(\alpha - k)(n + \theta)^{\alpha - k - 1},$$

with $0 < \theta < k + 1$. Thus,

$$0 < \Delta^{k+1} n^\alpha = \frac{\alpha(\alpha - 1) \cdots (\alpha - k)}{(n + \theta)^{k+1-\alpha}} < \frac{\alpha^{k+1}}{n^{k+1-\alpha}} < 1$$

provided $n > \alpha^{\frac{k+1}{k+1-\alpha}}$. This is a contradiction because $\Delta^{k+1} n^\alpha$ is an integer. $\square$

Actually, SIEGEL proved the following theorem:

(7.19) If α is a real positive number and if there exist three distinct prime numbers p_1, p_2, p_3 such that p_1^α, p_2^α, p_3^α are algebraic numbers, then $\alpha \in \mathbb{Q}$.

Or, equivalently, if $\alpha > 0$, and if α is not rational, then there are at most two primes p_1, p_2 such that p_1^α, p_2^α, are algebraic numbers.

E.   Hilbert's 7th problem

As already mentioned in the historical survey, Hilbert's 7th problem was solved independently and simultaneously by GEL'FOND and by SCHNEIDER in 1934.

First, I note the following equivalent formulations:

(7.20) The following statements are equivalent:

(1) If α, β are algebraic numbers, $\alpha \neq 0$, and $\log \alpha \neq 0$, and β is irrational, then $\alpha^\beta = \exp(\beta \log \alpha)$ is transcendental.

(2) If α, $\beta \in \mathbb{Q}^{\mathrm{alg}}$, α, $\beta \neq 0$, and if $\log \alpha$, $\log \beta$ are linearly independent over $\mathbb{Q}$, then $\log \alpha$, $\log \beta$ are linearly independent over $\mathbb{Q}^{\mathrm{alg}}$.

(3) If β, $\lambda \in \mathbb{C}$, $\lambda \neq 0$, $\beta \notin \mathbb{Q}$, then one of the numbers e^λ, β, $e^{\beta\lambda}$ is transcendental.

PROOF.   (1) $\Rightarrow$ (2). Let α, $\beta \in \mathbb{Q}^{\mathrm{alg}}$, with α, $\beta \neq 0$, 1, and assume that $\log \alpha$, $\log \beta$ are are linearly independent over $\mathbb{Q}$. If there exists γ, $\delta \in \mathbb{Q}^{\mathrm{alg}}$, with $\delta \neq 0$ (for example) such that $\gamma \log \alpha + \delta \log \beta = 0$, then $\gamma \neq 0$ and $\frac{\log \alpha}{\log \beta} = -\frac{\gamma}{\delta} = \mu \in \mathbb{Q}^{\mathrm{alg}}$ with $\mu \notin \mathbb{Q}$. By (1), α^μ is transcendental. However, $\alpha^\mu = e^{\mu \log \alpha} = e^{\log \beta} = \beta$, which is a contradiction.

(2) $\Rightarrow$ (3). Let β, $\lambda \in \mathbb{C}$, with $\lambda \neq 0$, $\beta \notin \mathbb{Q}$, and assume that e^λ, β, $e^{\beta\lambda}$ are algebraic numbers. Note that λ, $\beta\lambda$ are linearly independent over $\mathbb{Q}$. By (2), λ, $\beta\lambda$ are also linearly independent over $\mathbb{Q}^{\mathrm{alg}}$, hence $\beta = \frac{\beta\lambda}{\beta}$ is transcendental, which is a contradiction.

(3) $\Rightarrow$ (1). Let $\alpha \in \mathbb{Q}^{\mathrm{alg}}$, with $\alpha \neq 0$, 1, and let $\beta \in \mathbb{Q}^{\mathrm{alg}} \setminus \mathbb{Q}$. Let $\lambda = \log \alpha \neq 0$. Since $\alpha = e^\lambda$ and β are algebraic numbers, then by (3), $e^{\beta\lambda} = e^{\beta \log \alpha} = \alpha^\beta$ is transcendental.   □

The theorem of GEL'FOND and SCHNEIDER is the following:

(7.21) If $\alpha \in \mathbb{Q}^{\mathrm{alg}}$, $\alpha \neq 0$, $\log \alpha \neq 0$, and if $\beta \in \mathbb{Q}^{\mathrm{alg}} \setminus \mathbb{Q}$, then α^β is a transcendental number.

With this theorem one deduces that $\sqrt{2}^{\sqrt{2}}$ is a transcendental number, thus the question which motivated this chapter is solved. But one sees also that if a, b are integers such that $a^m \neq b^n$ (for all non-zero integers m, n), then $\frac{\log a}{\log b}$ is transcendental (this had been conjectured already by EULER).

Similarly,

(7.22) e^π is a transcendental number.

PROOF.   Since i, $e^{i\pi} = -1$ are algebraic numbers, by (3) above, e^π is transcendental.   □

Already in 1932, KOKSMA and POPKEN had shown that e^π is transcendental. The methods of GEL'FOND and of SCHNEIDER have been used to prove that other numbers are transcendental.

F.   The work of Baker

BAKER began in 1968 to publish a series of penetrating papers on effective lower bounds of linear forms in logarithms. Here I shall be content to quote, among his results, those which are most directly relevant to the theory of transcendental numbers. BAKER's own book (1975) is highly recommended for more results and proofs.

(7.23) Let α_1, ..., α_n be non-zero algebraic numbers. If $\log \alpha_1$, ..., $\log \alpha_n$ are linearly independent over $\mathbb{Q}$, then $1, \log \alpha_1, \ldots, \log \alpha_n$ are linearly independent over $\mathbb{Q}^{\text{alg}}$.

This result contains many important corollaries, which are quite easy to derive.

(7.24) If α_1, ..., α_n, β_1, ..., β_n are algebraic numbers, α_1, ..., α_n non-zero, and if $\theta = \sum_{i=1}^n \beta_i \log \alpha_i \neq 0$, then θ is transcendental.

PROOF.   If θ were an algebraic number, then it follows that $(-\theta) \times 1 + \sum_{i=1}^n \beta_i \log \alpha_i = 0$, so $\log \alpha_1, \ldots, \log \alpha_n$ would be linearly dependent over $\mathbb{Q}^{\text{alg}}$, hence also over $\mathbb{Q}$. So, there exist $r_1, \ldots, r_n \in \mathbb{Q}$, not all equal to 0, such that $\sum_{i=1}^n r_i \log \alpha_i = 0$. Suppose, for example, $r_n \neq 0$; then

$$0 = r_n(-\theta + \sum_{i=1}^n \beta_i \log \alpha_i)$$
$$= r_n(-\theta) + (r_n\beta_1 - r_1\beta_n)\log \alpha_1 + (r_n\beta_2 - r_2\beta_n)\log \alpha_2$$
$$+ \cdots + (r_n\beta_{n-1} - r_{n-1}\beta_n)\log \alpha_{n-1}.$$

Proceeding by induction on n, it follows that $r_n\theta$ is transcendental and θ could not be algebraic.   □

(7.25) If $n \geq 0$ and α_1, ..., α_n, β_1, ..., β_n are non-zero algebraic numbers, then $e^{\beta_0}\alpha_1^{\beta_1} \cdots \alpha_n^{\beta_n}$ is transcendental.

PROOF.   If $\theta = e^{\beta_0}\alpha_1^{\beta_1} \cdots \alpha_n^{\beta_n}$ is algebraic, then $\beta_1 \log \alpha_1 + \cdots + \beta_n \log \alpha_n - \log \theta = -\beta_0 \neq 0$ is algebraic, which contradicts the preceding proposition.   □

The following lemma will be required in the next proposition:

Lemma. *If $\gamma_1, \ldots, \gamma_m, \delta_1, \ldots, \delta_m$, for $n \geq 1$, are algebraic numbers such that each $\gamma_i \neq 0$, 1, and $\delta_1, \ldots, \delta_m$ are linearly independent over $\mathbb{Q}$, then $\sum_{i=1}^{m} \delta_i \log \gamma_i \neq 0$.*

PROOF.   If $m = 1$ it is true. Proceeding by induction on m, if $\sum_{i=1}^{m} \delta_i \log \gamma_i = 0$, then $\log \gamma_1, \ldots, \log \gamma_m$ are linearly dependent over $\mathbb{Q}^{\mathrm{alg}}$, hence by **(7.22)** they are linearly dependent over $\mathbb{Q}$. So there exist $r_1, \ldots, r_m \in \mathbb{Q}$, not all equal to 0, such that $\sum_{i=1}^{m} r_i \log \gamma_i = 0$. For example, $r_m \neq 0$, hence

$$\sum_{i=1}^{m-1} r_m \delta_i \log \gamma_i = -r_m \delta_m \log \gamma_m = \delta_m \left(\sum_{i=1}^{m-1} r_i \log \gamma_i \right),$$

thus

$$\sum_{i=1}^{m-1} (r_m \delta_i - \delta_m r_i) \log \gamma_i = 0.$$

But $r_m \delta_1 - \delta_m r_1, \ldots, r_m \delta_{m-1} - \delta_m r_{m-1}$ are linearly independent over $\mathbb{Q}$, as easily seen.

By induction, this is a contradiction.   □

(7.26) If $\alpha_1, \ldots, \alpha_n, \beta_1, \ldots, \beta_n$ are algebraic numbers, such that each $\alpha_i \neq 0$, 1, and 1, $\beta_1, \ldots, \beta_n$ are linearly independent over $\mathbb{Q}$, then $\alpha_1^{\beta_1} \cdots \alpha_n^{\beta_n}$ is transcendental.

PROOF.   If $\theta = \alpha_1^{\beta_1} \cdots \alpha_n^{\beta_n}$ is algebraic, then $\theta \neq 0$ and also $\theta \neq 1$, otherwise $\sum_{i=1}^{n} \beta_i \log \alpha_i = 0$, which is contrary to the lemma. So, $\sum_{i=1}^{n} \beta_i \log \alpha_i - \log \theta = 0$.

But 1, $\beta_1, \ldots, \beta_n$ are linearly independent over $\mathbb{Q}$, which contradicts the lemma.   □

(7.27) If α is any non-zero algebraic number, then $\pi + \log \alpha$ is transcendental.

PROOF.   From $e^{i\pi} = -1$ it follows that $\pi = -i \log(-1)$. If $\pi + \log \alpha = \beta \in \mathbb{Q}^{\mathrm{alg}}$, then $-i \log(-1) + \log \alpha - \beta$. So, $\log(-1)$, $\log \alpha$,

1 are linearly dependent over $\mathbb{Q}^{\mathrm{alg}}$, hence $\log(-1)$, $\log \alpha$ are linearly dependent over $\mathbb{Q}$, by **(7.22)**. So, there exist integers m, n, not both 0, such that $m \log(-1) + n \log \alpha = 0$. If $n = 0$, then $m \neq 0$, and so $\log(-1) = 0$, hence $\pi = 0$, a contradiction. So $n \neq 0$, and $(-1)^m \alpha^n = 1$ implies $\alpha^{2n} = 1$, thus $2n \log \alpha = 2ki\pi$ (for some integer k), hence $\beta = \pi + \pi i k$, so π would be an algebraic number. Thus $i\pi$ is algebraic, hence by **(7.21)** it follows that $-1 = e^{i\pi}$ would be transcendental, which is absurd.   □

I remark that **(7.24)** contains the theorem of LINDEMANN and HERMITE, **(7.25)** contains the theorem of GEL'FOND and SCHNEIDER, while **(7.26)** contains as a special case the transcendence of π. All this shows the strength of BAKER's theorem.

Another important fact of Baker's theorem concerns the *effective* determination of lower bounds for the linear forms in logarithms and the applications to effective determination of solutions of wide classes of diophantine equations.

Regretably, I will not treat this connection here.

G.   The conjecture of Schanuel

The proof that specific transcendental numbers are algebraically independent over $\mathbb{Q}$ is rarely a simple task. So, it is advantageous to imagine what *should* be true.

In his book of 1966, LANG ennunciated an interesting conjecture of SCHANUEL. First, I recall some terminology.

Let L be a field extension of the field K (I shall be mostly concerned with the case where $K = \mathbb{Q}$ (or $\mathbb{Q}^{\mathrm{alg}}$).

Suppose that there exist n elements $\alpha_1, \ldots, \alpha_n \in L$ with the following properties:

(1) $\alpha_1, \ldots, \alpha_n$ are algebraically independent over K; that is, if $f \in K[X_1, \ldots, X_n]$ and $f(\alpha_1, \ldots, \alpha_n) = 0$, then f is the zero polynomial;

(2) if $\beta \in L$, then β is algebraic over the field $K(\alpha_1, \ldots, \alpha_n)$, generated by $\alpha_1, \ldots, \alpha_n$.

In this case, $\{\alpha_1, \ldots, \alpha_n\}$ is a *transcendence basis* of $L|K$. It may be shown that any other transcendence basis has the same number n of elements. This number n is called the *transcendence degree* of $L|K$ and denoted by tr $\deg(L|K)$.

If $L = K(\alpha_1, \ldots, \alpha_n)$, then tr $\deg(L|K) \leq n$, and there exists a subset of $\{\alpha_1, \ldots, \alpha_n\}$ which is a transcendence basis of $L|K$.

Moreover, if tr $\deg(L|K) = n$, then $\{\alpha_1, \ldots, \alpha_n\}$ is a transcendence basis.

I note also that

$$\text{tr } \deg(\mathbb{Q}(\alpha_1, \ldots, \alpha_n)|\mathbb{Q}) = \text{tr } \deg(\mathbb{Q}^{\text{alg}}(\alpha_1, \ldots, \alpha_n)|\mathbb{Q}^{\text{alg}})$$

(where $\alpha_1, \ldots, \alpha_n$ are any complex numbers).

Schanuel's conjecture is the following:

Conjecture (S). *If $\alpha_1, \ldots, \alpha_n \in \mathbb{C}$ are linearly independent over $\mathbb{Q}$, then* tr $\deg(\mathbb{Q}(\alpha_1, \ldots, \alpha_n, e^{\alpha_1}, \ldots, e^{\alpha_n})|\mathbb{Q}) \geq n$.

This conjecture is true, for example, when $\alpha_1, \ldots, \alpha_n \in \mathbb{Q}^{\text{alg}}$. Indeed, under this additional hypothesis, (S) becomes the theorem of LINDEMANN and WEIERSTRASS.

There are many interesting conjectures about transcendental numbers which follow more or less readily from the all-embracing Schanuel's conjecture.

GEL'FOND proposed:

Conjecture (S_1). *If $\alpha_1, \ldots, \alpha_n, \beta_1, \ldots, \beta_n \in \mathbb{Q}^{\text{alg}}$, with each $\beta_i \neq 0$, if $\alpha_1, \ldots, \alpha_n$ are linearly independent over $\mathbb{Q}$ and $\log \beta_1, \ldots, \log \beta_n$ are also linearly independent over $\mathbb{Q}$, then $e^{\alpha_1}, \ldots, e^{\alpha_n}, \log \beta_1, \ldots, \log \beta_n$ are algebraically independent over $\mathbb{Q}^{\text{alg}}$.*

Indeed, (S) implies (S_1) because

$$\text{tr } \deg(\mathbb{Q}^{\text{alg}}(e^{\alpha_1}, \ldots, e^{\alpha_n}, \log \beta_1, \ldots, \log \beta_n)|\mathbb{Q}^{\text{alg}})$$
$$= \text{tr } \deg(\mathbb{Q}^{\text{alg}}(\alpha_1, \ldots, \alpha_n, e^{\alpha_1}, \ldots, e^{\alpha_n}, \log \beta_1, \ldots, \log \beta_n,$$
$$\beta_1, \ldots, \beta_n)|\mathbb{Q}^{\text{alg}})$$
$$= 2n,$$

hence $e^{\alpha_1}, \ldots, e^{\alpha_n}, \log \beta_1, \ldots, \log \beta_n$ are algebraically independent over $\mathbb{Q}^{\text{alg}}$.

A special case of conjecture (S_1) is the following:

Conjecture (S_2). *If $\beta_1, \ldots, \beta_n \in \mathbb{Q}^{\text{alg}}$ with each $\beta_i \neq 0$, and if $\log \beta_1, \ldots, \log \beta_n$ are linearly independent over $\mathbb{Q}$, then $\log \beta_1, \ldots, \log \beta_n$ are algebraically independent over $\mathbb{Q}^{\text{alg}}$.*

I recall that BAKER proved Proposition **(7.21)** which is weaker than conjecture (S_2).

The following conjecture is also a consequence of (S):

Conjecture (S_3). *If* $\alpha, \beta_1, \ldots, \beta_n \in \mathbb{Q}^{\mathrm{alg}}$, $\alpha \neq 0, 1$, *and* $1, \beta_1,$ $\ldots, \beta_n$ *are linearly independent over* $\mathbb{Q}$, *then* $\log \alpha, \alpha^{\beta_1}, \ldots, \alpha^{\beta_n}$ *are algebraically independent over* $\mathbb{Q}^{\mathrm{alg}}$.

Indeed, $\log \alpha, \beta_1 \log \alpha, \ldots, \beta_n \log \alpha$ are linearly independent over $\mathbb{Q}$, hence,

$$\mathrm{tr} \deg(\mathbb{Q}^{\mathrm{alg}}(\log \alpha, \beta_1 \log \alpha, \ldots, \beta_n \log \alpha, \alpha, \alpha^{\beta_1}, \ldots, \alpha^{\beta_n})|\mathbb{Q}^{\mathrm{alg}}) \geq n+1.$$

Since $\alpha, \beta_1, \ldots, \beta_n \in \mathbb{Q}^{\mathrm{alg}}$, then necessarily $\log \alpha, \alpha^{\beta_1}, \ldots, \alpha^{\beta_n}$ are algebraically independent over $\mathbb{Q}$.

The special case of (S_3) when $n = 1$ is the conjecture:

Conjecture (S_4). *If* $\alpha, \beta \in \mathbb{Q}^{\mathrm{alg}}$, $\alpha \neq 0, 1$, *and* $\beta \notin \mathbb{Q}$, *then* $\log \alpha, \alpha^{\beta}$ *are algebraically independent over* $\mathbb{Q}^{\mathrm{alg}}$.

The following special case of (S_3) is a conjecture of GEL'FOND:

Conjecture (S_5). *If* $\alpha, \beta \in \mathbb{Q}^{\mathrm{alg}}$, *and if* β *has degree* $d \geq 2$, *then* $\mathrm{tr} \deg(\mathbb{Q}(\alpha^{\beta}, \ldots, \alpha^{\beta^{d-1}})|\mathbb{Q}) = d - 1$.

Now, $1, \beta, \beta^2, \ldots, \beta^{d-1}$ are linearly independent over $\mathbb{Q}$; by (S_3), $\log \alpha, \alpha^{\beta}, \ldots, \alpha^{\beta^{d-1}}$ are algebraically independent over $\mathbb{Q}$, hence $\mathrm{tr} \deg(\mathbb{Q}(\alpha^{\beta}, \ldots, \alpha^{\beta^{d-1}})|\mathbb{Q}) = d - 1$.

The following conjecture, which follows also from (S), was stated in special cases by LANG and RAMACHANDRA:

Conjecture (S_6). *If* $\alpha_1, \ldots, \alpha_n$ *are linearly independent over* $\mathbb{Q}$, *and* β *is a transcendental number, then*

$$\mathrm{tr} \deg(\mathbb{Q}(e^{\alpha_1}, \ldots, e^{\alpha_n}, e^{\alpha_1 \beta}, \ldots, e^{\alpha_n \beta})|\mathbb{Q}) \geq n - 1.$$

I show that (S_6) follows from (S): order the numbers $\alpha_1, \ldots, \alpha_n$ in such a way that a basis of the $\mathbb{Q}$-vector space generated by $\{\alpha_1, \ldots, \alpha_n, \beta \alpha_1, \ldots, \beta \alpha_n\}$ is $\{\alpha_1, \ldots, \alpha_n, \beta \alpha_1, \ldots, \beta \alpha_m\}$ where $0 \leq m \leq n$. Then $\mathrm{tr} \deg(\mathbb{Q}(\alpha_1, \ldots, \alpha_n, \beta)|\mathbb{Q}) \leq m + 1$. Indeed, since β is transcendental, there is a transcendence basis of $\mathbb{Q}(\alpha_1, \ldots, \alpha_n, \beta)|\mathbb{Q}$ which is $\{\alpha_{i_1}, \ldots, \alpha_{i_s}, \beta\}$ (with $1 \leq i_1 < i_2 < \cdots < i_s \leq n$); then $\alpha_1, \ldots, \alpha_n, \beta \alpha_{i_1}, \ldots, \beta \alpha_{i_s}$ are linearly independent over $\mathbb{Q}$, so $s + n \leq m + n$, hence $s \leq m$, as required.

On the other hand, from (S) one deduces that

$$\mathrm{tr} \deg(\mathbb{Q}(\alpha_1, \ldots, \alpha_n, \beta \alpha_1, \ldots, \beta \alpha_m, e^{\alpha_1}, \ldots, e^{\alpha_n}, e^{\beta \alpha_1}, \ldots, e^{\beta \alpha_m})|\mathbb{Q})$$
$$\geq n + m,$$

hence also

$$\text{tr } \deg(\mathbb{Q}(\alpha_1, \ldots, \alpha_n, \beta\alpha_1, \ldots, \beta\alpha_n, e^{\alpha_1}, \ldots, e^{\alpha_n}, e^{\beta\alpha_1}, \ldots, e^{\beta\alpha_n})|\mathbb{Q})$$
$$\geq n + m.$$

Comparing with the transcendence degree of $\mathbb{Q}(\alpha_1, \ldots, \alpha_n, \beta)|\mathbb{Q}$, it follows that at least $n - 1$ of the numbers e^{α_i}, $e^{\alpha_i\beta}$ $(i = 1, \ldots, n)$ are algebraically independent.

Here is another interesting consequence of (S) (log denotes the principle value of the logarithm).

Conjecture (S_7). *The numbers e, e^π, e^e, e^i, π, π^π, π^e, π^i, 2^π, 2^e, 2^i, $\log\pi$, $\log 2$, $\log 3$, $\log\log 2$, $(\log 2)^{\log 3}$, $2^{\sqrt{2}}$ are algebraically independent over $\mathbb{Q}$ (and, in particular, they are transcendental).*

PROOF. I begin by noting that $i\pi$, $\log 2$ are linearly independent over $\mathbb{Q}$. By (S), tr $\deg(\mathbb{Q}(i\pi, \log 2, -1, 2|\mathbb{Q})) = 2$, so $i\pi$, $\log 2$ are algebraically independent over $\mathbb{Q}$; hence so are π and $\log 2$. Therefore, 2, 3, π, $\log 2$ are multiplicatively independent: if $2^a 3^b \pi^c (\log 2)^d = 1$ (with a, b, c, $d \in \mathbb{Z}$), then $a = b = c = d = 0$. Thus, $\log\pi$, $\log 2$, $\log 3$, $\log\log 2$ are linearly independent over $\mathbb{Q}$. Hence, also $i\pi$, $\log\pi$, $\log 2$, $\log 3$, $\log\log 2$ are linearly independent over $\mathbb{Q}$. By (S),

$$\text{tr } \deg(\mathbb{Q}(i\pi, \log\pi, \log 2, \log 3, \log\log 2, -1, \pi, 2, 3, \log 2)|\mathbb{Q}) = 5,$$

thus π, $\log\pi$, $\log 2$, $\log 3$, $\log\log 2$ are algebraically independent over $\mathbb{Q}$. Hence 1, $i\pi$, $\log\pi$, $\log 2$, $\log 3$, $\log\log 2$ are linearly independent over $\mathbb{Q}$. By (S),

$$\text{tr } \deg(\mathbb{Q}(1, i\pi, \log\pi, \log 2, \log 3, \log\log 2, e, -1, \pi, 2, 3, \log 2)|\mathbb{Q}) = 6,$$

hence e, π, $\log\pi$, $\log 2$, $\log 3$, $\log\log 2$, are algebraically independent. So 1, $i\pi$, π, $\log\pi$, e, $e\log\pi$, $\pi\log\pi$, $\log 2$, $\pi\log 2$, $e\log 2$, $i\log 2$, i, $i\log\pi$, $\log 3$, $\log\log 2$, $(\log 3)(\log\log 2)$, $\sqrt{2}\log 2$ are linearly independent over $\mathbb{Q}$. By (S),

$$\text{tr } \deg(\mathbb{Q}(i\pi, \pi, \log\pi, e, e\log\pi, \pi\log\pi, \log 2, \pi\log 2, e\log 2, i\log 2,$$
$$i, i\log\pi, \log 3, \log\log 2, (\log 3)(\log\log 2), \sqrt{2}\log 2, -1, e^\pi, \pi,$$
$$e^e, \pi^e, \pi^\pi, 2, 2^\pi, 2^e, 2^i, e^i, \pi^i, 3, \log 2, (\log 2)^{\log 3}, 2^{\sqrt{2}})|\mathbb{Q}) = 17.$$

Hence, π, $\log\pi$, e, $\log 2$, $\log 3$, $\log\log 2$, e^π, e^e, π^e, π^π, 2^π, 2^e, 2^i, e^i, π^i, $(\log 2)^{\log 3}$, $2^{\sqrt{2}}$ are algebraically independent over $\mathbb{Q}$. $\square$

LANG considered the following conjecture. Let K_1 be the field of all numbers which are algebraic over the field $\mathbb{Q}^{\text{alg}}(e^\alpha)$. Let K_2 be the field of all numbers which are algebraic over the field $K_1(e^\alpha)_{\alpha \in K_1}$. In the same way, define the fields K_3, K_4, ..., and let $K = \bigcup_{n \geq 1} K_n$.

Conjecture (S_8). $\pi \notin K$.

LANG sketched a proof of how this conjecture is a consequence of (S).

There have been the classical results of algebraic independence, by HERMITE, WEIERSTRASS and the more recent results of BAKER, all involving the experimental and the logarithmic function. For a long time it was desirable to obtain an algebraic independence result involving the gamma function $\Gamma(x)$.

In a *tour de force*, culminating deep research on algebraic independence, NESTERENKO established the result below 1997 (see also GRAMAIN (1998)):

(7.28) The numbers π, e^π, $\Gamma(\frac{1}{4})$ are algebraically independent over $\mathbb{Q}$.

This theorem was hailed by experts in the field and by broadly informed mathematicians. In other circles of mathematicians—and good ones at that—one wondered why time and energy should be spent on questions of no practical importance, such as this one. And for that matter, about transcendental numbers like $\sqrt{2}^{\sqrt{2}}$ ....

Mathematics has the unique character of being a scientific discipline with applications to all kinds of other sciences and to practical life. But mathematics is also an art, the beauty lying in the symmetries, patterns, and intricately deep relationships which enchant the beholder. Discoveries that require the invention of new methods and great ingenuity are indeed to be hailed as important—at least from one point of view. Will these be of any practical use some day? Is it a legitimate question? Indeed, numerous are the examples when theories seemed for centuries to be gratuitous speculations, like the study of prime numbers, but today a mainstay of crucial applications in communications. It is the intrinsic quality of a new result which confers its importance.

H. Transcendence measure and the classification of Mahler

To classify the complex numbers, MAHLER considered the values of polynomial expressions and measured how close to zero they may become.

Let $n \geq 1$, $H \geq 1$ be integers, let $\mathbb{Z}_{n,H}[X]$ be the set of all polynomials $f(X) \in \mathbb{Z}[X]$ of degree at most n and height at most H; i.e., $f(X) = \sum_{i=0}^{n} a_i X^i$, with $a_i \in \mathbb{Z}$ and $\max\{|a_i|\} \leq H$. The set $\mathbb{Z}_{n,H}[X]$ is clearly finite.

If $\alpha \in \mathbb{C}$, let

$$w_{n,H}(\alpha) = \min\{|f(\alpha)| : f \in \mathbb{Z}_{n,H}[X] \text{ and } f(\alpha) \neq 0\}.$$

Taking $f(X) = 1$, one has $0 < w_{n,H}(\alpha) \leq 1$. Also, if $n \leq n'$, $H \leq H'$, then $w_{n,H}(\alpha) \geq w_{n',H'}(\alpha)$.

Let $w_n(\alpha) = \limsup_{H \to \infty} \frac{-\log w_{n,H}(\alpha)}{\log H}$ for all $n \geq 1$, and let $w(\alpha) = \limsup_{n \to \infty} \frac{w_n(\alpha)}{n}$.

Thus $0 \leq w_n(\alpha) \leq \infty$ and $w_n(\alpha) \leq w_{n+1}(\alpha)$ for $n \geq 1$. Hence $0 \leq w(\alpha) \leq \infty$.

Let $\mu(\alpha) = \inf\{n \mid w_n(\alpha) = \infty\}$, so $1 \leq \mu(\alpha) \leq \infty$, and if $\mu(\alpha) < \infty$, then $w(\alpha) = \infty$.

This leads to the following partition of complex numbers into four disjoint classes, proposed by MAHLER (1930, 1932b):

(1) α is an *A-number* when $w(\alpha) = 0$, $\mu(\alpha) = \infty$;
(2) α is an *S-number* when $0 < w(\alpha) < \infty$, $\mu(\alpha) = \infty$;
(3) α is a *T-number* when $w(\alpha) = \infty$, $\mu(\alpha) = \infty$;
(4) α is a *U-number* when $w(\alpha) = \infty$, $\mu(\alpha) < \infty$.

MAHLER proved:

(7.29) α is an *A-number* if and only if it is an algebraic number.

Moreover:

(7.30) If α, β are numbers in different classes, then α, β are algebraically independent.

The *S*-numbers may be classified according to their type, which I define now. From $w(\alpha) < \infty$ it follows that the sequence $\frac{w_n(\alpha)}{n}$ is bounded above, so there exists $t > 0$ such that

$$\limsup_{H \to \infty} \frac{-\log w_{n,H}(\alpha)}{\log H} = w_n(\alpha) < tn$$

for every $n \geq 1$. Hence, for every $\epsilon > 0$ there exists $H_0 \geq 1$ (depending on n, t, ϵ) such that $\frac{-\log w_{n,H}(\alpha)}{\log H} < n(t + \epsilon)$ for all $H > H_0$. Hence, $w_{n,H}(\alpha) > H^{-n(t+\epsilon)}$ for $H > H_0$. Choosing $c_n = \min_{1 \leq H \leq H_0}\{1, \frac{1}{2}w_{n,H}(\alpha)H^{n(t+\epsilon)}\}$, then $w_{n+1}(\alpha) > \frac{c_n}{H^{n(t+\epsilon)}}$ for all $H \geq 1$. Thus, there exists $\theta > 0$ such that for every $n \geq 1$ there exists $c_n > 0$ satisfying $w_{n,H}(\alpha) > \frac{c_n}{H^{n-\theta}}$ for all $H \geq 1$.

The *type* α is defined to be the infimum of all θ with the above property. It may be shown that $\theta(\alpha) = \sup_{n \geq 1}\{\frac{w_n(\alpha)}{n}\}$.

Now I investigate the cardinality and measure of the sets of S-numbers, T-numbers and U-numbers.

In 1932, MAHLER showed that the sets of real, respectively complex, S-numbers have measure 1 (in the sense of linear, respectively plane Lebesgue measure). The following more precise statement was conjectured by MAHLER in the same paper.

Using a classification given by KOKSMA, in analogy to MAHLER's classification, SPRINDŽUK proved Mahler's conjecture in 1965:

(7.31) (1) All real numbers (with the exception of a subset with measure 0 in $\mathbb{R}$) are S-numbers of type 1.

(2) All complex numbers (with the exception of a subset with measure 0 in $\mathbb{C}$) are S-numbers of type $\frac{1}{2}$.

So, the set of S-numbers is uncountable.

Yet, it is not generally straightforward to give examples of S-numbers and *a fortiori* to compute their type.

MAHLER showed that $\alpha = 0.123456789101112\ldots$ (already considered in subsection B., Example 5) is an S-number.

For many years it was not known whether the set of T-numbers was empty or not. SCHMIDT showed in 1968 (without exhibiting an explicit example) that the set of T-numbers is not empty. He gave a simpler proof in 1969.

As for the U-numbers, an easy characterization implies readily:

(7.32) Every Liouville number is a U-number.

Moreover, LEVEQUE showed (1953):

(7.33) For every integer $\mu \geq 1$, there exists a U-number α such that $\mu(\alpha) = \mu$.

It follows that the set of U-numbers is uncountable, even though it has measure zero (by **(7.30)**).

In 1971, 1972, MAHLER modified his classification of transcendental numbers; various problems arising from the new set-up were solved by DURAND in 1974.

Now I introduce the concept of transcendence measure of a transcendental number.

The function $T(n, H)$ (with real positive values), defined for integers $n \geq 1$, $H \geq 1$, is a *transcendence measure* for the transcendental number α if $|f(\alpha)| \geq T(n, H)$ for every $f \in \mathbb{Z}_{n,H}[X]$.

The best transcendence measure is, of course, $w_{n,H}(\alpha)$ as defined above by MAHLER. However, it is usually very difficult to calculate.

I indicate some results about transcendence measures for numbers like e, π, $\log r$ (r rational, $r \neq 1$, $r > 0$).

BOREL (1899) and POPKEN (1929a) gave transcendence measures for e.

In particular, POPKEN's result implied that e is not a Liouville number. It should be noted that this may be also proved from the continued fraction expansion of e, which implies that

$$\left| e - \frac{a}{b} \right| \geq \frac{\log \log(4b)}{18 \log(4b) b^2}$$

for all rational numbers $\frac{a}{b}$, $b > 0$ (see also BUNDSCHUH (1971)).

MAHLER proved in 1932:

(7.34) For every $n \geq 1$ there exists $H_0(n) \geq 1$ such that if $H > H_0(n)$, then

$$|f(e)| > \frac{1}{H^{n+Cn^2 \frac{\log(n+1)}{\log \log H}}}$$

for every $f(X) \in \mathbb{Z}_{n,H}[X]$, where $C > 0$ is a constant, independent of n and H.

It follows that

(7.35) e is an S-number of type $\theta(e) = 1$; hence e is not a Liouville number.

Next, MAHLER showed (1932):

(7.36) Let $\alpha = \pi$ or $\alpha = \log r$ where r is a positive rational number, $r \neq 1$. Then, for every $n \geq 1$ and $H \geq 1$, $|f(\alpha)| > \frac{C(n)}{H^{s^n}}$ for every $f(X) \in \mathbb{Z}_{n,H}[X]$, where $C(n) > 0$ and $s > 0$ is a constant, independent of n, H.

It follows that

(7.37) π and $\log r$ (r rational, $r > 0$, $r \neq 1$) are not U-numbers, hence they are not Liouville numbers.

For more information on transcendence measures, the reader may consult the paper of WALDSCHMIDT (1978).

8   Final comments

It is preferable now to interrupt this survey lest it become too tiring for the reader (but never for me). Apart from the fact that many topics evoked in the survey were no more than evoked, there are many aspects which were completely ignored: metric problems concerning continued fractions, normal numbers, uniform distribution modulo 1, questions of irrationality and transcendence of values of entire function, of certain meromorphic functions, or of functions which are solutions of certain types of differential equations. Nor did I touch on questions of simultaneous approximations, nor did I ....

Fortunately, there are many books and surveys on various aspects of the theory (some of these have already been cited): MAILLET (1906) (the first book devoted to transcendental numbers), MINKOWSKI (1907), PERRON (1910, 1913), KHINTCHINE (1935), KOKSMA (1936), SIEGEL (1949), GEL'FOND (1952), NIVEN (1956), CASSELS (1957), SCHNEIDER (1957), MAHLER (1961, 1976a), NIVEN (1963), LANG (1966), LIPMAN (1966), FEL'DMAN (1967), RAMACHANDRA (1969), SCHMIDT (1972), WALDSCHMIDT (1974, 1979), BAKER (1975), and MIGNOTTE (1976).

I hope that the reader derived some enjoyment and that the present survey has stimulated the desire for further studies of numbers.

References

1572 R. Bombelli. *L'Algebra, Parte Maggiore dell'Aritimetica Divisa in Tre Libri*. Bologna. Reprinted by Feltrinelli, Milano, 1966.

1655 J. Wallis. *Arithmetica Infinitorum*. Reprinted in *Opera Mathematica*, Vol. I, Oxford, 1695.

1685 J. Wallis. *Tractatus de Algebra.* Reprinted in *Opera Mathematica,* Vol. II, Oxford, 1695.

1737 L. Euler. De fractionibus continuis dissertation. *Comm. Acad. Sci. Petr.,* 9:98–137. Reprinted in *Opera Omnia,* Ser. I, Vol. 14, *Commentationes Analyticae,* 187–215. B. G. Teubner, Leipzig, 1924.

1748 L. Euler. *Introductio in Analysin Infinitorum,* Vol. I, Chapter VI, §105. Lausanne. Reprinted in *Opera Omnia,* Ser. I, Vol. 8, 108–109. B. G. Teubner, Leipzig, 1922.

1755 L. Euler. De relatione inter ternas pluresve quantitates instituenda. *Opuscula Analytica,* 2:91–101. Reprinted in *Opera Omnia,* Ser. I, Vol. 4, *Commentationes Arithmeticae,* 136–146. B. G. Teubner, Leipzig, 1941.

1761 J. H. Lambert. Mémoire sur quelques propriétés remarquables des quantités transcendantes circulaires et logarithmiques. *Mém. Acad. Sci. Berlin,* 17:265–322.

1769a J. L. Lagrange. Solution d'un probléme d'arithmétique. *Miscellanea Taurinensia, 1769–79,* 4. Reprinted in *Oeuvres,* Vol. I, 671–731. Gauthier-Villars, Paris, 1867.

1769b J. L. Lagrange. Sur la solution des problèmes indéterminés du second degré. *Mém. Acad. Royale Sci. Belles-Lettres de Berlin,* 23. Reprinted in *Oeuvres,* Vol. II, 377–535. Gauthier-Villars, Paris, 1868.

1770 J. L. Lagrange. Additions au mémoire sur la résolution des équations numériques. *Mém. Acad. Royale Sci. Belles-Lettres de Berlin,* 24. Reprinted in *Oeuvres,* Vol. II, 581–652. Gauthier-Villars, Paris, 1868.

1770 J. H. Lambert. *Vorläufige Kenntnisse für die, so die Quadratur und Rectification des Circuls suchen.* Berlin. Beitrage zum Gebrauche der Mathematik und deren Anwendung (2. Teil), 144–169, Berlin, 1770.

1779 L. Euler. De formatione fractionum continuarum. *Acta. Acad. Sci. Imper. Petropolitane, I (1779),* 9:3–29. Reprinted in *Opera Omnia,* Vol. XV, 314–337. B. G. Teubner, Leipzig, 1927.

1794 A. M. Legendre. *Éléments de Géométrie* (12e édition), Note IV, 286–296. Firmin Didot, Paris. 1823 (12e édition), 1794 (1e édition).

1801 C. F. Gauss. *Disquisitiones Arithmeticae.* G. Fleischer,

Leipzig. Translated by A. A. Clarke, Yale Univ. Press, New Haven, 1966.

1808   A. M. Legendre. *Essai sur la Théorie des Nombres (Seconde Édition)*. Courcier, Paris.

1815   M. J. De Stainville. *Mélanges d'Analyse Algébrique et de Géométrie*. Courcier, Paris.

1829   E. Galois. Démonstration d'un théorème sur les fractions continues périodiques. *Ann. Math. Pures et Appl., de M. Gergonne*, 19:294–301. Reprinted in *Écrits et Mémoires Mathématiques*, (par R. Bourgne et J. P. Azra), 365–377. Gauthier-Villars, Paris, 1962.

1832   F. Richelot. De resolutione algebraica aequationes $X^{257} = 1$, sive de divisione circuli per bisectionem anguli septies repetitam in partes 257 inter se aequalis commentatio coronata. *J. reine u. angew. Math.*, 9:1–26, 146–161, 209–230, 337–358.

1836   G. L. Dirichlet. Sur les intégrales eulériennes. *J. reine u. angew. Math.*, 15:258–263. Reprinted in *Werke*, Vol. I, 273–282. G. Reimer, Berlin, 1889.

1840a   J. Liouville. Additif à la note sur l'irrationalité du nombre e. *J. Math. Pures et Appl.*, 5(1):193.

1840b   J. Liouville. Sur l'irrationalité du nombre e. *J. Math. Pures et Appl.*, 5(1):192.

1842   G. L. Dirichlet. Verallgemeinerung eines Satzes aus der Lehre von den Kettenbrüchen nebst einigen Anwendungen auf die Theorie der Zahlen. *Sitzungsber. Preuss. Akad. d. Wiss.*, Berlin, 93–95. Reprinted in *Werke*, Vol. I, 633–638. G. Reimer, Berlin, 1889. Reprinted by Chelsea Publ. Co., New York, 1969.

1844a   J. Liouville. Nouvelle démonstration d'un théorème sur les irrationalles algébriques, inséré dans le compte rendu de la dernière séance. *C. R. Acad. Sci. Paris*, 18:910–911.

1844b   J. Liouville. Sur des classes très étendues de quantités dont la valeur n'est ni algébrique, ni même réductible à des irrationnelles algébriques. *C. R. Acad. Sci. Paris*, 18: 883–885.

1851   J. Liouville. Sur des classes très étendues de quantités dont la valeur n'est ni algébrique, ni même réductible à des irrationnelles algébriques. *J. Math. Pures et Appl.*, 16(1):

133–142.

1869 G. Cantor. Über die einfachen Zahlensysteme. *Zeitsch. f. Math. u. Physik*, 14:121–128. Reprinted in *Gesammelte Abhandlungen*, 35–42. Springer-Verlag, Berlin, 1932.

1873 C. Hermite. Sur la fonction exponentielle. *C. R. Acad. Sci. Paris*, 77:18–24, 74–79, 226–233, 285–293. Reprinted in *Oeuvres*, Vol. III, 150–181. Gauthier-Villars, Paris, 1912.

1874 G. Cantor. Über eine Eigenschaft der Inbegriffes aller reellen algebraischen Zahlen. *J. reine u. angew. Math.*, 77:258–262. Reprinted in *Gesammelte Abhandlungen*, 115–118. Springer-Verlag, Berlin, 1932.

1874 F. Mertens. Ein Beitrag zur analytischen Zahlentheorie Über die Verheilung der Primzahlen. *J. reine u. angew. Math.*, 78:46–63.

1874 T. Muir. *The expression of a quadratic surd as a continued fraction.* Glasgow.

1876 J. W. L. Glaisher. Three theorems in arithmetics. *Messenger of Math.*, 5:21–22.

1878 G. Cantor. Ein Beitrag zur Mannigfaltigkeitslehre. *J. reine u. angew. Math.*, 84:242–258. Reprinted in *Gesammelte Abhandlungen*, 119–133. Springer-Verlag, Berlin, 1932.

1878 J. W. L. Glaisher. Series and products for π and powers of π. *Messenger of Math.*, 7:75–80.

1879 A. Markoff. Sur les formes quadratiques binaires indéfinies. *Math. Annalen*, 15:381–409.

1880 J. J. Sylvester. On a point in the theory of vulgar fractions. *Amer. J. of Math.*, 3:332–335. Reprinted in *Mathematical Papers*, Vol. III, 440–445. University Press, Cambridge, 1909.

1882a F. Lindemann. Sur le rapport de la circonférence au diamètre, et sur les logarithmes népériens des nombres commensurables ou des irrationnelles algébriques. *C. R. Acad. Sci. Paris*, 95:72–74.

1882b F. Lindemann. Über die Ludolph'sche Zahl. *Sitzungsber. Preuß. Akad. Wiss. zu Berlin*, 679–682.

1882c F. Lindemann. Über die Zahl π. *Math. Annalen*, 20:213–225.

1883 J. Lüroth. Über die eindeutige Entwicklung von Zahlen in eine unendliche Reihe. *Math. Annalen*, 21:411–423.

1884 L. Kronecker. Näherungsweise ganzzahlige Auflösung linearer Gleichungen. *Sitzungsber. Preuß. Akad. d. Wiss. zu Berlin*, 1179–1193 and 1271–1299. Reprinted in *Werke*, Vol. III, 47–110. B. G. Teubner, Leipzig, 1930.

1885 K. Weierstrass. Zu Lindemann's Abhandlung "Über die Ludolph'sche Zahl". *Sitzungsber. Preuß. Akad. Wiss. zu Berlin*, 1067–1085. Reprinted in *Mathematische Werke*, Vol. 11, 341–362. Mayer & Müller, Berlin, 1895.

1891a A. Hurwitz. Über die angenäherte Darstellung der Irrationalzahlen durch rationale Brüche. *Math. Annalen*, 39:279–284. Reprinted in *Mathematische Werke*, Vol. II, 122–128. Birkhäuser, Basel, 1963.

1891b A. Hurwitz. Über die Kettenbruch-Entwicklung der Zahl e. *Schriften phys. ökon. Gesellschaft zu Königsberg*, 32 Jahrg.: 59–62. Reprinted in *Mathematische Werke*, Vol. II, 129–133. Birkhäuser, Basel, 1933.

1893 A. Hurwitz. Beweis der Transzendenz der Zahl e. *Math. Annalen*, 43:220–222. Reprinted in *Mathematische Werke*, Vol. II, 134–135. Birkhäuser, Basel, 1933.

1899 E. Borel. Sue la nature arithme'tique du nombre e. *C. R. Acad. Sci. Paris*, 128:596–599.

1899 E. Landau. Sur la série des inverses des nombres de Fibonacci. *Bull. Soc. Math. France*, 27:298–300.

1900 D. Hilbert. Mathematische Probleme. Göttinger Nachrichten, 253–297 and also *Archiv d. Math. u. Physik, Ser. 3*, 1, 1901, 44–63 and 213–237. English translation appeared in *Bull. Amer. Math. Soc.*, 8, 1902, 437–479, and also in *Mathematical Developments arising from Hilbert Problems. Proc. Symp. Pure Math.*, 28, 1976, 1–34.

1901 M. Lerch. Sur la fonction $\zeta(s)$ pour valeurs impaires de l'argument. *Jornal Ciencias Mat. e Astron.*, 14:65–69. Published by F. Gomes Teixeira, Coimbra.

1904 O. Veblen. The transcendence of π and e. *Amer. Math. Monthly*, 11:219–223.

1906 E. Maillet. *Introduction à la Théorie des Nombres Transcendants et des Propriétés Arithmétiques des Fonctions.* Gauthier-Villars, Paris.

1907 H. Minkowski. *Diophantische Approximationen.* B. G. Teubner, Leipzig.

1909 A. Thue. Über Annäherungswerte algebraische Zahlen. *J. reine u. angew. Math.*, 135:284–305. Reprinted in *Selected Mathematical Papers*, 232–253. Universitetsforlaget, Oslo, 1982.

1910 O. Perron. *Irrationalzahlen.* W. de Gruyter, Berlin. Reprinted by Chelsea Publ. Co., New York, 1951.

1910 W. Sierpiński. Sur la valeur asymptotique d'une certaine somme. *Bull. Intern. Acad. Sci. Cracovie*, 9–11. Reprinted in *Oeuvres Choisies*, Vol. I, 158–160. Warszawa, 1974.

1912 A. Thue. Über eine Eigenschaft, die keine transcendente Grössen haben kann. *Kristiania Vidensk. Selskab Skr., I, Mat. Nat. Kl.*, No. 20. Reprinted in *Selected Mathematical Papers*, 479–492. Universiteteforlaget, Oslo, 1982.

1913 F. Engel. *Verhandl. d. 52. Versammlung deutsche Philologen u. Schulmänner*, 190–191. Marburg.

1913 O. Perron. *Die Lehne von den Kettenbrüche.* B. G. Teubnen, Leipzig. Reprinted by Chelsea Publ. Co., New York, 1950.

1914 S. Kakeya. On the partial sums of an infinite series. *Science Reports Tôhoku Imp. Univ.*, 3(1):159–163.

1919 F. Hausdorff. Dimension und äußeres Maß. *Math. Annalen*, 79:157–179.

1921 O. Perron. Über die Approximation irrationaler Zahlen durch rationale, I, II. *Sitzungsber. Heidelberg Akad. d. Wiss.*, Abh. 4, 17 pages and Abh. 8, 12 pages.

1921 C. L. Siegel. Über den Thueschen Satz. *Norske Vidensk. Selskab Skrifter, Kristiania, Ser. 1.*, No. 16, 12 pages. Reprinted in *Gesammelte Abhandlungen*, Vol. I, 103–112. Springer-Verlag, Berlin, 1966.

1924 A. J. Khintchine. Einige Sätze über Kettenbrüche mit Anwendungen auf die Theorie der Diophantióchen Approximationen. *Math. Annalen*, 92:115–125.

1924 G. Pólya and G. Szegö. *Aufgaben und Lehrsätze der Analysis, I.* Springer-Verlag, Berlin.

1925 S. L. Malurkar. On the application of Herr Mellin's integrals to some series. *J. Indian Math. Soc.*, 16:130–138.

1929 K. Mahler. Arithmetische Eigenshaften def Lösungen einer klasse von Funktionalgleichungen. *Math. Annalen*, 101: 342–366.

⸌1929a J. Popken. Zur Transzendenz von *e*. *Math. Z.*, 29:525–541.

1929b J. Popken. Zur Transzendenz von π. *Math. Z.*, 29:542–548.

1929 K. Shibata. On the order of approximation of irrational numbers by rational numbers. *Tôhoku Math. J*, 30:22–50.

1930 K. Mahler. Über Beziehungen zwischen der Zahl *e* und den Liouvilleschen Zahlen. *Math. Z.*, 31:729–732.

1932 J. F. Koksma and J. Popken. Zur transzendenz von e^π. *J. reine u. angew. Math.*, 168:211–230.

1932a K. Mahler. Über das Mass der Menge aller *S*-Zahlen. *Math. Annalen*, 106:131–139.

1932b K. Mahler. Zur Approximation der Exponentialfunktion und des Logarithmus, I, II. *J. reine u. angew. Math.*, 166: 118–136 and 137–150.

1932 O. Perron. Über mehrfach transzendente Erweiterungen des natürlichen Rationalitäts bereiches. *Sitzungsber. Bayer Akad. Wiss.*, H2, 79–86.

1933 B. L. van der Waerden. Die Seltenheit der Gleichungen mit Affekt. *Math. Annalen*, 109:13–16.

1934a A. O. Gel'fond. Sur le septième problème de Hilbert. *Dokl. Akad. Nauk SSSR*, 2:1–6.

1934b A. O. Gel'fond. Sur le septième problème de Hilbert. *Izv. Akad. Nauk SSSR*, 7:623–630.

1934 T. Schneider. Tranzendenzuntersuchungen periodischer Funktionen. *J. reine u. angew. Math.*, 172:65–74.

1935 A. J. Khintchine. *Continued Fractions*. Moscow. Translation of the 3rd edition by P. Wynn, Noordhoff, Gröningen, 1963.

1936 J. F. Koksma. *Diophantische Approximationen*. Springer-Verlag, Berlin. Reprinted by Chelsea Publ. Co., New York, 1953.

1937 K. Mahler. Eigenshaften eines Klasse von Dezimalbrüchen. *Nederl. Akad. Wetensch., Proc. Ser. A*, 40:421–428.

1938 G. H. Hardy and E. M. Wright. *An Introduction to the Theory of Numbers*. Clarendon Press, Oxford, 5th (1979) edition.

1939 J. F. Koksma. Über die Mahlersche Klasseneinteilung der transzendente Zahlen und die Approximation komplexer Zahlen durch algebraischen Zahlen. *Monatshefte Math. Phys.*, 48:176–189.

1943 F. J. Dyson. On the order of magnitude of the partial quotients of a continued fraction. *J. London Math. Soc.*, 18:40–43.

1947 F. J. Dyson. The approximation of algebraic numbers by rationals. *Acta Arith.*, 79:225–240.

1949 T. Schneider. Über eine Dysonsche Verschärfung des Siegel-Thuesche Satzes. *Arch. Math.*, 1:288–295.

1949 C. L. Siegel. *Transcendental Numbers.* Annals of Math. Studies, 16, Princeton, N.J.

1952 A. O. Gel'fond. *Transcendental and Algebraic Numbers* (in Russian). G.I.T.T.L., Moscow. English translation at Dover, New York, 1960.

1953a W. J. LeVeque. Note on S-numbers. *Proc. Amer. Math. Soc.*, 4:189–190.

1953b W. J. LeVeque. On Mahler's U-numbers. *J. London Math. Soc.*, 28:220–229.

1953 K. Mahler. On the approximation of π. *Indag. Math.*, 15: 30–42.

1954 M. M. Hjortnaes. Overføng av rekken $\sum_{k=1}^{\infty} \frac{1}{k^3}$ til et bestemt integral. In *Proc. 12th Congr. Scand. Math., Lund, 1953*. Lund Univ.

1955 K. F. Roth. Rational approximations to algebraic numbers. *Mathematika*, 2:1–20. Corrigendum, p. 168.

1956 I. Niven. *Irrational Numbers.* Math. Assoc. of America, Washington.

1957 J. W. S. Cassels. *An Introduction to Diophantine Approximation.* Cambridge Univ. Press, Cambridge.

1957 S. Ramanujan. *Notebooks of Srinivasan Ramanujan (2 volumes).* Tata Institute of Fund. Res., Bombay.

1957 T. Schneider. *Einführung in die Transzendenten Zahlen.* Springer-Verlag, Berlin. Translated into French by F. Eymard. Gauthier-Villars, Paris, 1959.

1959 G. H. Hardy and E. M. Wright. *The Theory of Numbers, 4th ed.* Clarendon Press, Oxford.

1961 K. Mahler. *Lectures on Diophantine Approximations.* Notre Dame Univ., South Bend, IN.

1962 P. Erdős. Representation of real numbers as sums and products of Liouville numbers. *Michigan Math. J.*, 9:59–60.

1962 W. M. Schmidt. Simultaneous approximation and algebraic independence of numbers. *Bull. Amer. Math. Soc.*, 68:475–478.

1963 I. Niven. *Diophantine Approximations.* Wiley-Interscience, New York.

1963 C. D. Olds. *Continued Fractions.* Math. Assoc. of America, Washington.

1964a A. Baker. Approximations to the logarithms of certain rational numbers. *Acta Arith.*, 10:315–323.

1964b A. Baker. Rational approximations to $\sqrt[3]{2}$ and other algebraic numbers. *Quart. J. Math. Oxford*, 15:375–383.

1964 D. Knuth. Transcendental numbers based on the Fibonacci sequence. *Fibonacci Q.*, 2:43–44.

1965 V. G. Sprindžuk. A proof of Mahler's conjecture on the measure of the set of S-numbers. *Izv. Akad. Nauk SSSR, Ser. Mat.*, 29:379–436. Translated into English in *Amer. Math. Soc. Transl. (2)*, 51, 1960, 215–272.

1966 S. Lang. *Introduction to Transcendental Numbers.* Addison-Wesley, Reading, MA.

1966 J. N. Lipman. *Transcendental Numbers.* Queen's Papers in Pure and Applied Mathematics, No. 7. Queen's University, Kingston, Ont., 1966.

1967 N. I. Fel'dman and A. B. Shidlovskii. The development and present state of the theory of transcendental numbers. *Russian Math. Surveys*, 22:1–79. Translated from Uspehi Mat. Nauk SSSR, 22:3–81.

1968 W. M. Schmidt. T-numbers do exist. In *Symp. Math., IV, 1st. Naz. di Alta Mat., Roma*, 3–26. Academic Press, London, 1970.

1969 K. Ramachandra. *Lectures in Transcendental Numbers.* Ramanujan Institute, Madras.

1970 P. Bundschuh. Ein Satz über ganze Funktionen und Irrationalitätsaussagen. *Invent. Math.*, 9:175–184.

1970 E. Grosswald. Die Werte der Riemannschen Zetafunktion an ungeraden Argumentstellen. *Nachr. der Akad. Wiss. Göttingen*, 9–13.

1970 E. Schenkman. The independence of some exponential values. *Amer. Math. Monthly*, 81:46–49.

1971 J. L. Brown. On generalized bases for real numbers. *Fibonacci Q.*, 9:477–496.

1971 P. Bundschuh. Irrationalitätsmasse für e^a, $a \neq 0$, rational oder Liouville Zahl. *Math. Annalen*, 192:229–242.

1971 L. Carlitz. Reduction formulas for Fibonacci summations. *Fibonacci Q.*, 9:449–466 and 510.

1971 S. Lang. Transcendental numbers and diophantine approximation. *Bull. Amer. Math. Soc.*, 77:635–677.

1971 K. Mahler. On the order function of a transcendental number. *Acta Arith.*, 18:63–76.

1971 W. M. Schmidt. Mahler's T-numbers. *Proc. Sympos. Pure Math., Vol. XX*, 275–286.

1972 E. Grosswald. Comments on some formulae of Ramanujan. *Acta Arith.*, 21:25–34.

1972 W. M. Schmidt. *Approximation to Algebraic Numbers.* Enseign. Math., Monograph #19, Génève.

1973 K. Katayama. On Ramanujan's formula for values of Riemann zeta function at positive odd integers. *Acta Arith.*, 22:149–155.

1973 K. Mahler. The classification of transcendental numbers. In *Analytic Number Theory (Proc. Sympos. Pure Math., Vol. XXIV, St. Louis Univ., St. Louis, MO, 1972)*, 175–179. Amer. Math. Soc., Providence, R.I.

1973 Z. A. Melzak. *Companion to Concrete Mathematics, 2 volumes.* John Wiley & Sons, New York. 1973, 1976.

1973 M. Mignotte. Construction de nombres transcendants, grâce aux théorèmes de Liouville, Thue, Siegel et Roth. In *Sém. Waldschmidt*, chapter 3. Orsay.

1973 L. Nový. *Origins of Modern Algebra.* P. Noordhoff, Leyden.

1973 J. R. Smart. On the values of the Epstein zeta function. *Glasgow Math. J.*, 14:1–12.

1973 M. Waldschmidt. La conjecture de Schanuel. In *Sém. Waldschmidt*, Orsay.

1974 B. C. Berndt. Ramanujan's formula for $\zeta(2n+1)$. In *Professor Srinivasan Ramanujan Commemoration Volume*, 2–9. Jupiter Press, Madras.

1974 L. Comtet. *Advanced Combinatorial Analysis.* D. Reidel, Dordrecht.

1974 A. Durand. Quatre problèmes de Mahler sur la fonction ordre d'un nombre transcendant. *Bull. Soc. Math. France,* 102:365–377.

1974 I. J. Good. A reciprocal series of Fibonacci numbers. *Fibonacci Q.,* 12:346.

1974 H. Halberstam. Transcendental numbers. *Math. Gaz.,* 58: 276–284.

1974 M. Mignotte. Approximations rationnelles de π et de quelques autres nombres. *Bull. Soc. Math. France,* 37: 121–132.

1974 D. Shanks. Incredible identities. *Fibonacci Q.,* 12:271 and 280.

1974 M. Waldschmidt. *Nombres Transcendants.* Lecture Notes in Mathematics #402. Springer-Verlag, Berlin.

1975 A. Baker. *Transcendental Number Theory.* Cambridge Univ. Press, London.

1976 V. E. Hoggatt and M. Bicknell. A reciprocal series of Fibonacci numbers with subscripts $2^n k$. *Fibonacci Q.,* 14: 453–455.

1976a K. Mahler. *Lectures on Transcendental Numbers.* Lecture Notes in Mathematics #546. Springer-Verlag, New York.

1976b Kurt Mahler. On a class of transcendental decimal fractions. *Comm. Pure Appl. Math.,* 29:717–725.

1976 M. Mignotte. *Approximation des Nombres Algébriques.* Publ. Math. Orsay, no. 77–74, Orsay, France.

1977 B. C. Berndt. Modular transformations and generalizations of some formulae of Ramanujan. *Rocky Mt. J. Math.,* 7: 147–189.

1978/9 A. J. van der Poorten. Some wonderful formulae... footnotes to Apéry's proof of the irrationality of $\zeta(3)$. *Sém. Delange-Pisot-Poitou,* 20e année(29):7 pages.

1978 M. Waldschmidt. Transcendence measures for exponentials and logarithms. *J. Austral. Math. Soc. Ser. A,* 25(4):445–465.

1979 K. Alladi. Legendre polynomials and irrational numbers. Matscience report no. 100, Inst. Math. Sciences, Madras. 83 pages.

1979 R. Apéry. Irrationalité de $\zeta(2)$ et $\zeta(3)$. *Astérisque,* 61: 11–13. Société Math. France.

1979 F. Beukers. A note on the irrationality of $\zeta(2)$ and $\zeta(3)$. *Bull. London Math. Soc.*, 11:268–272.

1979 M. Waldschmidt. *Transcendence Methods.* Queen's Papers in Pure and Applied Mathematics, No. 52. Queen's University, Kingston, Ont.

1979 A. J. van der Poorten. A proof that Euler missed. . . Apéry's proof of the irrationality of $\zeta(3)$. *Math. Intelligencer*, 1: 193–203.

1980 ———. *Numéro Spécial π.* Supplément au "Petit Archimède".

1981 H. Cohen. Généralisation d'une construction de R. Apéry. *Bull. Soc. Math. France*, 109:269–281.

1982 D. Zagier. On the number of Markoff numbers below a given bound. *Math. of Comp.*, 39:709–723.

1983 M. Waldschmidt. Les débuts de la théorie des nombres transcendants. *Cahiers Sém. Histoire Math.*, 4:93–115.

1984 F. Gramain. Les nombres transcendants. *Pour la Science*, 80:70–79.

1985 B. C. Berndt. *Ramanujan's Notebooks, Part I.* Springer-Verlag, New York.

1986 P. Ribenboim. Some fundamental methods in the theory of Diophantine equations. In *Aspects of mathematics and its applications*, 635–663. North-Holland, Amsterdam.

1988 D. Castellanos. The ubiquity of π. *Math. Mag.*, 61:67–98 and 148–163.

1989 B. C. Berndt. *Ramanujan's Notebooks, Part II.* Springer-Verlag, New York.

1989 D. V. Chudnovsky and G. V. Chudnovsky. Transcendental methods and theta-functions. In *Theta functions—Bowdoin 1987, Part 2 (Brunswick, ME, 1987)*, 167–232. Amer. Math. Soc., Providence, RI.

1991 R. André-Jeannin. A note on the irrationality of certain Lucas infinite series. *Fibonacci Q.*, 29:132–136.

1991 K. Nishioka. *Mahler Functions and Transcendence (Springer Lect. Notes in Math. #1631).* Springer-Verlag, Berlin.

1993 D. V. Chudnovsky and G. V. Chudnovsky. Hypergeometric and modular function identities, and new rational approximations to and continued fraction expansions of classical constants and functions. *Contemp. Math.*, 143:117–162.

1993 M. Hata. Rational approximations to π and some other numbers. *Acta Arith.*, 63:335–349.

1994 P. G. Becker and T. Töpfer. Irrationality results for reciprocal sums of certain Lucas numbers. *Arch. Math. (Basel)*, 62:300–305.

1994 S. Landau. How to tangle with a nested radical. *Math. Intelligencer*, 16(2):49–55.

1996 R. P. Brent, A. J. van der Poorten, and H. J. J. te Riele. A comparative study of algorithms for computing continued fractions of algebraic numbers. In *Algorithmic Number Theory (Talence, 1996)*, Lecture Notes in Computer Science #1122, 35–47. Springer-Verlag, Berlin.

1997 Yu. V. Nesterenko. On the measure of algebraic independence of values of Ramanujan functions (in Russian). *Tr. Mat. Inst. Steklova*, 218:299–334.

1998 F. Gramain. Quelques résultats d'indépendance algébrique. In *Proceedings of the International Congress of Mathematicians, Vol. II (Berlin, 1998)*, 173–182.

11

Galimatias Arithmeticae*

You may read in the *Oxford English Dictionary* that *galimatias* means confused language, meaningless talk. This is what you must expect in this talk. As a token of admiration to GAUSS, I dare to append the word *Arithmeticae* to my title. I mean no offense to the Prince, who, at age 24, published *Disquisitiones Arithmeticae*, that imperishable masterwork.

As I retire (or am hit by retirement), it is time to look back at events in my career. Unlike what most people do, I would rather talk about mathematical properties and problems of some numbers connected with highlights of my life. I leave for the end the most striking conjunction.

I will begin with the hopeful number 11 and end with the ominous number 65.

---

*This chapter is a modified version of a talk at the University of Munich, given in November 1994 at a festive colloquium in honor of Professor Sibylla Priess-Crampe.

11

- At age 11 I learned how to use x to represent an unknown quantity in order to solve problems like this one: "Three brothers, born two years apart, had sums of ages equal to 33. What are their ages?" The power of the method was immediately clear to me and determined that I would be interested in numbers, even after my age would surpass the double of the sum of the ages of the three brothers.

  But 11 is interesting for many better reasons.

- 11 is the smallest prime repunit. A number with n digits all equal to 1 is called a *repunit* and denoted by R_n. So, $11 = R_2$. The following repunits are known to be prime: R_n with $n = 2$, 19, 23, 317, and 1031. It is not known whether there are infinitely many prime repunits.

- If $n > 11$, there exists a prime $p > 11$ such that

$$p \text{ divides } n(n+1)(n+2)(n+3).$$

  A curiosity? Not quite. A good theorem (by MAHLER states that if $f(x)$ is a polynomial with integral coefficients of degree two or more (for two, the theorem is PÓLYA'S), and if H is a finite set of primes (such as $\{2, 3, 5, 7, 11\}$), then there exists n_0 such that if all primes factors of $f(n)$ are in H, then $n \le n_0$.

  Another way of saying this is as follows: $\lim_{n \to \infty} P[f(n)] = \infty$, where $P[f(n)]$ denotes the largest prime factor of $f(n)$. With the theory of BAKER on linear forms in logarithms, COATES gave an effective bound for n_0. For the particular polynomial $f(x) = x(x+1)(x+2)(x+3)$, the proof is elementary.

- 11 is the largest positive integer d that is square-free and such that $\mathbb{Q}(\sqrt{-d})$ has a euclidean ring of integers. The other such fields are those with $d = 1$, 2, 3, and 7. This means that if α, $\beta \in \mathbb{Z}[\sqrt{-d}]$, there exist γ, $\delta \in \mathbb{Z}[\sqrt{-d}]$ such that $\alpha = \beta\gamma + \delta$ where $\delta = 0$ or $N(\delta) < N(\beta)$. (Here, for $\alpha = a + b\sqrt{-d}$, $N(\alpha) = a^2 + db^2$. The situation is just like that for euclidean division in the ring $\mathbb{Z}$ of ordinary integers.)

- It is not known whether there exists a cuboid with sides a, b, and c measured in integers, as well as all diagonals measured in integers. In other words, it is not known whether the following system has a solution in non-zero integers:

$$\begin{cases} a^2 + b^2 = d^2 \\ b^2 + c^2 = e^2 \\ c^2 + a^2 = f^2 \\ a^2 + b^2 + c^2 = g^2 \end{cases}$$

If such integers exist, then 11 divides abc.

- 11 is the smallest integer that is not a *numerus idoneus*.

  You do not know what a *numerus idoneus* is? I too needed to reach 65 before realizing how this age and *idoneus* numbers are connected with each other. So be patient.

- According to the theory of supersymmetry, the world has 11 dimensions: 3 for space position, 1 for time, and 7 to describe the various possible superstrings and their different vibrating patterns, so explaining subatomic particles' behavior.

  Is this a joke or a new theory to explain the world?

- The Mersenne numbers are the integers $M_q = 2^q - 1$ where q is a prime. Big deal: some are prime, some are composite. Bigger deal: how many of each kind? Total mystery!

  $M_{11} = 2^{11} - 1 = 2047 = 23 \cdot 28$. It is the smallest composite Mersenne number. The largest known composite Mersenne number is M_q with $q = 72021 \times 2^{23630} - 1$.

19

- One of my favorite numbers has always been 19. At this age Napoleon was winning battles—this we should forget. At the same age, GAUSS discovered the law of quadratic reciprocity—this you cannot forget, once you have known it.

- First a curiosity concerning the number 19. It is the largest integer n such that

$$n! - (n-1)! + (n-2)! - \cdots \pm 1!$$

is a prime number. The other integers n with this property are

$$n = 3, 4, 5, 6, 7, 8, 9, 10, \text{ and } 15.$$

- Both the repunit R_{19} and the Mersenne number M_{19} are prime numbers.

- Let $U_0 = 0$, $U_1 = 1$, and $U_n = U_{n-1} + U_{n-2}$ for $n \geq 2$; these are the Fibonacci numbers. If U_n is prime, then n must also be prime, but not conversely. 19 is the smallest prime index that provides a counterexample: $U_{19} = 4181 = 37 \cdot 113$.

- The fields $\mathbb{Q}(\sqrt{-19})$, $\mathbb{Q}(\sqrt{19})$ have class number 1. (The class number is a natural number which one associates to every number field. It is 1 for the field of rationals; it is also 1 for the field of Gaussian numbers, and for any field whose arithmetical properties resemble those of the rational numbers. The larger the class number of a number field, the more its arithmetical properties "deviate" from those of the rationals. For more on these concepts, see RIBENBOIM (2000).) The ring of integers of $\mathbb{Q}(\sqrt{19})$ is euclidean, while the ring of integers $\mathbb{Q}(\sqrt{-19})$ is not euclidean.

- Let $n > 2$, $n \not\equiv 2 \pmod 4$, and let $\zeta_n = e^{2\pi i/n}$ denote a primitive nth root of 1. 19 is the largest prime p such that $\mathbb{Q}(\zeta_p)$ has class number 1. This was important in connection with KUMMER's research on Fermat's Last Theorem.

MASLEY and MONTGOMERY determined in 1976 all integers n, $n \not\equiv 2 \pmod 4$, such that $\mathbb{Q}(\zeta_n)$ has class number 1, namely:

$$n = 1, 3, 4, 5, 7, 8, 9, 11, 12, 13, 15, 16, 17, 19, 20, 21, 24, 25,$$
$$27, 28, 32, 33, 35, 36, 40, 44, 45, 48, 60, \text{ and } 84.$$

- BALASUBRAMANIAN, DRESS, and DESHOUILLERS showed in 1986 that every natural number is the sum of at most 19 fourth powers. DAVENPORT had shown in 1939 that every sufficiently large natural number is the sum of at most 16 fourth powers. This provided a complete solution of the two forms of Waring's problem for fourth powers.

29

- Twin primes, such as 29 and 31, are not like the ages of twins— their difference is 2. Why? There are many twin persons and many twin primes, but in both cases, it is not known whether there are infinitely many ....

EULER showed that

$$\sum_{p \text{ prime}} \frac{1}{p} = \infty.$$

On the other hand, BRUN showed that

$$\sum_{p,\, p+2 \text{ primes}} \frac{1}{p} < \infty$$

BRUN's result says that either there are only finitely many twin primes, or, if there are infinitely many twin primes, their size must increase so rapidly that the sum above remains bounded. All of this is amply discussed in my book on prime numbers RIBENBOIM (1996).

- A curiosity observed by EULER: If 29 divides the sum $a^4+b^4+c^4$, then 29 divides $\gcd(a, b, c)$.

- Let p be a prime. The *primorial* of p is

$$p\sharp = \prod_{q \leq p,\, q \text{ \{prime\}}} q;$$

$29 = 5\sharp - 1$. The expressions $p\sharp + 1$ and $p\sharp - 1$ have been considered in connection with variants of the EUCLID's proof that there exist infinitely many primes. The following primes p are the only ones less than or equal to 35000 such that $p\sharp - 1$ is prime:

$$p = 3,\, 5,\, 11,\, 13,\, 41,\, 89,\, 317,\, 991,\, 1873,\, 2053,\, 2371,\, 4093,\, 4297,$$
$$4583,\, 6569,\, 13033,\, 15877.$$

For this and similar sequences, see RIBENBOIM (1996).

- $2 \cdot 29^2 - 1 = \square$ (a square), similarly $2 \cdot 1^2 - 1 = \square$, $2 \cdot 5^2 - 1 = \square$. In fact, there are infinitely many natural numbers x such that $2x^2 - 1 = \square$. Here is how to obtain all pairs of natural numbers (t, x) such that $t^2 - 2x^2 = -1$. From $(t + \sqrt{2}x)(t - \sqrt{2}x) = -1$, it follows that $t + \sqrt{2}x$ is a unit in the field $\mathbb{Q}(\sqrt{2})$. The fundamental unit is $1+\sqrt{2}$ with the norm $(1+\sqrt{2})(1-\sqrt{2}) = -1$, so $t + \sqrt{2}x = (1 + \sqrt{2})^n$ with n odd. Thus we have

$$(1 + \sqrt{2})^2 = 3 + 2\sqrt{2},$$
$$(1 + \sqrt{2})^3 = 7 + 5\sqrt{2},$$
$$(1 + \sqrt{2})^5 = 41 + 29\sqrt{2}.$$

The next solution is obtained from

$$(1 + \sqrt{2})^7 = 239 + 169\sqrt{2},$$

namely $2 \cdot 169^2 - 1 = 239^2$.

- The ring of integers of $\mathbb{Q}(\sqrt{29})$ is euclidean. There are 16 real quadratic fields $\mathbb{Q}(\sqrt{d})$ with a euclidean ring of integers, namely

$$d = 2, 3, 5, 6, 7, 11, 13, 17, 19, 21, 29, 33, 37, 41, 57, 73.$$

- $2X^2 + 29$ is an *optimal prime-producing polynomial*. Such polynomials were first considered by EULER—they are polynomials $f \in \mathbb{Z}[X]$ that assume as many initial prime values as they possibly can. More precisely, let $f \in \mathbb{Z}[X]$, with positive leading coefficient and $f(0) = q$, a prime. There exists the smallest $r > 0$ such that $f(r) > q$ and $q \mid f(r)$. The polynomial is *optimal prime-producing* if $f(k)$ is prime for $k = 0, 1, \ldots, r - 1$.

EULER observed that $X^2 + X + 41$ is optimal prime-producing, since it assumes prime values at $k = 0, 1, \ldots, 39$, while $40^2 + 40 + 41 = 41^2$.

In 1912, RABINOVITCH showed that the polynomial $f(X) = X^2 + X + q$ (with q prime) is optimal prime-producing if and only if the field $\mathbb{Q}(\sqrt{1 - 4q})$ has class number 1.

HEEGNER, STARK, and BAKER determined all the imaginary quadratic fields $\mathbb{Q}(\sqrt{d})$ (with $d < 0$ and d square-free) with class number 1:

$$d = -1, -2, -5, -7, -11, -19, -43, -67, -163.$$

These correspond to the only optimal prime-producing polynomials of the form $X^2 + X + q$, namely $q = 2, 3, 5, 11, 17, 41$. $X^2 + X + 41$ is the record prime-producing polynomial of the form $X^2 + X + q$.

FROBENIUS (1912) and HENDY (1974) studied optimal prime-producing polynomials in relation to imaginary quadratic fields having class number 2. There are three types of such fields:

(i) $\mathbb{Q}(\sqrt{-2p})$, where p is an odd prime;

(ii) $\mathbb{Q}(\sqrt{-p})$, where p is a prime, and $p \equiv 1 \pmod 4$;

(iii) $\mathbb{Q}(\sqrt{-pq})$ where p, q are odd primes with $p < q$ and $pq \equiv 3 \pmod{4}$.

For the types of fields above, the following theorem holds:

(i) $\mathbb{Q}(\sqrt{-2p})$ has class number 2 if and only if $2X^2+p$ assumes prime values at $k = 0, 1, \ldots, p-1$.

(ii) $\mathbb{Q}(\sqrt{-p})$ has class number 2 if and only if $2X^2 + 2X + \frac{p+1}{2}$ assumes prime values at $k = 0, 1, \ldots, \frac{p-3}{2}$.

(iii) $\mathbb{Q}(\sqrt{-pq})$ has class number 2 if and only if $pX^2+pX+\frac{p+q}{4}$ assumes prime values at $k = 0, 1, \ldots, \frac{p+q}{4} - 2$.

STARK and BAKER determined the imaginary quadratic fields $\mathbb{Q}(\sqrt{d})$ (with $d < 0$ and d square-free) that have class number 2. According to their types, they are:

(i) $d = -6, -10, -22, -58$.

(ii) $d = -5, -13, -37$.

(iii) $d = -15, -35, -51, -91, -115, -123, -187, -235, -267, -403, -427$.

With these values of d one obtains optimal prime-producing polynomials.

In particular, $2X^2 + 29$ is an optimal prime-producing polynomial, with prime values at $k = 0, 1, \ldots, 28$; it corresponds to the field $\mathbb{Q}(\sqrt{-58})$, which has class number 2.

- 29 is the number of distinct topologies on a set with 3 elements. Let τ_n denote the number of topologies on a set with n elements; thus $\tau_1 = 1$ and $\tau_2 = 2$. One knows the values of τ_n for $n \leq 9$ (RADOUX (1975)).

Approaching the thirties, the age of confidence, life was smiling. 29 was the first twin prime age I reached since I became a mathematician by profession, so I select the number

30

- At this age I was in Bahia Blanca, Argentina, preparing a book which has, I believe, the distinction of being the southern-most published mathematical book. (At least this is true for books on ordered groups—but mine is not the northern-most published book on the subject.)

- There is only one primitive pythagorean triangle with area equal to its perimeter, namely $(5, 12, 13)$, with perimeter 30.

- 30 is the largest integer d such that if $1 < a < d$ and $\gcd(a, d) = 1$, then a is a prime. Other numbers with this property are: 3, 4, 6, 8, 12, 18, and 24. This was first proved by SCHATUNOWSKY in 1893 and, independently, by WOLFSKEHL in 1901. (WOLFSKEHL is the rich mathematician who donated 100,000 golden marks to be given to the author of the first proof of Fermat's Last Theorem to be published in a recognized mathematical journal.)

This result has an interpretation as follows. Given $d > 1$ and a, $1 \leq a < d$, $\gcd(a, d) = 1$, by Dirichlet's theorem, there exist infinitely many primes of the form $a + kd$ $(k \geq 0)$. Let $p(a, d)$ be the smallest such prime, and let

$$p(d) = \max\{p(a, d) \mid 1 \leq a < d, \gcd(a, d) = 1\}.$$

If $d > 30$, then $p(d) > d + 1$. In particular,

$$\liminf \frac{p(d)}{d + 1} > 1.$$

POMERANCE has shown:

$$\liminf \frac{p(d)}{\varphi(d) \log d} \geq e^{\gamma}$$

where $\varphi(d)$ is EULER's totient of d and γ is the Euler-Mascheroni constant.

On the other hand, as shown by LINNIK, for d sufficiently large, $p(d) \leq d^L$, where L is a constant. HEATH-BROWN showed that $L \leq 5.5$.

32

- 32 is the smallest integer n such that the number γ_n of groups of order n (up to isomorphism) is greater than n: $\gamma_{32} = 51$.

I hate the number 32. At 32 degrees Fahrenheit, water becomes ice and snow begins to fall. Let us change the subject!

Older people remember best the events of their youth and those of the more recent past. I haven't forgotten anything I did not want to forget, so I could let you know about all the years 33, 34, ... But I would rather concentrate on the 60's.

60

- 60 was the base of numeration in the counting system of the Sumerians (ca. 3500 B.C.). Today we still use the sexagesimal system in astronomy and in the subdivisions of the hour.

- 60 is a *highly composite number*. Such numbers were introduced and studied by RAMANUJAN (1915): The natural number n is *highly composite* if $d(n) > d(m)$ for every m, $1 \le m < n$, where $d(n)$ = number of divisors of n. Thus $d(60) = d(2^2 \cdot 3 \cdot 5) = 3 \cdot 2 \cdot 2 = 12$. The smallest highly composite numbers are

$$2, 4, 6, 12, 24, 32, 48, 60, 120, 180, 240, 360, 720, 840, \dots .$$

- 60 is a *unitarily perfect number*, which I now define. A number d is a *unitary divisor* of n if $d \mid n$ and $\gcd(d, n/d) = 1$; n is *unitarily perfect* if

$$n = \sum \{d \mid 1 \le d < n, \ d \text{ unitary divisor of } n\}.$$

Unitary divisors of 60 are 1, 3, 4, 5, 12, 15, 20 and their sum is indeed 60.

Conjecture: There exist only finitely many unitarily perfect numbers.

The only known unitarily perfect numbers are

$$6, 60, 90, 87360 \quad \text{and} \quad 2^{18} \cdot 3 \cdot 7 \cdot 11 \cdot 13 \cdot 19 \cdot 37 \cdot 79 \cdot 109 \cdot 157 \cdot 313.$$

- 60 is the number of straight lines that are intersections of the pairs of planes of the faces of a dodecahedron.

- 60 is the order of the group of isometries of the icosahedron. This is the alternating group on 5 letters. It is the non-abelian simple group with the smallest order. The simple groups have been classified—a great achievement! There are 18 infinite families:

- cyclic groups of prime order;

- alternating groups A_n, with $n \geq 5$;

- six families associated to the classical groups;

- ten families associated to Lie algebras (discovered by DICKSON, CHEVALLEY, SUZUKI, REE, and STEINBERG.

There are also 26 "sporadic" groups, which do not belong to the above families. The sporadic groups with the largest order is FISCHER's monster, which has

$$2^{46} \cdot 3^{20} \cdot 5^9 \cdot 7^6 \cdot 11^2 \cdot 13^3 \cdot 17 \cdot 19 \cdot 23 \cdot 29 \cdot 31 \cdot 41 \cdot 47 \cdot 59 \cdot 71 \geq 8 \cdot 10^{53}$$

elements.

61

- A curiosity: Let $k \geq 0$, and let $a_1, \ldots, a_k$, x, y be digits. If the number (in decimal notation)

$$a_1 a_2 \ldots a_k xyxyxyxyxyxy$$

is a square, then $xy = 21$, 61, or 84. Examples:

$$17392885161616161616 = 1318820881^2;$$
$$258932382121212121 = 508853989^2.$$

- The Mersenne number $M_{61} = 2^{61} - 1$ is a prime. Today there are 37 known prime Mersenne numbers $M_p = 2^p - 1$, namely, those with

$$p = 2, 3, 5, 7, 13, 17, 19, 31, 61, 89, 107, 127, 521, 607, 1279, 2203,$$
$$2281, 3217, 4253, 4423, 9689, 9941, 11213, 19937, 21701, 23209,$$
$$44497, 86243, 110503, 132049, 216091, 756839, 859433, 1257787,$$
$$1398269, 2976221, \text{ and } 3021227.$$

$2^{3021227} - 1$ is also the largest prime known today.

62

This number is remarkable for being so uninteresting. As a matter of fact, suppose that, for some reason or another, there is some number that is not remarkable. Then there is the smallest non-remarkable number, which is therefore remarkable for being the smallest non-remarkable number.

But this is just another example of Russell's paradox ....

63

- This number appears in a cycle associated with *Kaprekar's algorithm* for numbers with 2 digits. This algorithm, for numbers with k digits, goes as follows: Given k digits $a_1 \ldots a_k$, not all equal, with $a_1 \geq a_2 \geq \ldots \geq a_k \geq 0$, consider two numbers formed using these digits: $a_1 a_2 \ldots a_k$, and $a_k a_{k-1} \ldots a_1$. Compute their difference, and repeat the process with the k digits so obtained.

  Kaprekar's algorithm for 2, 3, 4, and 5 digits leads to the following fixed points or cycles.

  2 digits $\rightarrow$ cycle 63 - 27 - 45 - 09 - 81

  3 digits $\rightarrow$ 495

  4 digits $\rightarrow$ 6174

  5 digits $\rightarrow$ one of the 3 cycles: 99954 - 95553

  98532 - 97443 - 96642 - 97731

  98622 - 97533 - 96543 - 97641

  Example: $\{3, 5\}$: $53 - 35 = 18$, $81 - 18 = 63$, $63 - 36 = 27$, $72 - 27 = 45$, $54 - 45 = 09$, $90 - 09 = 81$.

- 63 is the unique integer $n > 1$ such that $2^n - 1$ does not have a *primitive prime factor*. Explanation: If $1 \leq b < a$, with $\gcd(a, b) = 1$, consider the sequence of binomials $a^n - b^n$ for $n \geq 1$. The prime p is a *primitive prime factor* of $a^n - b^n$ if $p \mid a^n - b^n$, but $p \nmid a^m - b^m$ if $1 \leq m < n$.

  ZSIGMONDY proved, under the above assumptions, that every binomial $a^n - b^n$ has a primitive prime factor, except in the following cases:

(i) $n = 1$, $a - b = 1$;

(ii) $n = 2$, a and b odd, and $(a + b)$ a power of 2;

(iii) $n = 6$, $a = 2$, $b = 1$.

This theorem has many applications in the study of exponential diophantine equations; see RIBENBOIM (1994). Explicitly, when $a = 2$ and $b = 1$, the sequence is:

$$1, 3, 7, 15 = 3 \cdot 5, 31, 63 = 3^2 \cdot 7, 127, 257, 511, 1023 = 3 \cdot 11 \cdot 31, \ldots$$

64

64 is almost 65, a number I hated to reach, but which nevertheless has many interesting features.

65

- 65 is the smallest number that is the sum of 2 squares of natural numbers in 2 different ways (except for the order of summands):

$$65 = 8^2 + 1^2 = 7^2 + 4^2.$$

Recall Fermat's result: n is a sum of 2 squares if and only if for every prime $p \equiv 3 \pmod 4$, $v_p(n)$ is even. (Here $v_p(n)$ denotes the p-adic value of n, that is, $p^{v_p(n)} \mid n$ but $p^{v_p(n)+1}$ does not divide n.) The following formula gives the number

$$r(n) = \#\{(a, b) \mid 0 \le b \le a \text{ and } n = a^2 + b^2\}.$$

For each $d \ge 1$, let

$$\chi(d) = \begin{cases} (-1)^{\frac{d-1}{2}} & \text{if } d \text{ is odd,} \\ 0 & \text{if } d \text{ is even.} \end{cases}$$

Let $R(n) = \sum_{d \mid n} \chi(d)$. Then

$$r(n) = \begin{cases} \dfrac{R(n)}{2} & \text{if } R(n) \text{ is even,} \\ \dfrac{1 + R(n)}{2} & \text{if } R(n) \text{ is odd.} \end{cases}$$

Example: $65 = 5 \cdot 13$ has divisors 1, 5, 13, 65, and $R(65) = \sum_{d|65} \chi(d) = 4$, so $r(65) = 2$.

- 65 is the smallest hypotenuse common to two pythagorean triangles. This follows from the parameterization of the sides of pythagorean triangles: If $0 < x$, y, z, with y even and $x^2 + y^2 = z^2$, then there exist a and b, $1 \le b < a$, such that

$$x = a^2 - b^2; \quad y = 2ab; \quad z = a^2 + b^2.$$

Moreover, the triangle is primitive (i.e. $\gcd(x, y, z) = 1$) if and only if $\gcd(a, b) = 1$. From $65 = 8^2 + 1^2 = 7^2 + 4^2$ one gets the pythagorean triangles $(63, 16, 65)$ and $(33, 56, 65)$.

- A curiosity: 65 is the only number with 2 digits d, e, $0 \le e < d \le 9$, such that $(de)^2 - (ed)^2 = \square$, a square. Indeed, $65^2 - 56^2 = 33^2$, and the uniqueness follows from the parameterization indicated above.

- 65 is also a remarkable number of the *second kind,* that is, it counts the number of remarkable numbers satisfying some given property. In the present case, 65 is perhaps the number of EU-LER's *numeri idonei.* I say "perhaps" because there is still an open problem, and instead of 65 there may eventually exist 66 such numbers.

Numeri idonei

What are these *numeri idonei* of EULER? Also called *convenient numbers,* they were used conveniently by EULER to produce prime numbers.

Now I will explain what the *numeri idonei* are. Let $n \ge 1$. If q is an odd prime and there exist integers x, $y \ge 0$ such that $q = x^2 + ny^2$, then:

(i) $\gcd(x, ny) = 1$;

(ii) if $q = x_1^2 + ny_1^2$ with integers $x_1, y_1 \ge 0$, then $x = x_1$ and $y = y_1$.

We may ask the following question. *Assume that q is an odd integer, and that $q = x^2 + ny^2$, with integers x, $y \ge 0$, such that conditions (i) and (ii) above are satisfied. Is q a prime number?*

The answer depends on n. If $n = 1$, the answer is "yes", as Fermat knew. For $n = 11$, the answer is "no": $15 = 2^2 + 11 \cdot 1^2$ and conditions (i) and (ii) hold, but 15 is composite. EULER called n a *numerus idoneus* if the answer to the above question is "yes".

EULER gave a criterion to verify in a finite number of steps whether a given number is convenient, but his proof was flawed. Later, in 1874, GRUBE found the following criterion, using in his proof results of GAUSS, which I will mention soon. Thus, n is a convenient number if and only if for every $x \geq 0$ such that $q = n + x^2 \leq \frac{4n}{3}$, if $q = rs$ and $2x \leq r \leq s$, then $r = s$ or $r = 2x$.

For example, 60 is a convenient number, because

$$60 + 1^2 = 61 \, (\star),$$
$$60 + 2^2 = 64 = 4 \cdot 16 = 8 \cdot 8,$$
$$60 + 3^2 = 69 \, (\star),$$
$$60 + 4^2 = 76 \, (\star)$$

and the numbers marked with a $(\star)$ do not have a factorization of the form indicated.

EULER showed, for example, that 1848 is a convenient number, and that

$$q = 18518809 = 197^2 + 1848 \cdot 100^2$$

is a prime number. At EULER's time, this was quite a feat.

GAUSS understood convenient numbers in terms of his theory of binary quadratic forms. The number n is convenient if and only if each genus of the form $x^2 + ny^2$ has only one class.

Here is a list of the 65 convenient numbers found by EULER:

1, 2, 3, 4, 5, 6, 7, 8, 9, 10, 12, 13, 15, 16, 18, 21, 22, 24, 25, 28, 30, 33, 37, 40, 42, 45, 48, 57, 58, 60, 70, 72, 78, 85, 88, 93, 102, 105, 112, 120, 130, 133, 165, 168, 177, 190, 210, 232, 240, 253, 273, 280, 312, 330, 345, 357, 385, 408, 462, 520, 760, 840, 1320, 1365, 1848.

Are there other convenient numbers? CHOWLA showed that there are only finitely many convenient numbers; later, finer analytical work (for example, by BRIGGS, GROSSWALD, and WEINBERGER) implied that there are at most 66 convenient numbers.

The problem is difficult. The exclusion of an additional *numerus idoneus* is of a kind similar to the exclusion of a hypothetical tenth imaginary quadratic field (by HEEGNER, STARK, and BAKER), which I have already mentioned.

An extraordinary conjunction

If your curiosity has not yet subsided, I was struck in 1989, in Athens, at the occasion of my "Greek Lectures on Fermat's Last Theorem" by an extraordinary conjunction of numbers. Once in a lifetime, and not to be repeated before....

At that year, my wife's age and my age were 59 and 61—twin primes (but we are not twins); at that same year, we had been married 37 years—the smallest irregular prime. If you are still interested, KUMMER had proved that Fermat's Last Theorem is true for all odd prime exponents p that are regular primes. These are the primes p that do not divide the class number of the cyclotomic field generated by the pth root of 1. KUMMER also discovered that 37 is the smallest irregular prime. Pity that 1989 (the year of my Athens lecture) is not a prime.

So you are challenged to find the next occurrence of numbers like 37, 59, 61, but in a prime numbered year.

Notes. This paper on remarkable numbers would not have been possible were it not for the very original book by F. LE LIONNAIS, *Les Nombres Remarquables*, published in 1983 by Hermann, in Paris.

François LE LIONNAIS was not a mathematician by profession, but rather a scientific writer, and as such, very well informed. His book *Les Grands Courants de la Pensée Mathématique* is very engrossing to read even today. Just after the war he gathered in this book the ideas of several young French mathematicians—still little known at that time—who would soon rise to the pinnacle. An English translation and the original are available in good libraries. I have an autographed copy of the book on remarkable numbers, where Le Lionnais thanked me for calling his attention to the number 1093. You may read about this number in Chapter 8 of this book.

Another book of the same kind, which served me well, is: D. WELLS, *The Penguin Dictionary of Curious and Interesting Numbers*, Penguin, London, UK, 1986.

For results on algebraic numbers, nothing is easier for me than to quote my own book RIBENBOIM (2000), to appear in a new edition at Springer-Verlag. For *numeri idonei*, see FREI (1984). Concerning primitive factors of binomials, see RIBENBOIM (1994). On prime numbers, Fibonacci numbers and similar topics, see RIBENBOIM (1996). For further reference, see GUY (1994).

The following list of references is, it goes without saying, incomplete.

References

1984 G. Frei. Les nombres convenables de Leonhard Euler. In *Number theory (Besançon), 1983–1984*, Exp. No. 1, 58. Univ. Franche-Comté, Besançon.

1994 R. K. Guy. *Unsolved Problems in Number Theory*. Springer-Verlag, New York, 2nd edition.

1994 P. Ribenboim. *Catalan's Conjecture*. Academic Press, Boston.

1996 P. Ribenboim. *The New Book of Prime Number Records*. Springer-Verlag, New York.

2000 P. Ribenboim. *The Classical Theory of Algebraic Numbers*. Springer-Verlag, New York.

Index of Names

Index of Subjects

DATE DUE

DEC 0 4 2000		
DEC 1 2 2001		
12/25/03		
OCT 1 4 2005		
MAY 1 8 2006		